Methods and Protocols in Food Science

Series Editor
Anderson S. Sant'Ana
University of Campinas
Campinas, Brazil

For further volumes:
http://www.springer.com/series/16556

Methods and Protocols in Food Science series is devoted to the publication of research protocols and methodologies in all fields of food science.

Volumes and chapters will be organized by field and presented in such way that the readers will be able to reproduce the experiments in a step-by-step style. Each protocol will be characterized by a brief introductory section, followed by a short aims section, in which the precise purpose of the protocol will be clarified.

Phytochemicals

Extraction, Quantification, Isolation, and Characterization

Edited by

Ashwani Kumar

Institute of Food Technology, Bundelkhand University, Jhansi, Uttar Pradesh, India

Vidisha Tomer

Department of Horticulture and Food Science, VIT School of Agricultural Innovations and Advanced Learning, Vellore Institute of Technology, Vellore, Tamil Nadu, India

Piyoosh Kumar Babele

Bioprocess Department, Hindustan Petroleum Green R&D Centre, Bengaluru, Karnataka, India

Humana Press

Editors
Ashwani Kumar
Institute of Food Technology
Bundelkhand University
Jhansi, Uttar Pradesh, India

Vidisha Tomer
Department of Horticulture and Food Science
VIT School of Agricultural Innovations and Advanced Learning
Vellore Institute of Technology
Vellore, Tamil Nadu, India

Piyoosh Kumar Babele
Bioprocess Department
Hindustan Petroleum Green R&D Centre
Bengaluru, Karnataka, India

ISSN 2662-950X ISSN 2662-9518 (electronic)
Methods and Protocols in Food Science
ISBN 978-1-0716-4839-1 ISBN 978-1-0716-4840-7 (eBook)
https://doi.org/10.1007/978-1-0716-4840-7

This Humana imprint is published by the registered company Springer Science+Business Media, LLC, part of Springer Nature.
The registered company address is: 1 New York Plaza, New York, NY 10004, U.S.A.

Foreword

Plants have long been a primary source of food and medicine since the dawn of humankind. Every major civilization has developed its own traditional system of medicine, using plant-based bioactive compounds to treat ailments. Some of the most well-known traditional systems include Indian, Chinese, Greco-Arab, and African medicine. In India, *Ayurveda* is the traditional medicinal system, which focuses on balancing three elements: *Vata*, *Pitta*, and *Kapha*. Chinese medicine, on the other hand, emphasizes the balance between yin (negative forces) and yang (positive forces). The Greco-Arab system, known as the *Unani* system, is based on the concept of four humors—blood, phlegm, yellow bile, and black bile—which are linked to the physical, mental, emotional, and spiritual aspects of health. In Africa, traditional medicine is still widely practiced, especially in Guinea, where about 80% of the population adheres to it. All of these systems rely on plants and herbs for healing.

However, a limitation of traditional medicine is that the effectiveness of herbs can vary depending on their source and the concentration of bioactive compounds (phytochemicals) present. This variability has made traditional medicines less popular than allopathic (modern) medicines. Yet, the growing incidence of side effects from allopathic medicines has reignited interest in herbal remedies. This has prompted extensive research into the phytochemicals found in medicinal plants and their effects on human health when used as a part of food or consumed in concentrated, purified forms. These phytochemicals can influence metabolism, gene expression, and cellular signaling, while some also reduce oxidative stress and inflammation by enhancing the expression of antioxidant enzymes. Purified and concentrated phytochemicals, when administered in known dosages, have proven to be more effective than their natural counterparts. As a result, the development of functional foods (foods containing bioactive compounds, particularly phytochemicals) and nutraceuticals (concentrated and purified forms of these compounds) has gained significant traction.

Phytochemicals with established health benefits include isoprenoids, polyphenols, alkaloids, dietary fibers, and peptides. Some of the most studied are carotenoids, phytosterols, phytostanols, saponins, phenolic acids, flavonoids, stilbenes, lignans, and alkaloids like morphine and nicotine. Carotenoids, such as beta-carotene, alpha-carotene, and beta-cryptoxanthin, are precursors to vitamin A, while lycopene, lutein, etc. are antioxidants. Phytosterols and phytostanols help reduce cardiovascular diseases by limiting cholesterol absorption. Flavonoids possess antioxidative, anti-inflammatory, antimutagenic, and anticarcinogenic properties. Saponins have surfactant properties and are found in foods like quinoa and ginseng and possess antioxidant and anti-inflammatory properties. Stilbenes like resveratrol, found in grapes, improve cardiovascular health. Lignans, found in flaxseed and sesame, offer antioxidant and anti-inflammatory effects. Alkaloids like morphine and nicotine exhibit anticancer, analgesic, and anti-inflammatory properties.

To harness the potential of these phytochemicals in creating functional foods and nutraceuticals, it is crucial to extract, quantify, purify, and characterize them effectively. The book *Phytochemicals: Extraction, Quantification, Isolation, and Characterization* provides comprehensive information on the sources and health benefits of different phytochemicals, along with detailed, step-by-step methodologies for their extraction, quantification, isolation, and characterization. This book will be invaluable to researchers working with phytochemicals. I congratulate the editors for selecting such an important topic and for presenting the material in an accessible and understandable way.

Vice Chancellor, Bundelkhand University, Jhansi, Uttar Pradesh, India ***Mukesh Pandey***

Preface to the Series

The Methods and Protocols in Food Science series is devoted to the publication of research protocols and methodologies in all fields of food science. The series is unique as it includes protocols developed, validated, and used by food and related scientists as well as provides theoretical basis for each protocol. Aspects related to improvements in the protocols and adaptations and further developments in the protocols may also be approached.

The Methods and Protocols in Food Science series aims to bring the most recent developments in research protocols in the field as well as very well-established methods. As such the series targets undergraduates, graduates, and researchers in the field of food science and correlated areas. The protocols documented in the series will be highly useful for scientific inquiries in the field of food sciences and presented in such a way that the readers will be able to reproduce the experiments in a step-by-step style.

Each protocol will be characterized by a brief introductory section, followed by a short aims section, in which the precise purpose of the protocol is clarified. Then, an in-depth list of materials and reagents required for employing the protocol is presented, followed by comprehensive and step-by-step procedures on how to perform that experiment. The next section brings the dos and don'ts when carrying out the protocol, followed by the main pitfalls faced and how to troubleshoot them. Finally, template results will be presented and their meaning/conclusions addressed.

The Methods and Protocols in Food Science series will fill an important gap, addressing a common complain of food scientists, regarding the difficulties in repeating experiments detailed in scientific papers. With this, the series has a potential to become a reference material in food science laboratories of research centers and universities throughout the world.

About the Book

Plants produce a range of specialized metabolites, and they are key players in complex interactions between plants and their environment. These specialized metabolites, often referred to as secondary metabolites, include a wide range of phytochemicals like phenolic acids, carotenoids, alkaloids, flavonoids, terpenoids, phytosterols, saponins, etc. These biochemicals play vital roles in plants, such as defending against herbivores, pathogens, environmental stress, and contribute to the plant's color, flavor, and aroma. Phytochemistry is an important field because many of these plant-derived secondary metabolites have pharmaceutical and nutraceutical properties. The book titled *Phytochemicals: Extraction, Quantification, Isolation, and Characterization* delves into various techniques and methodologies employed in the research and study of plant-based bioactive compounds, which are believed to possess health-promoting properties.

This book is written with the primary objective of providing a comprehensive understanding of methods and protocols involved in the extraction, quantification, isolation, and characterization of phytochemicals. It covers various extraction techniques, ranging from traditional solvent extraction methods to advanced techniques like supercritical fluid extraction. Analytical techniques used for the quantification and characterization of phytochemicals, including UV-VIS spectroscopy, Fourier-transform infrared (FTIR) spectroscopy, thin-layer chromatography (TLC), high-performance liquid chromatography (HPLC), gas chromatography (GC), and liquid chromatography (LC) coupled with mass spectrometry (MS) are also outlined. A key aspect of the book is the validation of the health benefits of phytochemicals. This is supported by a dedicated chapter on in vitro (laboratory-based) methods for evaluating the effectiveness of phytochemicals in promoting health.

Overall, this book serves as a comprehensive guide for researchers, students, and professionals interested in plant-based bioactive compounds. It provides an in-depth understanding of the entire process, from extraction and quantification to the validation of health-promoting properties. By offering both theoretical knowledge and practical techniques, it equips the reader with the tools necessary to explore the therapeutic potential of phytochemicals and contribute to the advancement of natural product research.

About the Book

Contents

Contributors

PIYOOSH KUMAR BABELE • *Bioprocess Department, Hindustan Petroleum Green R&D Centre, Bengaluru, Karnataka, India*
DEVENDRA KUMAR BHATT • *Institute of Food Technology, Bundelkhand University, Jhansi, Uttar Pradesh, India*
GARIMA CHOUHAN • *Department of Biotechnology, Sharda School of Bioscience & Technology (SSBT), Sharda University, Greater Noida, Uttar Pradesh, India*
PINTU CHOUDHARY • *Department of Food Technology, College of Agricultural Engineering and Technology, Dr. Rajendra Prasad Central Agricultural University, Pusa, Samastipur, Bihar, India*
NIDHI DANGI • *School of Health and Allied Sciences (Nutrition & Dietetics), Geeta University, Panipat, Haryana, India*
IMMANUEL DHANASINGH • *Centre for Bio Separation Technology, Vellore Institute of Technology, Vellore, Tamil Nadu, India*
YOGESH GAT • *Department of Food Engineering and Technology, Institute of Chemical Technology, Mumbai, India*
PRAGATI GODARA • *Department of Nutrition & Dietetics, Chandigarh School of Business, Mohali, Punjab, India*
ESTHER GÓMEZ-MEJÍA • *Department of Analytical Chemistry, Faculty of Chemistry, Complutense University of Madrid, Madrid, Spain*
PRIXIT GULERIA • *Department of Food Science and Nutrition, Kanya Mahavidyalaya Kharkhoda, Sonipat, Haryana, India*
ADITI JAIN • *Institute of Food Technology, Bundelkhand University, Jhansi, Uttar Pradesh, India*
ARYA KADAM • *Department of Food Engineering and Technology, Institute of Chemical Technology, Mumbai, India*
SHIVA KANT • *Department of Biology and Environmental Science, Auburn University at Montgomery, Montgomery, AL, USA*
PIYUSH KASHYAP • *Department of Food Technology and Nutrition, School of Agriculture, Lovely Professional University, Phagwara, Punjab, India*
AMARJEET KAUR • *Department of Food and Nutrition, Punjab Agricultural University, Ludhiana, Punjab, India*
SIMARDEEP KAUR • *Division of Crop Science, ICAR-Research Complex for North Eastern Hill Region, Umiam, Meghalaya, India*
BABABODE KEHINDE • *Department of Biosystems and Agricultural Engineering, University of Kentucky, Lexington, KY, USA*
KHUSHI KESHARI • *Department of Biotechnology, Sharda School of Bioscience & Technology (SSBT), Sharda University, Greater Noida, Uttar Pradesh, India*
KIRTAN KOTAK • *Department of Food Engineering and Technology, Institute of Chemical Technology, Mumbai, India*
AJAY KUMAR • *Department of Food Science and Technology, Dr. Yashwant Singh Parmar, University of Horticulture and Forestry, Nauni, Solan, Himachal Pradesh, India*
ASHWANI KUMAR • *Institute of Food Technology, Bundelkhand University, Jhansi, Uttar Pradesh, India*

MUKUL KUMAR • *Department of Food Technology and Nutrition, Lovely Professional University, Phagwara, Punjab, India*
SATISH KUMAR • *Department of Food Science and Technology, Dr. Yashwant Singh Parmar, University of Horticulture and Forestry, Nauni, Solan, Himachal Pradesh, India*
MANOHAR LAL • *Chandigarh Group of Colleges Jhanjeri, Mohali, Punjab, India*
AMIT LATHER • *Geeta Institute of Pharmacy, Geeta University, Panipat, Haryana, India*
MARÍA EUGENIA LEÓN-GONZÁLEZ • *Department of Analytical Chemistry, Faculty of Chemistry, Complutense University of Madrid, Madrid, Spain*
MEGHANA MASKE • *Department of Food Science and Technology, University of Wisconsin-Stout, Menomonie, WI, USA*
BHAVNA NEGI • *Department of Food Engineering and Technology, Institute of Chemical Technology, Mumbai, India*
OROOFA • *Division of Food Science and Technology, Sher-e-Kashmir University of Agricultural Sciences and Technology, Srinagar, Jammu and Kashmir, India*
KULDEEP PANDEY • *Department of Fruit Science, Acharya Narendra Deva University of Agriculture and Technology, Ayodhya, Uttar Pradesh, India*
PRAGYA PANDEY • *Department of Food and Nutrition, Acharya Narendra Deva University of Agriculture and Technology (ANDUAT), Ayodhya, Uttar Pradesh, India; Department of Food Science and Nutrition, Acharya Narendra Deva University of Agriculture & Technology (NDUAT), Ayodhya, Uttar Pradesh, India*
SHALINI PANDEY • *Dr. B.R. Choudhary Agricultural Research Station, Mandor, Agriculture University, Jodhpur, India*
SUBHASH CHANDRA PANI • *Centre for Bio Separation Technology, Vellore Institute of Technology, Vellore, Tamil Nadu, India*
ANIKA PANWAR • *Department of Food Science and Technology, Dr. Yashwant Singh Parmar, University of Horticulture and Forestry, Nauni, Solan, Himachal Pradesh, India*
AASIMA RAFIQ • *Krishi Vigyan Kendra Srinagar, Sher-e-Kashmir University of Agricultural Sciences and Technology, Srinagar, Jammu and Kashmir, India*
PRATIBHA RATHEE • *Department of Pharmacy, Maharishi Markandeshwar (Deemed to be University), Mullana, Haryana, India*
SANDRA RODRÍGUEZ-BLÁZQUEZ • *Department of Analytical Chemistry, Faculty of Chemistry, Complutense University of Madrid, Madrid, Spain*
NOELIA ROSALES-CONRADO • *Department of Analytical Chemistry, Faculty of Chemistry, Complutense University of Madrid, Madrid, Spain*
ADITI SACHDEVA • *Department of Biotechnology, Sharda School of Bioscience & Technology (SSBT), Sharda University, Greater Noida, Uttar Pradesh, India*
PRASHANT SAHNI • *College of Dairy & Food Technology, Agriculture University, Jodhpur, Rajasthan, India*
NEHA SHARMA • *School of Life Sciences, Devi Ahilya University, Indore, Madhya Pradesh, India*
POORVA SHARMA • *Department of Food and Bioproduct Sciences, University of Saskatchewan, Saskatoon, SK, Canada*
VISHANT SHARMA • *Department of Food Science and Technology, Dr. Yashwant Singh Parmar, University of Horticulture and Forestry, Nauni, Solan, Himachal Pradesh, India*
PRERANA SHERE • *MIT School of Food Technology, MIT Art, Design & Technology University, Pune, India*
SHABNUM SHOWKAT • *Division of Food Science and Technology, Sher-e-Kashmir University of Agricultural Sciences and Technology, Srinagar, Jammu and Kashmir, India*

ANAND KRISHNA SINGH • *Department of Life Science/Biotechnology, Shri Vaishnav Vidyapeeth Vishwavidyalaya, Indore, Madhya Pradesh, India*

JAGVEER SINGH • *Department of Fruit Science, Acharya Narendra Deva University of Agriculture and Technology, Ayodhya, Uttar Pradesh, India*

VIGNESHWARAN SUBRAMANI • *Department of Horticulture and Food science, VIT School of Agricultural Innovations and Advanced Learning, Vellore Institute of Technology, Vellore, Tamil Nadu, India*

PRIYANKA SUTHAR • *Department of Food Science and Technology, Dr. Yashwant Singh Parmar, University of Horticulture and Forestry, Nauni, Solan, Himachal Pradesh, India*

ABHAY TIWARI • *Department of Biotechnology, Sharda School of Bioscience & Technology (SSBT), Sharda University, Greater Noida, Uttar Pradesh, India*

VIDISHA TOMER • *Department of Horticulture and Food Science, VIT School of Agricultural Innovations and Advanced Learning, Vellore Institute of Technology, Vellore, Tamil Nadu, India*

ANKUR TRIPATHI • *College of Dairy & Food Technology, Agriculture University, Jodhpur, Rajasthan, India*

PURNIMA TRIPATHI • *Department of Life Science/Biotechnology, Shri Vaishnav Vidyapeeth Vishwavidyalaya, Indore, Madhya Pradesh, India*

PRAVEEN YADAV • *Department of Biotechnology, Sharda School of Bioscience & Technology (SSBT), Sharda University, Greater Noida, Uttar Pradesh, India*

Chapter 1

Extraction and Quantification of Polyphenols

Sandra Rodríguez-Blázquez, Esther Gómez-Mejía, Noelia Rosales-Conrado, and María Eugenia León-González

Abstract

Polyphenols are one of the main classes of secondary metabolites derived from plants offering several health benefits resulting in their use as functional ingredients. Their structural diversity makes them very different from one another in terms of their chemical and physical properties, which makes analysis challenging. The main sources of phenolic compounds are fruits, vegetables, nuts, and processed products, to name but a few. However, attention is now turning to alternative sources of phenolic compounds, namely bio-residues and by-products generated by the agri-food industry. To this end, it is essential to design analytical strategies, thereby allowing the extraction and characterization of phenolic compounds recovered from agri-food residues. Therefore, this chapter aims to give a broad view of the developed methods for the extraction and characterization of polyphenols, including the different workflows and steps involving in the analysis of polyphenols from food stuff and bio-residues.

Key words Polyphenols, Sources of polyphenols, Agri-food by-products, Conventional extraction methods, Non-conventional extraction methods, Experimental design, Response surface methodology, Chromatographic methods, Spectrophotometric methods

1 Introduction

Polyphenols, also known as phenolic compounds, are a group of natural compounds with phenolic structural features. These compounds originate in plants, which synthesize them in large quantities as a product of their secondary metabolism [1]. Some are essential for plant physiological functions, others are involved in abiotic and biotic stress defense functions, that is, salinity, heavy metal, water scarcity, temperature, UV lights, and disease progression [2]. In addition, polyphenols are widely acknowledged for contributing not only to the natural color of plants (blue, purple, red, and yellow pigments), but also to the organoleptic characteristics of fruits and vegetables [3]. Furthermore, phenolic compounds exhibit excellent biological properties, which include anticancer, antioxidant, anti-inflammatory, anti-aging,

Ashwani Kumar, Vidisha Tomer and Piyoosh Kumar Babele (eds.), *Phytochemicals: Extraction, Quantification, Isolation, and Characterization*, Methods and Protocols in Food Science, https://doi.org/10.1007/978-1-0716-4840-7_1,

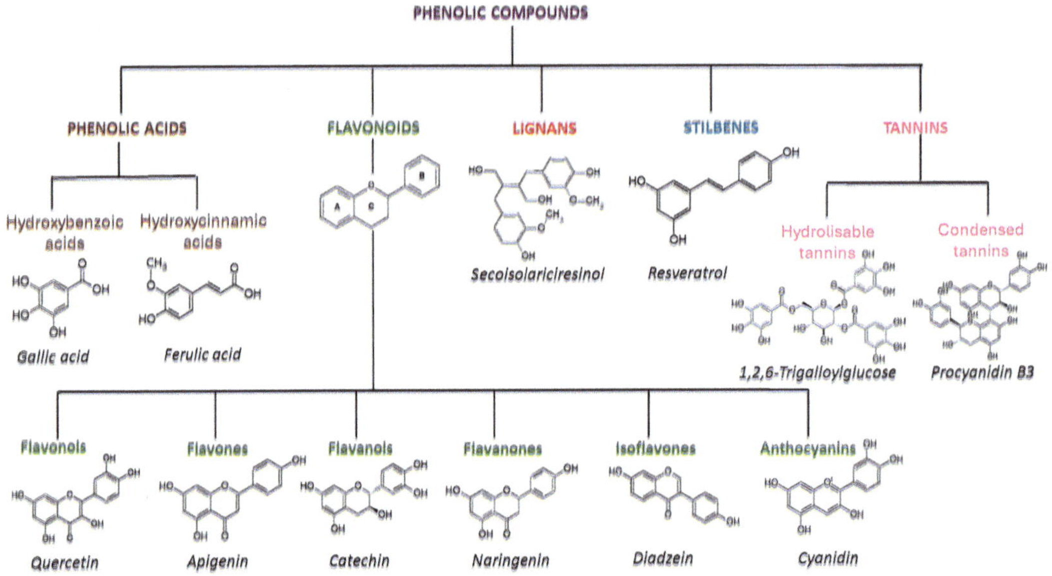

Fig. 1 Main classification of phenolic compounds. The name in italics denotes an example of a phenolic compound of each family group

antihypertensive and antimicrobial effects. These bioactive compounds are even an important source of anti-infective elements against antibiotic-resistant human pathogens [3, 4].

At present, 8000 different types of phenolic compounds have been identified in almost all edible and medicinal plants [4]. Polyphenols can be classified in different ways including by their origin, biological function, and chemical structure. This last one is the most adopted classification. Regarding their chemical structure, polyphenols are a well-known group of phenolic systems characterized by at least two phenyl rings and one or more hydroxyl substituents. This description covers a large number of heterogeneous compounds due to their complexity. Therefore, phenolic compounds, as shown in Fig. 1, can be subcategorized into five major groups: phenolics acids (hydroxybenzoic and hydroxycinnamic acids); flavonoids (flavones, flavonols, isoflavones, flavanones, anthocyanins); stilbenes (resveratrol, piceatannol); lignans (sesamol, pinoresinol, sinol, enterodiol); and tannins (hydrolyzable, non-hydrolyzable, and condensed tannins) [3, 5].

Phenolic acids can be further classified as hydroxybenzoic acids (represented by gallic acid, protocatechuic acid, etc.) and hydroxycinnamic acids (represented by chlorogenic acid, ferulic acid, caffeic acid, etc.) based on the presence of a carboxylic acid group linked to the phenolic ring. Hydroxybenzoic acid, derived from benzoic acid, has a C6-C1 parent nucleus, whereas hydroxycinnamic acid, derived from cinnamic acid, occurs in plants as simple esters with glucose or quinic acid. Taken together, the specific structure of phenolic acids can cause H-atom donation due to their phenolic moiety and stabilized resonance structure and thus exhibit antioxidant activity through the mechanism of radical scavenging, electron donation, or simple oxygen quenching [6]. On the other hand, as shown in Fig. 1, flavonoids are composed of two phenolic rings (rings A and C) connected by an oxygenated heterocycle containing three carbon atoms (ring B). These are the most abundant class of phenolic compounds, which consists of a series of phenolic compounds with low molecular weight polyphenolic structures (including flavones, isoflavones, flavanones, flavonols, and anthocyanins). To date, almost 6000 flavonoids have been identified [7, 8].

In the meantime, one or more phenylpropane units create stilbenes and lignans, respectively [9]. As to lignans, they are a class of natural compounds produced by the polymerization of two phenylpropanoid derivatives (namely C6-C3 monomers) that show enormous structural diversity [10]. Stilbenes are a type of phenylpropanoids characterized by a 1,2-diphenylethylene core containing two phenolic rings linked by a methylene bridge [11]. Lastly, tannins are high molecules with a molecular weight range from 500 Da to 30,000 Da, which constitute the third important group of phenolics and may be subdivided into hydrolyzable tannins (500–3000 Da) that are glycoside esters of gallic and ellagic acid, and condensed tannins (1–30 kDa), also known as proanthocyanidins (polymeric flavonoids) [12, 13].

Polyphenols are widely distributed in plants, such as vegetables, fruits, nuts, cereals, tobacco, and manufacturer products (e.g., tea, olive oil, beer, chocolate, wine and others) [1]. However, in recent years, food biowaste has gained increasing recognition as a source of phenolic compounds [14]. Horticultural wastes, which designate all those parts of fruits and vegetables discarded during the processing, delivery, and consumption phases of the food supply chain, have been recognized as important sources [15, 16]. In addition, other non-horticultural agri-food wastes derived from food manufacturing, such as yeasts used in beer and sparkling wine production, are known to present significant amounts of phenolic compounds [17].

Currently, the recovery of phenolic compounds from agri-food waste has become the preferred alternative to overcome certain economic and technical limitations in relation to the chemical and biogenic production of polyphenols [18]. This approach to food waste valorization enables the recovery of high value-added compounds from underutilized raw materials, minimizing their environmental impact and contributing to the efficient management of natural resources, which is a priority of the United Nations [19].

For this purpose, it is necessary to design analytical strategies, thereby allowing the extraction and characterization of phenolic compounds recovered from agri-food by-products [20, 21]. Unfortunately, there is neither a single nor a standard method, and the

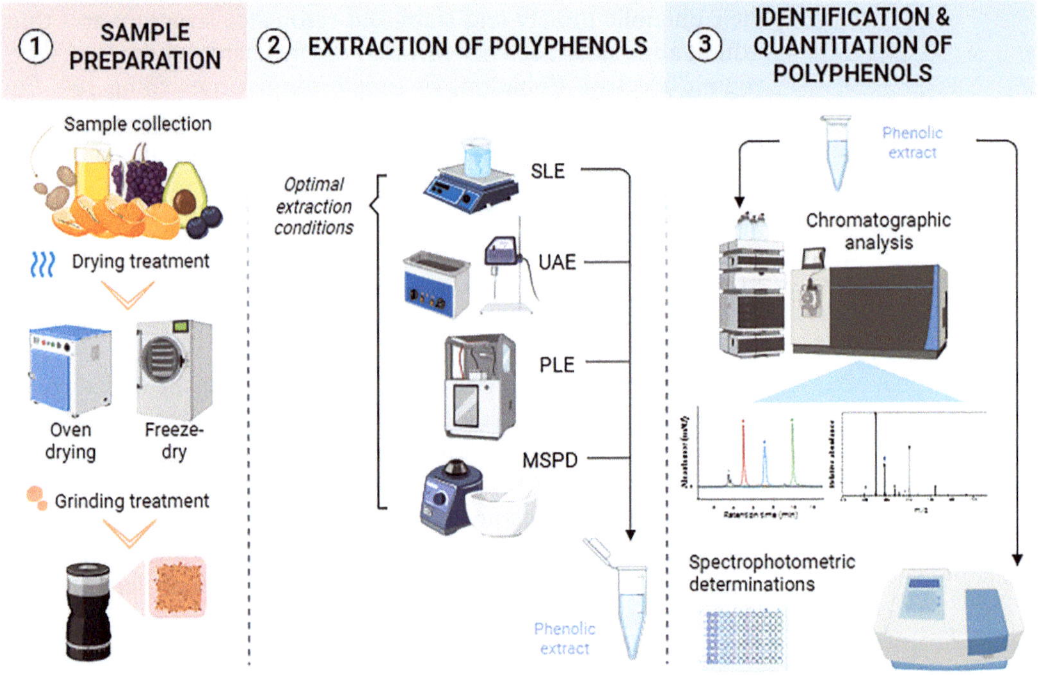

Fig. 2 Experimental flow chart for the analysis of polyphenols from food and agri-food residues

general outline to be followed for the extraction and analysis of phenolic compounds in samples from agri-food waste is as shown in Fig. 2.

Extraction is the most critical step toward obtaining phenolic compounds from bio-residues, seeking to combine the maximum recovery with the highest selectivity for the compounds of interest. The extraction of phenolic compounds has been performed by means of different techniques that can be categorized as conventional (e.g., maceration, solid-liquid extraction…) or non-conventional (namely, ultrasound assisted extraction, pressurized liquid extraction, matrix solid-phase dispersion extraction…) [20, 22, 23] (Fig. 2). Furthermore, recent methods increasingly use green solvents such as water or ethanol-water mixtures. SLE is a conventional technique based on the migration of analytes contained in a solid matrix into a contact solvent [24], usually accompanied by heating and stirring, either magnetic or mechanical, to further improve extraction speed and yield [25]. As such, temperature, solvent, sample-to-solvent ratio, and extraction time are some of the factors that must be considered when optimizing the performance of SLE extraction [5, 24]. Table 1 summarizes selected experimental conditions employed for the extraction of phenolic compounds by SLE. Overall, SLE is among the first extraction techniques chosen due to its ease of management, low cost, user-

Table 1
Extraction methods applied for the determination of phenolic compounds in agri-food matrices

Extraction technique	Sample amount	Solvent (volume)	Temperature	Time	Other factors	Extract processing	References
SLE	50 mg–1.0 g	Water MeOH-water Ethanol-water (1–100 mL)	25–98 °C	5 min–2 h	Stirring at 2–150 rpm	Centrifugation Filtration Freeze-drying SPE cleaning	[25–33]
UAE	0.50–5.0 g	EtOH Acetone-water MeOH-water EtOH-water At ratios from 80:20 to 50:50 (v/v) Choline chloride-lactic acid at 1:2 molar ratio (15–62.5 mL)	25–70 °C	15–40 min	56–800 W 20–25 kHz 30–50% amplitude	Centrifugation Filtration SPE cleaning	[33–40]
PLE	1.0–6.0 g	Water EtOH-Water 50:50 (v/v)	60–200 °C	5–30 min	580– 1500 psi 1–3 cycles 2–5 g dispersing agents (diatomaceous earth, sand, glass beads)	Centrifugation Filtration	[41–45]
MSPD	0.1–1.0 g	MeOH EtOH-water MeOH-water At ratio 80:20 or 70:30 (v/v) (0.5–15 mL)	–	5–20 min	0.5–3.0 g of solid support (C18, sea sand, diatomaceous earth, TiO_2 nanoparticles)	Centrifugation Filtration Solvent evaporation	[46–53]

friendly application on an industrial scale, and its efficiency, which are comparable to other non-conventional techniques [24, 41]. Hence, it has been widely exploited in tea leaves [27, 28], viticultural wastes (wine lees, grape pomace, seeds, etc.) [29, 30], spent coffee grounds [31, 54], citrus peels [25, 55], and brewer's spent yeast [32, 46], to name but a few.

However, SLE-based methodologies do present some drawbacks, including high solvent consumption (up to 100 mL) and long extraction times (up to 2 h), together with poor selectivity [24, 56]. To overcome the limitations of the SLE, non-conventional techniques like UAE and PLE are used.

UAE has emerged as an advanced and sustainable alternative technique for obtaining polyphenols from a variety of agri-food matrices, including bean coats [34], spent coffee grounds [35], citrus peels [36, 37], raspberries [38] and avocado peels [39]. It is grounded on the cavitation phenomenon caused by the propagation of ultrasound waves in the liquid medium, where small gas bubbles gradually grow until collapsing, releasing a large amount of energy, and triggering significant changes in pressure and temperature. This not only promotes the rupture of matrix tissues, but also enhances mass and heat transfer phenomena and, so, the extraction efficiency [56, 57]. UAE can be performed with either an ultrasound bath or a focused probe, although the latter is rather more versatile [56]. Apart from the nature of the solvent, time, solvent-to-solid ratio, pH and temperature, operational factors, namely ultrasound frequency, amplitude and power, play a key role in the effectiveness of UAE, given that the latter may induce changes in the native phenolic forms. Accordingly, definition of the optimal experimental conditions is paramount to maximize the recovery of polyphenols [58]. Compared to conventional SLE methods, generally, less solvent, temperature, and extraction time are required (Table 1), yet the use of UAE carries some associated shortcomings concerning the oxidation of the sample by the formation of free radicals and the critical demand for experimental optimization for each sample [24, 56, 58].

PLE is an advanced and non-conventional technique involving the use of liquid solvents at high temperature and pressure, so that the extraction solvent remains as a subcritical fluid. Thereby, deeper penetration of the solvent into the sample matrix and higher solubility of the target compounds occur, which promotes mass transfer and extraction yields [42, 56, 57]. In terms of working conditions, PLE is carried out in cycles, either with the extraction solvent pumped continuously through the sample vessels (dynamic configuration) or with several extraction cycles in which the solvent is replaced between cycles (static configuration) [42, 56]. The efficiency of PLE depends on both kinetic and thermodynamic parameters, including flow rate, pressure, temperature, and time, although the physical state of the sample and the solvent/sample ratio must also be evaluated when optimizing the process [42, 43]. This non-conventional extraction technology has great potential given the low solvent consumption and the widespread use of water, ethanol, or their mixtures (Table 1), as has been demonstrated in the recovery of phenolic compounds from olive pomace [41], wine lees [41], sweet cherry pomace [44], and avocado peel [45].

Despite claims that non-conventional techniques are more environmentally friendly and cost-effective than conventional techniques, the high temperatures and specific equipment requirements compromise the viability and profitability of the process, thus demanding a cost-benefit analysis for industrial scale-up [57]. At the same time, phenolic extracts are often centrifuged, filtered, lyophilized, and even purified by solid phase extraction (SPE) (Table 1), which may lead to alteration or loss of native phenolic compounds, as well as prolongation of extraction procedures. Thus, recent trends in analytical extraction methods are moving toward more environmentally friendly approaches, where extraction and purification of the extract are performed in a single step, increasing sensitivity and selectivity, while shortening protocols [59, 60].

Along these lines, matrix solid-phase dispersion extraction (MSPD) has been proposed as a simple, fast, efficient and versatile technique, which has aroused attention for the isolation of compounds susceptible to degradation [60, 61], such as phenolic compounds. As such it has been applied for the recovery of polyphenols from winery residues [47, 48, 50], oregano [51], pomegranate molasses [52], cherry seeds [49], plum seeds [53] and brewer's spent yeast [46]. The MSPD technique is based on grinding the sample with an abrasive and usually inert material also known as the dispersant (e.g., C18, sand, diatomaceous earth, silica), producing a semi-dry, homogeneous, and easy to handle material (the mixture). Additionally, the disruptive dispersant can be a functional material, the so-called sorbent or adjuvant, which can interact with the target analytes and prompt the selectivity and efficiency of this technique [60, 61]. Selecting a suitable solid support, sample-to-solid support ratio, solvent, and elution volume will directly determine the efficiency and selectivity of the MSPD method and should therefore be evaluated when optimizing the extraction conditions [60].

Particularly, chemometric tools have been pivotal for the development of extraction methods aimed at the recovery of polyphenols from agri-food waste, since allowing the identification of the main experimental factors that allow obtaining either high extraction yields or selective extractions of compounds of interest [58, 62–64]. Although in many cases univariate optimization has been used, the main problem of this approach is the high number of experiments that need to be carried out to know the interaction that may exist between the different variables (experimental factors), which can affect the extraction process. In order to minimize the number of experiments and obtain the maximum possible information, different designs of experiments have been used and the selection of the design will depend on the number of responses to be optimized (extraction yield, bioactive properties, specific polyphenols), the number of variables (experimental factors) to be studied, the relationship between the response to be assessed and the factor (linear, quadratic) and the possible interactions between the factors that may influence the extraction process [58, 65–67]. Most statistical designs (factorial designs, Box-Behnken designs, central composite designs, Taguchi designs, among others) provide polynomial equations that allow the calculation of optimal values for the evaluated responses and also allow the representation of the response or responses studied according to the factors under study, using the response surface methodology [46–49, 54, 55, 62–64, 68, 69]. The goodness of the polynomial fit method can be assessed using the coefficient of correlation (R^2). In some cases, it is necessary to optimize multiple responses, and, in these cases, the so-called desirability function is used to obtain the best possible conditions for all the answers evaluated [46–49, 55, 62, 63, 66–68].

As for analyzing polyphenols, it poses challenges due to the extensive variety of compounds that may be present in samples and their structural complexity. Moreover, this complexity is exacerbated by the frequent occurrence of these compounds in low concentrations [70]. As a result, the analysis of polyphenols necessitates the use of sensitive, selective, and precise analytical techniques capable of simultaneously identifying and quantifying these compounds in complex samples. Several analytical methods, mainly spectrophotometric and chromatographic, have been developed and commonly used to determine phenolic compounds in a wide range of samples [5].

Spectrophotometric methods provide versatile, simple, and low-cost approaches, with minimum analytical instrumentation requirements, for characterization of polyphenols. Among the range of spectrophotometric techniques available, UV-Vis spectroscopy is distinguished as a favored option due to its simplicity and sensitivity. It enables swift screening and accurate quantification of total polyphenol content [5, 71]. The Folin-Ciocalteu colorimetric method is frequently employed for the determination of Total Phenolic Content (TPC). This spectrophotometric assay is based on the redox reaction in alkaline conditions between the functional hydroxy groups of phenolic compounds and the Folin-Ciocalteu reagent, a mixture of phosphotungstic acid ($H_3PW_{12}O_{40}$) and phosphomolybdic acid ($H_3PMo_{12}O_{40}$). This reaction results in the formation of a blue chromophore, whose absorbance is measured at a wavelength of around 760 nm [71–73].

Total flavonoids, proanthocyanidin (condensed tannin), and hydrolyzable tannin can also be estimated by colorimetric methods [74]. For flavonols and flavones, various spectrophotometric assays are available; however, the most used is the aluminum chloride method. The Total Flavonoid Content (TFC) is determined by a sequence of reactions with $AlCl_3$ in alkaline medium and in the presence of $NaNO_2$, forming a pink-colored flavonoid-aluminum complex that can be quantified at specific wavelengths [5, 16, 75]. Anthocyanins, another main subclass of polyphenols, can be also quantified by spectrophotometric assays using the differential pH method. The quantification of Total Monomeric Anthocyanins (TMA) relies on the pH-dependent structural changes of anthocyanins: the red flavylium cation predominates under highly acidic conditions (pH <1), while the colorless hemiketal form is prevalent at pH 4.5. Therefore, by measuring the absorbance at the maximum wavelength within the red-visible spectrum in two solutions of anthocyanin samples, one dissolved in KCl solution (pH 1) and the other in sodium acetate buffer (pH 4.5), it is possible to determine the TMA [5, 71, 74].

These spectrophotometric techniques provide practical and effective approaches for estimating total content, offering valuable insights into a quantitative comparison of levels of polyphenols, flavonoids, and anthocyanins, while also serving as a convenient and cost-effective method for swift screening of multiple samples. Nevertheless, these methods do not distinguish and measure the compounds individually. Furthermore, they inherently carry a margin of error due to their limited specificity and are susceptible to matrix effects arising from variations in sample solubility, redox potential, and pH disparities [76]. In fact, other substances such as ascorbic acid, aromatic amines, reducing sugars, SO_2, tyrosine [74, 77], vitamins, amino acids, proteins, organic acids [78] and the presence of Fe(III) interfere with the results of the spectrophotometric assays [73]. For these various reasons, spectrophotometric methods serve primarily for rough approximations of phenolic compounds and should be supplemented with more selective and resilient techniques, such as chromatographic methods [5, 79].

Chromatographic methods have emerged as essential instruments for thoroughly characterizing polyphenols, offering robust tools for researchers to separate, quantify, and identify these compounds across a range of biological and food samples. They afford exceptional precision and specificity in both separation and identification processes. The selection of a chromatographic technique hinges on the particular polyphenols under scrutiny and the analytical objectives of the research. Among the array of chromatographic methods available, gas chromatography (GC) and liquid chromatography (LC), coupled with various detectors, have been widely adopted for discerning phenolic compounds in agricultural and food matrices [80]. Nevertheless, the polar and heat-sensitive characteristics of these analytes hinder direct analysis by GC, necessitating derivatization (typically through silylation) to reduce their polarity [81, 82]. Therefore, HPLC stands out as the leading separation technique for isolating polyphenols, thanks to its impressive resolution capabilities and remarkable versatility [5, 80, 83]. HPLC operates in various modes, including reversed phase (RP), normal phase, hydrophilic interaction liquid chromatography (HILIC), or ion exchange chromatography (IEC), among others. These diverse modes offer numerous analytical options for the separation of phenolic compounds from agri-food extracts [80]. However, RP-HPLC is commonly utilized to effectively separate phenolic compounds using non-polar stationary phases, typically silica particles modified with long-chain hydrocarbons (such as C8 and C18). Binary solvent systems, comprising acidified aqueous solutions and organic solvents such as methanol and acetonitrile, are frequently employed as mobile phases. Additionally, additives such as formic acid, acetic acid, or trifluoroacetic acid are often incorporated into the mobile phase to regulate both pH and ionic strength. Moreover, chromatographic separation of phenolic extracts typically involves gradient elution mode, although prior optimization of both composition and elution gradient is often necessary due to the complexity and diversity of phenolic compounds present in agri-food extracts [80, 83, 84].

Regarding the detection of polyphenols, several detectors have been employed combined with chromatographic methods for identifying and quantifying these compounds. UV-Vis or diode array detection (DAD), fluorescence, and mass spectrometry are extensively utilized for characterizing phenolic compounds in agri-food matrices [80, 85]. To enhance the identification and structural elucidation of polyphenol extracts, mass spectrometry (MS) is frequently coupled, enabling them to detect and characterize these compounds based on their mass-to-charge ratios. The combination of HPLC with MS allows the simultaneous screening, confirmation, and quantification of multiple phenolic compounds, especially through tandem mass spectrometry (MS/MS) [80, 83]. LC-MS/MS has the capability to improve noise reduction and sensitivity by employing multiple reaction monitoring (MRM) scan mode. Additionally, high resolution mass spectrometry (LC-HRMS) plays a pivotal role for the characterization and identification of new polyphenols [70].

Apart from chromatographic and spectrophotometric methods, several non-chromatographic and non-spectrophotometric methods contribute to the comprehensive characterization of polyphenols. These alternative approaches offer diverse insights into the structural and functional aspects of polyphenolic compounds. Thus, electroanalytical methods, particularly voltametric techniques, can be utilized to characterize polyphenols by examining their redox properties [86]. Furthermore, non-destructive infrared spectroscopy and electrochemical sensors analysis have been also applied for polyphenol determination [73, 87].

On the other hand, capillary electrophoresis (CE) can be paired with diverse detection methods, including UV absorbance, fluorescence, or MS, offering excellent resolution and effective separation of polyphenolic compounds based on their electrophoretic mobility [76, 88]. Additionally, in cases where mass spectral data are insufficient for determining a polyphenol definitive structure, NMR spectrometry emerges as a powerful complementary method for structural assignment. Therefore, in many instances, the combination of UV, MS, and ^{1}H-NMR is adequate for obtaining information necessary for structural elucidation [76].

Hence, considering all the above, the objective of this chapter is to show an overview of the most commonly used extraction and analysis methods for the characterization of polyphenols in agri-food matrices.

2 Materials

2.1 Raw Materials

Different fruits, vegetables [5], medicinal plants, beverages such as tea [87], wine [89], fruit juices and ciders [77], and agri-food residues, including citrus peels [55], brewing yeast [26], grape pomace [47], and/or seeds from fruits [49, 90] can be grinded and/or dried prior to the extraction and characterization protocols.

2.2 Instrumentation

1. Gas chromatography coupled with flame ionization detector (GC-FID).
2. Gas chromatography coupled with both flame ionization detector and mass spectrometry (GC-FID-MS).
3. High performance liquid chromatography coupled with UV-Vis (HPLC-UV-Vis).
4. High performance liquid chromatography coupled with mass spectrometry (HPLC-MS).
5. Fused silica capillary column (30 m × 0.25 mm I.D × 0.25 μm film thickness).
6. Reversed-phase C18 column.
7. UV-Vis spectrophotometer.
8. Diode array spectrophotometer.

2.3 Apparatus

1. Accelerated solvent extractor equipped with stainless steel extraction cells.
2. Analytical balance.
3. Centrifuge (≥15,000 rpm).

4. Freeze dryer.
5. Magnetic stirring hot plate.
6. Magnetic stirrer.
7. Solid-phase extraction cartridge.
8. Ultrasonic processor.
9. Nylon filters (0.45 μm).
10. Ceramic mortar.

2.4 Reagents

1. Distilled water.
2. Ethanol.
3. Glass beads.
4. Glass bottle with lid.
5. Methanol.
6. 2 N Folin-Ciocalteu's reagent (This chemical is supplied in 2 N).
7. Sodium carbonate (Na_2CO_3) (7.5%) (*see* **Note 1**).
8. Gallic acid (Standard) (*see* **Note 2**).
9. Sodium nitrite ($NaNO_2$) (5%) (*see* **Note 3**).
10. Aluminiumchloride 6-hydrate ($AlCl_3.6H_2O$) (10%) (*see* **Note 4**).
11. Sodium hydroxide (NaOH) (1 M) (*see* **Note 5**).
12. Quercetin (Standard) (*see* **Note 6**).
13. 0.025 M Potassium chloride buffer, pH 1 (*see* **Note** 7).
14. 0.025 M sodium acetate buffer, pH 4.5 (*see* **Note 8**).
15. Chlorohydric acid (To adjust the pH of buffer solutions).
16. Acetonitrile for HPLC gradient quality.
17. Acetonitrile for MS quality.
18. Methanol for HPLC gradient quality.
19. Methanol for MS quality.
20. Formic acid.
21. Acetic acid.
22. Trifluoroacetic acid.

2.5 Others

1. Beakers.
2. Diatomaceous earth.
3. Sand.
4. Spatula.

3 Methods

3.1 Extraction Procedure of Polyphenols

3.1.1 Solid-Liquid Extraction

1. Place a vessel/beaker on a heating and/or stirring plate.
2. Add necessary volume of solvent (typically hydroalcoholic mixtures with methanol or ethanol, in majority percentages of alcohol, for example, 70% and 80% (v), acidified or not) (*see* **Note 9**).
3. When the solvent achieves the desired temperature (e.g., 25–98 °C), the agri-food sample is added and kept under stirring.
4. Once the extraction time is over (e.g., 5 min–2 h), separate the phenolic extract from the solid residue by centrifugation (e.g., 15 min at 15000 rpm) or filtration using a paper and nylon membrane of 0.45 μm, and place in an opaque glass container at 4 °C until analyzed (*see* **Note 10**).

3.1.2 Ultrasound-Assisted Extraction

1. Determine the optimal experimental conditions for the UAE by conducting preliminary studies (Table 1).
2. Once the optimal conditions are defined, place the sample and solvent in the optimal ratio (*see* Table 1) in a container, for example, beaker or plastic tube, and start the extraction by putting in an ultrasonic bath or by inserting an ultrasonic probe. For the latter, it is important to ensure that the titanium tip does not touch any walls, nor the sample, to guarantee a proper operation.
3. The power, frequency, and amplitude of the ultrasound should then be programmed as needed, and the process should be maintained at the selected temperature.
4. Separate the phenolic extract by centrifugation (≥15,000 rpm) and/or filtration using Nylon-filters (0.45 μm) as indicated in Table 1.
5. Store the extract in hermetically sealed opaque vials at −20 °C until analysis.

3.1.3 Pressurized Liquid Extraction

1. Rinse the extraction cell with extraction solvent (e.g., water or ethanol-water mixture 50:50 (*v*/*v*)) to avoid contamination.
2. Program the cells to preheat to the desired temperature (Table 1).
3. Mix the sample with dispersing agent (sand, diatomaceous earth, glass beads, etc.) (e.g., sample-dispersant ratio of 1:1 or 1:2 (*w*/*w*)) and place in a stainless-steel extraction cell having a filter at the bottom.

4. Next, select pressure, wash volume, time, and number of cycles [41–45]. To prevent polyphenolic oxidation, the cells are usually purged with nitrogen for a short period of time, for example, 60 s, during extraction.
5. Filter with Nylon-filter (0.45 μm) or centrifuge the extract (≥15,000 rpm).
6. Finally, collect the phenolic extracts in glass vials.
7. Measure the volume and store under refrigerated conditions.

3.1.4 Matrix Solid Phase Dispersion Extraction

1. Grind the sample for 1–5 min using pestle and mortar.
2. Transfer it to a cartridge provided with a frit.
3. Compress the sample and elute the analytes with a suitable solvent (e.g., methanol, ethanol-water mixture or methanol-water mixture at 80:20 or 70:30 (*v/v*)) [61, 91]. Alternatively, ultrasound- or vortex-assisted stirring of the blend-solvent mixture followed by centrifugation (4000–11,000 rpm) can be used as a time-saving approach to the conventional SPE elution step [60, 92].
4. Finally, the eluate obtained can be directly analyzed, otherwise concentrated on the subsequent analytical procedure (Table 1).

3.2 Optimization of Extraction Conditions

Optimization methods can be carried out in two stages, when the number of factors to be optimized is high, so that in a first stage an experimental design can be chosen that, with a minimum number of experiments, provides information on the most significant variables about the evaluated response, and a second design in which more levels of the variables with greater importance are considered [62]. When the number of variables to be evaluated is not very high, a single design is usually used, Table 2 indicates the designs used for the optimization of polyphenol extraction in different samples, as well as the factors and responses studied.

The procedure to be followed would be:

1. Select the independent factors that can affect the extraction and the different levels at which they are going to be evaluated, usually two or three levels.
2. Select the responses that allow the evaluation of the effectiveness of the extraction.
3. Select the experimental design.
4. Conduct the experiments and record the results.
5. Obtain an equation that relates the answer or answers to the variables to be optimized.
6. In the case of using response surface analysis, represent the surface or surfaces that allow to visualize the representation in three dimensions of the variation of the responses as a function of two factors.

Table 2
Experimental design used for extraction of polyphenols from plant sources

Source	Compounds	Extraction method	Design of experiments (DOE)	Responses	Factors	Number of runs	References
Purple corn pericarp	Phenolic profiles	Maceration	Plackett-Burman design (PBD)	TAC TPC CT	Time (h), ethanol concentration (*v/v*), stirring speed (rpm) and acetic acid concentration (*v/v*)	13	[62]
			Box-Behnken design (BBD)		Time (h), ethanol concentration (*v/v*) and acetic acid concentration (*v/v*)	15	
			Facial central composite design (FCCD)		Time (h), ethanol concentration (*v/v*) and acetic acid concentration (*v/v*)	18	
Olive leaves	Phenolic profiles	Ultrasound-assisted extraction	BBD	TPC TFC Antioxidant activity (DPPH, ABTS, FRAP)	Extraction time (min, sample to solvent ratio ($g{\cdot}mL^{-1}$), and ethanol Concentration (*v/v*)	15	[68]
Corn cob (*Zea mays* L.)	Phenolic profiles	Solvent extraction	Full factorial design (FFD) with three replicates in the center	TMC	Solid to solvent ratio ($g{\cdot}mL^{-1}$), temperature, percentage of ethanol in the hydroalcoholic mixtures and extraction time	16 + 3	[64]
		MAE	BBD		Natural eutectic solvents (NADES) Percentage of ethanol in the hydroalcoholic mixtures and extraction time	15	

(continued)

Table 2 (continued)

Source	Compounds	Extraction method	Design of experiments (DOE)	Responses	Factors	Number of runs	References
Spent Coffee grounds	Gallic acid 3,4-Dihydroxybenzoic acid Chlorogenic acid *p*-Coumaric acid *Trans*-Ferulic acid Rutin Naringin Quercetin Kaempferol	Solid-liquid extraction	Factorial design (two levels) with three replicates at the center point	Amounts of polyphenols evaluated	Ethanol-water proportion (*v/v*), extraction time and temperature	8 + 3	[54]
Citrus peels (*Citrus sinensis* L.) Osbeck (*citrus lemon* L.) Citrusclementine	*p*-Coumaric acid *Trans*-Ferulic acid Rutin Hesperidin	Solid-liquid extraction	Multi-level factorial design with three replicates at the center point	Amounts of polyphenols evaluated	Extraction time, ethanol-water proportion and extraction temperature	18 + 3 orange (3 levels for ethanol-water proportion and temperature) 8 + 3 for lemon and clementine (2 levels for each factor)	[55]
Residual brewing yeast	*Trans*-Ferulic acid Naringin Kaempferol	MSPD extraction assisted by titanium dioxide nanoparticles	Factorial design (two levels) with three replicates at the center point	Amounts of polyphenols evaluated	Diatomaceous earth (g) anatase TiO_2 NPs (mg) and ethanol-water proportion (*v/v*)	8 + 3	[46]

Grape residues	Gallic acid 3,4-Dihydroxybenzoic acid Catechin Caffeic acid *p*-Coumaric acid Rutin Quercetin	MSPD extraction assisted by titanium dioxide nanoparticles	Factorial design (two levels) with three replicates at the center point	Amounts of polyphenols evaluated	Ethanol-water proportion (*v*/*v*), the grinding time (min)and the amount of anatase TiO_2 NPs (mg)	8 + 3	[47]
Grape seed residues	Gallic acid 3,4-Dihydroxybenzoic acid Catechin Resveratrol Quercetin Kaempferol	MSPD extraction	Factorial design (two levels) with three replicates at the center point	Amounts of polyphenols evaluated	Ethanol-water proportion (*v*/*v*), the stirring time (min) and the amount of diatomaceous earth (g)	8 + 3	[48]
Defatted cherry seed residues from liquor processing	Phenolic profiles	MSPD extraction	Factorial design (two levels) with three replicates at the center point	TPC TFC Antioxidant activity (DPPH)	Ethanol content of the extraction solvent (*v*/*v*), stirring time interval (min), and the sample-to-silica ratio (*w*/*w*)	8 + 3	[49]

TAC Total anthocyanin content, *TPC* Total phenolic content, *TFC* Total flavonoid content, *CT* Condensed tannins, *DPPH* (1,1-diphenyl-2-picrylhydrazyl) radical reduction test, *ABTS* 2,2'-azino-bis (3-ethylbenzothiazoline-6-sulphonic acid) Radical cation decolorization test, *FRAP* Ferric reducing antioxidant power test, *MAE* Microwave-assisted extraction, *MSPD* Matrix solid-phase dispersion, *TMC* Total metabolite content

7. Validate the model using analysis of variance (ANOVA) that will allow us to establish its goodness of fit of the model (R^2) and the factors that are statistically most significant.
8. Determine the optimal conditions, in case of multiple answers, use the desirability function to choose the optimal one. Perform new experiments in these conditions to evaluate the reliability of the model.

3.3 Spectrophotometric Determination of Polyphenols

3.3.1 Total Phenolic Content

1. Take 0–125μL aliquots of standards and extract (as obtained in Subheading 3).
2. To this add 175 μL of Folin-Ciocalteu (2 N) reagent.
3. To this add 150 μL of sodium carbonate (Na_2CO_3) (7.5% w/v).
4. To this add Milli-Q water to a volume of 25 mL.
5. Measure, the absorbance of the mixtures at 720 nm using a UV-Vis spectrophotometer.
6. Quantification is done using gallic acid as standard. Calibration graphs are obtained in the concentration range 0–40 μM, and TPC is determined using the following equations, and the data included in Table 3. The TPC content is expressed as mg of gallic acid equivalents per gram of the extract (mg GAE·g^{-1}) (*See* **Note 4**).

$$\mathrm{TPC}\left(\frac{\mu\mathrm{mol}}{\mathrm{L}}\right) = \frac{\mathrm{Absorbance} - \mathrm{intercept}}{\mathrm{slope}}$$

$$\mathrm{TPC}\left(\frac{\mathrm{mg\ GAE}}{\mathrm{g}}\right) = \mathrm{TPC}\left(\frac{\mu\mathrm{mol}}{\mathrm{L}}\right) \times 10^{-3}\left(\frac{\mathrm{mmol}}{\mu\mathrm{mol}}\right) \times \mathrm{MW}\left(\frac{\mathrm{mg}}{\mathrm{mmol}}\right) \times \mathrm{DF} \times V_{\mathrm{extraction}}(\mathrm{L}) \times \frac{1}{\mathrm{mass\ extract\ (g)}}$$

3.3.2 Total Flavonoid Content

1. Take a 10 mL volumetric flask.
2. To this add 2 mL of Milli-Q water and 0.15 mL of 5% (*w*/*v*) $NaNO_2$, and 500 μL of sample extracts (as obtained in Subheading 3).

Table 3
Magnitudes for determining the total phenolic content

MW(GAE)	Molecular weight of Gallic acid	170.12 g·mol^{-1}
DF	Dilution factor of the extract	–
$V_{extraction}$	Volume used in the extraction	–

Table 4
Magnitudes for determining the total flavonoid content

MW(QE)	Molecular weight of gallic acid	302.236 g·mol^{-1}
DF	Dilution factor of the extract	–
$V_{extraction}$	Volume used in the extraction	–

3. Let it react for 5 min and then add 0.15 mL of 10% (*w*/*v*) $AlCl_3$.
4. After another 5 min, 1 mL of 1 M Sodium hydroxide (NaOH) is added.
5. Mix the reaction solution manually and keep at room temperature for 15 min.
6. Next, add Milli-Q water to a final volume of 10 mL.
7. Measure the absorbance at 415 nm.
8. Quantification is done using quercetin polyphenol as standard. Calibration graphs are obtained in the concentration range 0–40 μM, and TFC is calculated using the following formulas as well as the data included in Table 4 and is expressed as mg of quercetin equivalents per gram of the extract (mg QE·g^{-1}) [55]. (*See* **Note 8**).

$$\mathrm{TFC}\left(\frac{\mu\mathrm{mol}}{\mathrm{L}}\right)=\frac{\mathrm{Absorbance}-\mathrm{intercept}}{\mathrm{slope}}$$

$$\mathrm{TFC}\left(\frac{\mathrm{mg\ QE}}{\mathrm{g}}\right)=\mathrm{TFC}\left(\frac{\mu\mathrm{mol}}{\mathrm{L}}\right)\times 10^{-3}\left(\frac{\mathrm{mmol}}{\mu\mathrm{mol}}\right)\times \mathrm{MW}\left(\frac{\mathrm{mg}}{\mathrm{mmol}}\right)\times \mathrm{DF}\times V_{\mathrm{extraction}}(\mathrm{L})\times\frac{1}{\mathrm{mass\ extract\ (g)}}$$

3.3.3 Total Monomeric Anthocyanin

Monomeric anthocyanins pigments reversibly change color with a change in pH. The colored oxonium form (Fig. 3) exists at pH 1.0; and the colorless hemiketal form (Fig. 3) predominates at pH 4.5.

The difference in the absorbance of the pigments at λ_{max} = 500 nm is proportional to the pigment concentration. Thus, the pigment content should be calculated by using the molecular weight (MW) and molar extinction coefficient of the major anthocyanin in the matrix according to the following equations and data included in Table 5 (*See* **Notes 9** and **10**).

$$\mathrm{Anthocyaninpigment}\left(\mathrm{CydEQ}\frac{\mathrm{mg}}{\mathrm{L}}\right)=\frac{A_{\mathrm{neta}}\times \mathrm{MW}\left(\frac{\mathrm{g}}{\mathrm{mol}}\right)\times \mathrm{DF}\times 10^{3}\left(\frac{\mathrm{mg}}{\mathrm{g}}\right)}{\varepsilon\left(\frac{\mathrm{L}}{\mathrm{mol\cdot cm}}\right)\times 1\ (\mathrm{cm})}$$

$$A_{\mathrm{neta}}=(A_{500\ \mathrm{nm}}-A_{700\ \mathrm{nm}})_{\mathrm{pH}\ 1.0}-(A_{500\ \mathrm{nm}}-A_{700\ \mathrm{nm}})_{\mathrm{pH}\ 4.5}$$

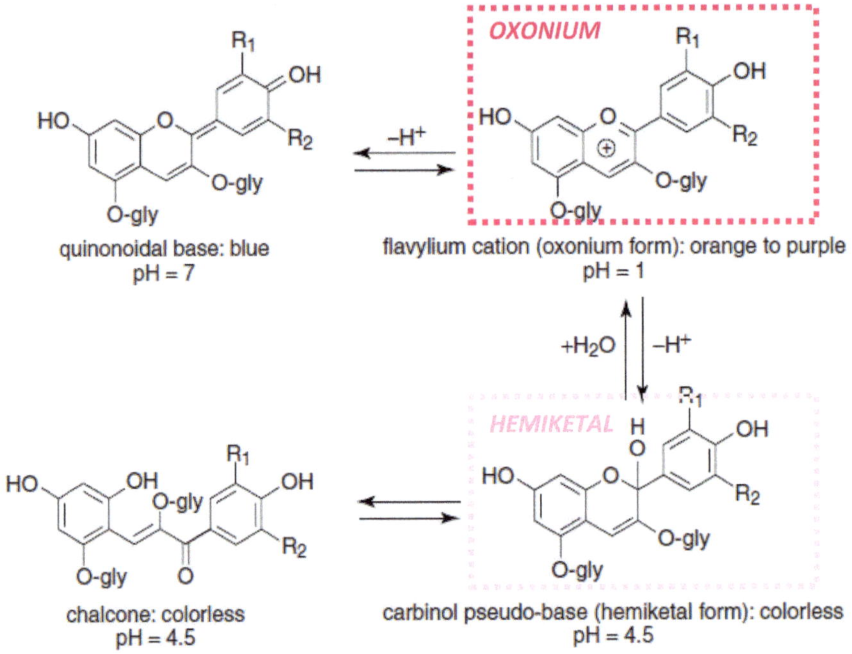

Fig. 3 Predominant form of anthocyanins at different pH values

Table 5
Magnitudes for determining the total monomeric anthocyanin content

A_{neta}	Differential absorbance	–
MW(Cyd – 3 – glu)	Molecular weight Cyd – 3 – glu	449.2 g·mol^{-1}
DF	Dilution factor of the extract	–
ε(Cyd – 3 – glu)	Molar extinction coefficient	269,000 L·mol^{-1}·cm^{-1}
10^3	Conversion from g to mg	mg·g^{-1}

Since natural matrices normally contain a mixture of anthocyanins, and their proportions will vary naturally, results are commonly expressed as equivalents of cyanindin-3-glucoside, which is the most common one in nature. The procedure for the estimation of TMA is as follows:

1. For determining the dilution factor and the λ_{max} for the sample, dilute a certain volume (depending on the dilution factor of the sample, the absorbance range should be in range 0.2–1) of the extract with KCl pH 1 buffer in 5 mL flask (the volume of the sample cannot exceed 20% of the total volume to not exceed the buffer's capacity). The absorbance should be in the range 0.2–1 at λ_{max}.

2. To prepare the sample's solutions, using the dilution factor determined in 5 mL flasks, prepare 2 dilutions of the sample, one with KCl buffer at pH 1.0 and the other with sodium acetate buffer at pH 4.5.
3. Let the dilutions equilibrate for 15 min. Then, measure the absorbance at λ_{max} ($\approx$ 500 nm) and 700 nm by using distilled water as spectrophotometric blank. Measure the absorbance within 20–50 min after preparation. If the sample turbidity is excessive, clarify by centrifuging (e.g., 5000 rpm) or filtering using Nylon filters (0.45 μm) before measurement [93, 94].

3.4 Chromatographic Determination of Polyphenols

3.4.1 Gas Chromatographic Determination

The analysis of polyphenols in food and vegetable samples might entail several purification processes, such as eliminating lipids from the extracts, liberating phenolic compounds from glycoside and ester bonds through enzymatic, alkaline, or acidic means, and implementing chemical alterations to generate more volatile derivatives. There are different types of reagents used for derivatization of phenolic compounds and to produce methyl or ethyl esters of phenolics including ethyl and methyl chloroformate, diazomethane and dimethyl sulfoxide in combination with methyl iodate. Other reagents that can be employed are trifluoroacetymide, N-(tert-butyldimethylsilyl)-N-methyltrifluoroacetamide and trimethylsilyl derivative. The silylation reaction is straightforward, devoid of undesirable by-products, and yields volatile products that do not disrupt the analysis, making it an excellent choice for identifying phenolic compounds. The most prevalent columns employed for phenolic quantification in GC techniques are typically fused silica capillaries of 30 m lengths in length, featuring internal diameters ranging from 25 to 32 μm and stationary phase particle sizes of 0.25 μm. In addition, helium is frequently used as carrier gas [74]. Regarding detection, the use of a flame ionization detector (FID) is a common method to detect phenolics [82]. Nevertheless, mass spectrometry (MS) has gained widespread adoption in recent times, mainly working with electron impact ionization at 70 eV in both polarities [74].

Some of the gas chromatographic techniques for the analysis of phenolic compounds are presented in Table 6.

3.4.2 High Liquid Chromatographic Determination

HPLC is the preferred technique for both separation and quantification of phenolic compounds because it avoids derivatization to volatile derivates. Several factors influence the HPLC analysis of polyphenols, encompassing sample purification, mobile phase composition, column selection, and detector choice. Typically, reversed-phase C18 columns are employed combined with a diode array (DAD) or MS detector with electron impact ionization at 70 eV in both polarities. Polar organic solvents (methanol or acetonitrile) acidified with acetic, trifluoroacetic, phosphoric or

Table 6
Summary of GC conditions to detect molecules belonging to phenolic classes

Sample	Polyphenols	Derivatization	Conditions	Detection	References
Leaves and stems from *Pandiakaheudelotii*	28 flavonoids 6 hydroxycinnamates 12 benzoic acid derivatives	N,O-bis (trimethylsilyl) trifluoroacetamide	Capillary HP INNOWax (30 m × 0.25 mm × 0.25 μm film thickness). Inlet and detection temperatures 250 and 320 °C. Split injection. Nitrogen as carrier gas. Temperature program	GC-FID	[82]
Guarana	3-Hydroxybenzoic, benzoic, gallic, syringic, isovanillic, protocatechuic acids, catechin, caffeine, epicatechin, quercetin	Hexamethyldisiloxane and dimethylchlorosilane	Zebron ZB-5 ms fused silica capillary column (30 m × 0.25 mm I.D. × 0.25 μm film thickness). Injector temperature 300 °C. Helium as carrier gas. Ion source temperature: 200 °C. Temperature program	GC-MS	[95]
Plant extracts	Gallic, p-hydroxybenzoic, gentisic, p-coumaric, vanillic, ferulic, syringic acids, catechin	Trimethylchlorosilane and N, O-bis(trimethylsilyl)-trifluoroacetamide with dimethyldichlorosilane	CP-Sil 8 capillary column (30 m × 0.32 mm ID × 0.25 μm film thickness). Temperature program	GC-MS	[96]
Olive oils	23 phenolic compounds	Bis(trimethylsilyl)trifluoracetamide and trimethylchlorosilane (99:1)	DB-5MS column (30 m × 0.25 mm ID, 0.25 μm film thickness. Transfer-line 275 °C and ion source 200 °C. Split injection. Temperature program. Helium as carrier gas	GC-MS GC-MS/MS	[97]

formic acids and water solutions acidified are usually employed as mobile phase for optimal performance under gradient elution mode [74]. Due to their several advantages, including shorter analysis times, reduced mobile phase and sample consumption, simplified coupling with MS, and enhanced sensitivity, C18 capillary analytical columns (150 × 0.3 mm I.D., 4 μm) are also used [98].

Some of the liquid chromatographic techniques for the analysis of phenolic compounds are presented in Table 7.

Table 7
Summary of LC conditions to detect molecules belonging to phenolic classes

Sample	Polyphenols	Technique	Column	Mobile phase	References
Apple peel	Phenolic profiling	HILIC-ESI-UV-MS/MS	Luna$^{(R)}$ C18 (2.0 mm × 150 mm, 3.0 μm)	0.1% acetic acid in water-ACN 80:20 (v/v) and 0.1% acetic acid in ACN Gradient elution	[99]
Orange, lemon and mandarin peels	12 phenolic compounds	HPLC-DAD	Agilent eclipse XDB-C18 (4.6 mm × 150 mm, 5 μm)	2% formic acid in water and ACN Gradient elution	[100]
Residual brewer's yeast	9 polyphenols	cLC-DAD	Capillary synergy™ fusion C18 (0.3 mm × 150 mm, 4 μm)	0.1% trifluoroacetic acid at pH 3.2 and ACN Gradient elution	[26]
Green and black teas	11 polyphenols	Nano HPLC-UV Nano HPLC-MS^n	Capillary C18 column (0.1 mm × 100 mm, 2.7 μm)	0.5% formic acid in ACN-MeOH-water Gradient elution	[101]
Pomegranate peels	Phenolic profiling	HPLC-DAD HPLC-ESI-MS^n	Ascentis express C18 (3.0 mm × 150 mm, 2.7 μm)	2% formic acid in water and 0.5% formic acid in MeOH-water 90:10 (v/v) Gradient elution	[102]
Kiwi peel	Phenolic profiling	HPLC-DAD-ESI-MS^n	Waters Spherisorb S3 ODS-2 (4.6 mm × 150 mm, 3 μm)	0.1% formic acid and acetonitrile Gradient elution	[103]

4 Notes

1. To prepare 7.5% (*w*/*v*) Na_2CO_3, exactly weigh 7.5 g of anhydrous Na_2CO_3. Add it to a beaker. Dissolve it in required quantities of Milli-Q water, transfer to a 100 mL volumetric flask, and make final volume 100 mL by adding rinsings from the beaker in which solution was prepared. This solution should be stored at room temperature for up to 3 weeks [104].
2. The stock and standard solutions of gallic acid are prepared following the procedure explained below. Firstly, 500 mM of gallic acid stock solution is prepared by weighing 2 g of gallic acid standard and dissolved it into a beaker using methanol as a solvent. Then, transfer to a final volume of 25 mL in a volumetric flask. This solution should be stored in the refrigerator at 4 °C for up to 1 month. Subsequently, a work solution of 8 mM of gallic acid is prepared taking 80 μL of the above solution and dissolving it in a 5 mL volumetric flask with Milli-Q water. Then, to prepare the calibration standard solutions (0–40 μM), increasing volumes of 0 to 125 μL of an 8 mM gallic acid and added in 25 mL volumetric flasks. Then, 175 μL of the *Folin-Ciocalteu* reagent and 150 μL of Na_2CO_3 are added and the volume is made up using Milli-Q water as a solvent. These solutions should be prepared daily [104].
3. To prepare 5% (*w*/*v*) $NaNO_2$, weigh exactly 5 g of $NaNO_2$. Transfer the substance to a beaker and dissolve it in the necessary amount of distilled water. Then, pour the solution into a final volume of 100 mL in a volumetric flask. This solution should be maintained at room temperature for up to 3 weeks [104].
4. To prepare 10% Aluminum chloride 6-hydrate, weigh exactly 10 g of Aluminum chloride 6-hydrate and add it to a beaker. Dissolve it in required quantities of distilled water and then transfer to a 100 mL volumetric flask. This solution should be stored at room temperature for up to 3 weeks [104].
5. To prepare 1 M of NaOH, weigh exactly 2 g of NaOH into a beaker. Then, dissolve it in the required quantities of distilled water and transfer to a 100 mL volumetric flask. This solution should be stored in the refrigerator at 4 °C for up to 3 weeks [104].
6. The stock and standard solutions of quercetin are prepared following the procedure described below. Firstly, 0.81 mM of quercetin stock solution is prepared by weighing 6 mg of quercetin standard and dissolving it into a beaker using ethanol-water mixture (80:20 (*v*/*v*)) as a solvent. Then, transfer to a final volume of 25 mL in a volumetric flask. This

solution should be stored in the refrigerator at 4 °C for up to 3 weeks. Subsequently, to prepare the calibration standard solutions (0–45 μM), 2 mL of Milli-Q water and 150 μL of $NaNO_2$, variable volumes of quercetin stock solution. After 5 min, 150 μL of Aluminum chloride 6-hydrate are added. After an additional 5 min, 1 mL of NaOH is added and the reaction is allowed to continue for 15 min before topping up the volume of Milli-Q water [104].

7. To prepare 0.025 M potassium chloride buffer pH 1.0, dissolve 0.945 g of KCl in 450 mL of distilled water. Adjust the pH to 1.0 with concentrate HCl and fix the volume to 500 mL.
8. To prepare 0.4 sodium acetate buffer pH 4.5, dissolve 27.215 g of sodium acetate in 450 mL of distilled water. Adjust the pH to 4.5 with concentrate HCl and fix the volume to 500 L.
9. Although methanol provides superior yields, ethanol is preferred in the food industry due to its lower toxicity, holding the "generally regarded as safe" (GRAS) status as defined by the American Food and Drug Administration [5, 56].
10. It is advisable to analyze the phenolic extract immediately after collection. A subsequent removal of the solvent, either by evaporation or lyophilization, is reported in several investigations [25, 27, 32].

Acknowledgments

This work is part of the research project PID2023-149424OB-I00, funded by MCIU/AEI/10.13039/501100011033 and ESF+. Sandra Rodríguez-Blázquez also acknowledges the Complutense University of Madrid and the Santander Bank for the pre-doctoral grant funding (CT25-24).

References

1. Abbas M, Saeed F, Anjum FM, Afzaal M, Tufail T, Bashir MS, Ishtiaq A, Hussain S, Suleria HAR (2017) Spolyphenols: an overview. Int J Food Prop. Taylor and Francis Inc, pp 1689–1699
2. Ahlawat YK, Singh M, Manorama K, Lakra N, Zaid A, Zulfiqar F (2023) Plant phenolics: neglected secondary metabolites in plant stress tolerance. Rev Bras Bot. https://doi.org/10.1007/s40415-023-00949-x
3. Singla RK, Dubey AK, Garg A, Sharma RK, Fiorino M, Ameen SM, Haddad MA, Al-Hiary M (2019) Natural polyphenols: chemical classification, definition of classes, subcategories, and structures. J AOAC Int 102:1397–1400. https://doi.org/10.5740/jaoacint.19-0133
4. Mustafa SK, Oyouni AAWA, Aljohani MMH, Ayaz Ahmad M (2020) Polyphenols more than an antioxidant: role and scope. J Pure Appl Microbiol 14:47–61. https://doi.org/10.22207/JPAM.14.1.08
5. Ignat I, Volf I, Popa VI (2011) A critical review of methods for characterisation of polyphenolic compounds in fruits and vegetables. Food Chem 126:1821–1835. https://doi.org/10.1016/j.foodchem.2010.12.026
6. Belščak-Cvitanović A, Durgo K, Huđek A, Bačun-Družina V, Komes D (2018) Overview of polyphenols and their properties. In:

Polyphenols: properties, recovery, and applications. Elsevier, pp 3–44

7. Pietta PG (2000) Flavonoids as antioxidants. J Nat Prod 63:1035–1042. https://doi.org/10.1021/np9904509
8. Kumar S, Pandey AK (2013) Chemistry and biological activities of flavonoids: an overview. Sci World J 2013. https://doi.org/10.1155/2013/162750
9. Oluwole O, Fernando WB, Lumanlan J, Ademuyiwa O, Jayasena V (2022) Role of phenolic acid, tannins, stilbenes, lignans and flavonoids in human health – a review. Int J Food Sci Technol. https://doi.org/10.1111/ijfs.15936
10. Cui Q, Du R, Liu M, Rong L (2020) Lignans and their derivatives from plants as antivirals. Molecules 25:183. https://doi.org/10.3390/molecules25010183
11. ElKhawand T, Courtois A, Valls J, Richard T, Krisa S (2018) A review of dietary stilbenes: sources and bioavailability. Phytochem Rev 17:1007–1029. https://doi.org/10.1007/s11101-018-9578-9
12. Romani A, Ieri F, Turchetti B, Mulinacci N, Vincieri FF, Buzzini P (2006) Analysis of condensed and hydrolysable tannins from commercial plant extracts. J Pharm Biomed Anal 41:415–420. https://doi.org/10.1016/j.jpba.2005.11.031
13. Gonzalez-Hernandezi MP, Karchesy J, Starkey EE (2003) Research observation: Hydrolyzable and condensed tannins in plants of northwest Spain forests. J Range Manag 56:461–465
14. Ueda JM, Pedrosa MC, Heleno SA, Carocho M, Ferreira ICFR, Barros L (2022) Food additives from fruit and vegetable by-products and bio-residues: a comprehensive review focused on sustainability. Sustainability (Switzerland) 14. https://doi.org/10.3390/su14095212
15. Maravić N, Teslić N, Nikolić D, Dimić I, Šereš Z, Pavlić B (2022) From agricultural waste to antioxidant-rich extracts: green techniques in extraction of polyphenols from sugar beet leaves. Sustain Chem Pharm 28. https://doi.org/10.1016/j.scp.2022.100728
16. Vijayalaxmi S, Jayalakshmi SK, Sreeramulu K (2015) Polyphenols from different agricultural residues: extraction, identification and their antioxidant properties. J Food Sci Technol 52:2761–2769. https://doi.org/10.1007/s13197-014-1295-9
17. Shinohara T, Kubodera S, Yanagida F (2000) Distribution of phenolic yeasts and production of phenolic off-flavors in wine fermentation. J Biosci Bioeng 90:90–97
18. Mir-Cerdà A, Nuñez O, Granados M, Sentellas S, Saurina J (2023) An overview of the extraction and characterization of bioactive phenolic compounds from agri-food waste within the framework of circular bioeconomy. Trends Anal Chem 161. https://doi.org/10.1016/j.trac.2023.116994
19. Dini I (2021) Bio discarded from waste to resource. Foods 10. https://doi.org/10.3390/foods10112652
20. Pagano I, Campone L, Celano R, Piccinelli AL, Rastrelli L (2021) Green non-conventional techniques for the extraction of polyphenols from agricultural food by-products: a review. J Chromatogr A 1651. https://doi.org/10.1016/j.chroma.2021.462295
21. Carpentieri S, Soltanipour F, Ferrari G, Pataro G, Donsì F (2021) Emerging green techniques for the extraction of antioxidants from agri-food by-products as promising ingredients for the food industry. Antioxidants 10. https://doi.org/10.3390/antiox10091417
22. Caldas TW, Mazza KEL, Teles ASC, Mattos GN, Brígida AIS, Conte-Junior CA, Borguini RG, Godoy RLO, Cabral LMC, Tonon RV (2018) Phenolic compounds recovery from grape skin using conventional and non-conventional extraction methods. Ind Crop Prod 111:86–91. https://doi.org/10.1016/j.indcrop.2017.10.012
23. Osorio-Tobón JF (2020) Recent advances and comparisons of conventional and alternative extraction techniques of phenolic compounds. J Food Sci Technol 57:4299–4315. https://doi.org/10.1007/s13197-020-04433-2
24. Sagar NA, Pareek S, Sharma S, Yahia EM, Lobo MG (2018) Fruit and vegetable waste: bioactive compounds, their extraction, and possible utilization. Compr Rev Food Sci Food Saf 17:512–531. https://doi.org/10.1111/1541-4337.12330
25. Montenegro-Landívar MF, Tapia-Quirós P, Vecino X, Reig M, Valderrama C, Granados M, Cortina JL, Saurina J (2021) Recovery of added-value compounds from orange and spinach processing residues: green extraction of phenolic compounds and evaluation of antioxidant activity. Antioxidants 10:1800. https://doi.org/10.3390/ANTIOX10111800
26. León-González ME, Gómez-Mejía E, Rosales-Conrado N, Madrid-Albarrán Y (2018) Residual brewing yeast as a source of polyphenols: extraction, identification and quantification by chromatographic and chemometric tools. Food Chem 267:246–254. https://doi.org/10.1016/j.foodchem.2017.06.141

27. Zhao C-N, Tang G-Y, Cao S-Y, Xu X-Y, Gan R-Y, Liu Q, Mao Q-Q, Shang A, Li H-B (2019) Phenolic profiles and antioxidant activities of 30 tea infusions from green, black, oolong, white, yellow and dark teas. Antioxidants 8:215. https://doi.org/10.3390/antiox8070215
28. Gómez-Mejía E, Rosales-Conrado N, León-González ME, Valverde A, Madrid Y (2022) A combined analytical-chemometric approach for the in vitro determination of polyphenol bioaccessibility by simulated gastrointestinal digestion. Anal Bioanal Chem 414:2739–2755. https://doi.org/10.1007/s00216-022-03922-x
29. Gómez-Mejía E, Roriz CL, Heleno SA, Calhelha R, Dias MI, Pinela J, Rosales-Conrado N, León-González ME, Ferreira ICFR, Barros L (2021) Valorisation of black mulberry and grape seeds: chemical characterization and bioactive potential. Food Chem 337. https://doi.org/10.1016/j.foodchem.2020.127998
30. Peixoto CM, Dias MI, Alves MJ, Calhelha RC, Barros L, Pinho SP, Ferreira ICFR (2018) Grape pomace as a source of phenolic compounds and diverse bioactive properties. Food Chem 253:132–138. https://doi.org/10.1016/j.foodchem.2018.01.163
31. Chatzimitakos T, Athanasiadis V, Kotsou K, Palaiogiannis D, Bozinou E, Lalas SI (2023) Optimized isolation procedure for the extraction of bioactive compounds from spent coffee grounds. Appl Sci 13. https://doi.org/10.3390/app13052819
32. Vieira EF, Carvalho J, Pinto E, Cunha S, Almeida AA, Ferreira IMPLVO (2016) Nutritive value, antioxidant activity and phenolic compounds profile of brewer's spent yeast extract. J Food Compos Anal 52:44–51. https://doi.org/10.1016/J.JFCA.2016.07.006
33. Di Donato P, Taurisano V, Tommonaro G, Pasquale V, Jiménez JMS, de Pascual-Teresa S, Poli A, Nicolaus B (2018) Biological properties of polyphenols extracts from agro industry's wastes. Waste Biomass Valorization 9:1567–1578. https://doi.org/10.1007/S12649-017-9939-4/FIGURES/3
34. Zhou Y, Xu XY, Gan RY, Zheng J, Li Y, Zhang JJ, Xu DP, Li H, Bin (2019) Optimization of ultrasound-assisted extraction of antioxidant polyphenols from the seed coats of red sword bean (Canavalia gladiate (Jacq.) DC.). Antioxidants 8:200. https://doi.org/10.3390/ANTIOX8070200
35. Al-Dhabi NA, Ponmurugan K, Maran Jeganathan P (2017) Development and validation of ultrasound-assisted solid-liquid extraction of phenolic compounds from waste spent coffee grounds. Ultrason Sonochem 34:206–213. https://doi.org/10.1016/J.ULTSONCH.2016.05.005
36. Nipornram S, Tochampa W, Rattanatraiwong P, Singanusong R (2018) Optimization of low power ultrasound-assisted extraction of phenolic compounds from mandarin (Citrus reticulata Blanco cv. Sainampueng) peel. Food Chem 241:338–345. https://doi.org/10.1016/J.FOODCHEM.2017.08.114
37. Ben AH, Abbassi A, Trabelsi A, Krichen Y, Chekir-Ghedira L, Ghedira K (2023) Optimization of ultrasound-assisted extraction of polyphenols and flavonoids from Citrus aurantium L. var. amara Engl. fruit peel using response surface methodology. Biomass Convers Biorefinery 1:1–13. https://doi.org/10.1007/S13399-023-03947-X/FIGURES/4
38. Teixeira BA, Vidigal MCTR, de Castro Leite Júnior B, ÉNR V, EMF M, Stringheta PC (2023) Optimization, kinetic and phenomenological modeling of ultrasound-assisted extraction process of bioactive compounds from raspberries (Rubus idaeus L.). Food Anal Methods 16:759–770. https://doi.org/10.1007/s12161-023-02462-z
39. Della Posta S, Gallo V, Ascrizzi AM, Gentili A, De Gara L, Dugo L, Fanali C (2023) Development of a green ultrasound-assisted procedure for the extraction of phenolic compounds from avocado peel with deep eutectic solvents. Green Anal Chem 7:100083. https://doi.org/10.1016/J.GREEAC.2023.100083
40. Anticona M, Blesa J, Lopez-Malo D, Frigola A, Esteve MJ (2021) Effects of ultrasound-assisted extraction on physicochemical properties, bioactive compounds, and antioxidant capacity for the valorization of hybrid Mandarin peels. Food Biosci 42. https://doi.org/10.1016/j.fbio.2021.101185
41. Tapia-Quirós P, Montenegro-Landívar MF, Reig M, Vecino X, Alvarino T, Cortina JL, Saurina J, Granados M (2020) Olive mill and winery wastes as viable sources of bioactive compounds: a study on polyphenols recovery. Antioxidants 9:1074–1074. https://doi.org/10.3390/ANTIOX9111074
42. Mustafa A, Turner C (2011) Pressurized liquid extraction as a green approach in food and herbal plants extraction: a review. Anal Chim Acta 703:8–18
43. Alvarez-Rivera G, Bueno M, Ballesteros-Vivas D, Mendiola JA, Ibañez E (2019) Pressurized liquid extraction. In: Liquid-phase extraction. Elsevier, pp 375–398

44. Domínguez-Rodríguez G, García MC, Marina ML, Plaza M (2021) Pressurized liquid extraction combined with enzymatic-assisted extraction to obtain bioactive non-extractable polyphenols from sweet cherry (Prunus avium L.) pomace. Nutrients 13:3242. https://doi.org/10.3390/NU13093242/S1
45. Grisales-Mejía JF, Álvarez-Rivera G, Torres-Castañeda HG, Andrade-Mahecha MM, Martínez-Correa HA, Mendiola JA, Cifuentes A, Ibañez E (2024) Hass avocado (Persea americana mill.) residues as a new potential source of neuroprotective compounds using pressurized liquid extraction. J Supercrit Fluids 204:106117. https://doi.org/10.1016/J.SUPFLU.2023.106117
46. Gómez-Mejía E, Rosales-Conrado N, León-González ME, Madrid Y (2019) Determination of phenolic compounds in residual brewing yeast using matrix solid-phase dispersion extraction assisted by titanium dioxide nanoparticles. J Chromatogr A 1601:255–265. https://doi.org/10.1016/j.chroma.2019.05.009
47. Gómez-Mejía E, Mikkelsen LH, Rosales-Conrado N, León-González ME, Madrid Y (2021) A combined approach based on matrix solid-phase dispersion extraction assisted by titanium dioxide nanoparticles and liquid chromatography to determine polyphenols from grape residues. J Chromatogr A 1644. https://doi.org/10.1016/j.chroma.2021.462128
48. Gómez-Mejía E, Vicente-Zurdo D, Rosales-Conrado N, León-González ME, Madrid Y (2022) Screening the extraction process of phenolic compounds from pressed grape seed residue: towards an integrated and sustainable management of viticultural waste. LWT 169. https://doi.org/10.1016/j.lwt.2022.113988
49. Rodríguez-Blázquez S, Fernández-Ávila L, Gómez-Mejía E, Rosales-Conrado N, León-González ME, Miranda R (2023) Valorization of defatted cherry seed residues from liquor processing by matrix solid-phase dispersion extraction: a sustainable strategy for production of phenolic-rich extracts with antioxidant potential. Antioxidants 12. https://doi.org/10.3390/antiox12122041
50. Lores Aguín M, Iglesias Estévez M, Álvarez Casas M, Llompart Vizoso MP, García Jares CM (2012) Extraction of bioactive polyphenols from grape marc by a matrix solid-phase dispersion method. Recursos Rurais 8:39–47
51. Simeoni MC, Pellegrini M, Sergi M, Pittia P, Ricci A, Compagnone D (2018) Analysis of polyphenols in the Lamiaceae Family by matrix solid-phase dispersion extraction followed by ultra-high-performance liquid chromatography–tandem mass spectrometry determination. ACS Omega. https://doi.org/10.1021/acsomega.8b02239
52. Kamal YT, Alam P, Alqasoumi SI, Foudah AI, Alqarni MH, Yusufoglu HS (2018) Investigation of antioxidant compounds in commercial pomegranate molasses products using matrix-solid phase dispersion extraction coupled with HPLC. Saudi Pharm J 26:839–844. https://doi.org/10.1016/J.JSPS.2018.03.015
53. Rodríguez-Blázquez S, Pedrera-Cajas L, Gómez-Mejía E, Vicente-Zurdo D, Rosales-Conrado N, León-González ME, Rodríguez-Bencomo JJ, Miranda R (2024) The potential of plum seed residue: unraveling the effect of processing on phytochemical composition and bioactive properties. Int J Mol Sci 25:1236. https://doi.org/10.3390/IJMS25021236
54. Ramón-Gonçalves M, Gómez-Mejía E, Rosales-Conrado N, León-González ME, Madrid Y (2019) Extraction, identification and quantification of polyphenols from spent coffee grounds by chromatographic methods and chemometric analyses. Waste Manag 96:15–24. https://doi.org/10.1016/j.wasman.2019.07.009
55. Gómez-Mejía E, Rosales-Conrado N, León-González ME, Madrid Y (2019) Citrus peels waste as a source of value-added compounds: extraction and quantification of bioactive polyphenols. Food Chem 295:289–299. https://doi.org/10.1016/j.foodchem.2019.05.136
56. Ameer K, Shahbaz HM, Kwon JH (2017) Green extraction methods for polyphenols from plant matrices and their byproducts: a review. Compr Rev Food Sci Food Saf 16:295–315. https://doi.org/10.1111/1541-4337.12253
57. Roselló-Soto E, Koubaa M, Moubarik A, Lopes RP, Saraiva JA, Boussetta N, Grimi N, Barba FJ (2015) Emerging opportunities for the effective valorization of wastes and by-products generated during olive oil production process: non-conventional methods for the recovery of high-added value compounds. Trends Food Sci Technol 45:296–310. https://doi.org/10.1016/J.TIFS.2015.07.003
58. Anaya-Esparza LM, Aurora-Vigo EF, Villagrán Z, Rodríguez-Lafitte E, Ruvalcaba-Gómez JM, Solano-Cornejo MÁ, Zamora-Gasga VM, Montalvo-González E, Gómez-Rodríguez H, Aceves-Aldrete CE, González-Silva N (2023) Design of experiments for optimizing ultrasound-assisted extraction of

bioactive compounds from plant-based sources. Molecules 28. https://doi.org/10.3390/molecules28237752

59. Poveda JM, Loarce L, Alarcón M, Díaz-Maroto MC, Alañón ME (2018) Revalorization of winery by-products as source of natural preservatives obtained by means of green extraction techniques. Ind Crop Prod 112:617–625. https://doi.org/10.1016/J.INDCROP.2017.12.063
60. Wianowska D, Gil M (2019) New insights into the application of MSPD in various fields of analytical chemistry. TrAC Trends Anal Chem 112:29–51. https://doi.org/10.1016/J.TRAC.2018.12.028
61. Capriotti AL, Cavaliere C, Laganà A, Piovesana S, Samperi R (2013) Recent trends in matrix solid-phase dispersion. TrAC Trends Anal Chem 43:53–66. https://doi.org/10.1016/J.TRAC.2012.09.021
62. Boateng ID, Mustapha A, Kuehnel L, Daubert CR, Kumar R, Agliata J, Flint-Garcia S, Wan C, Somavat P (2023) From purple corn waste (pericarp) to polyphenol-rich extract with higher bioactive contents and superior product qualities using two-step optimization techniques. Ind Crop Prod 200. https://doi.org/10.1016/j.indcrop.2023.116871
63. Chamali S, Bendaoud H, Bouajila J, Camy S, Saadaoui E, Condoret JS, Romdhane M (2023) Optimization of accelerated solvent extraction of bioactive compounds from Eucalyptus intertexta using response surface methodology and evaluation of its phenolic composition and biological activities. J Appl Res Med Aromat Plants 35. https://doi.org/10.1016/j.jarmap.2023.100464
64. Frosi I, Balduzzi A, Colombo R, Milanese C, Papetti A (2024) Recovery of polyphenols from corn cob (Zea mays L.): optimization of different green extraction methods and efficiency comparison. Food Bioprod Process 143:212–220. https://doi.org/10.1016/j.fbp.2023.11.009
65. Jankovic A, Chaudhary G, Goia F (2021) Designing the design of experiments (DOE) – an investigation on the influence of different factorial designs on the characterization of complex systems. Energ Buildings 250. https://doi.org/10.1016/j.enbuild.2021.111298
66. Marrubini G, Dugheri S, Cappelli G, Arcangeli G, Mucci N, Appelblad P, Melzi C, Speltini A (2020) Experimental designs for solid-phase microextraction method development in bioanalysis: a review. Anal Chim Acta 1119:77–100. https://doi.org/10.1016/j.aca.2020.04.012
67. Vera Candioti L, De Zan MM, Cámara MS, Goicoechea HC (2014) Experimental design and multiple response optimization. Using the desirability function in analytical methods development. Talanta 124:123–138. https://doi.org/10.1016/j.talanta.2014.01.034
68. El Adnany EM, Elhadiri N, Mourjane A, Ouhammou M, Hidar N, Jaouad A, Bitar K, Mahrouz M (2023) Impact and optimization of the conditions of extraction of phenolic compounds and antioxidant activity of olive leaves (Moroccan picholine) using response surface methodology. Separations 10. https://doi.org/10.3390/separations10060326
69. Ahmad I, Narsa A, Mr R, Zamruddin N, Iswahyudi I, Hajrah H, Indriyanti N, Arifuddin M, Siska S, Supandi S, Ambarwati NS (2023) Optimization of microwave-assisted extraction on polyphenol metabolite from Eleutherine bulbosa (Mill.) urb. bulbs using response surface methodology. J Adv Pharm Technol Res 14:113. https://doi.org/10.4103/japtr.japtr_613_22
70. Lucci P, Saurina J, Núñez O (2017) Trends in LC-MS and LC-HRMS analysis and characterization of polyphenols in food. Trends Anal Chem 88:1–24. https://doi.org/10.1016/j.trac.2016.12.006
71. Aleixandre-Tudo JL, Buica A, Nieuwoudt H, Aleixandre JL, Du Toit W (2017) Spectrophotometric analysis of phenolic compounds in grapes and wines. J Agric Food Chem 65:4009–4026. https://doi.org/10.1021/acs.jafc.7b01724
72. Sánchez-Rangel JC, Benavides J, Heredia JB, Cisneros-Zevallos L, Jacobo-Velázquez DA (2013) The Folin-Ciocalteu assay revisited: improvement of its specificity for total phenolic content determination. Anal Methods 5:5990–5999. https://doi.org/10.1039/c3ay41125g
73. Razem M, Ding Y, Morozova K, Mazzetto F, Scampicchio M (2022) Analysis of phenolic compounds in food by coulometric array detector: a review. Sensors 22. https://doi.org/10.3390/s22197498
74. Khoddami A, Wilkes MA, Roberts TH (2013) Techniques for analysis of plant phenolic compounds. Molecules 18:2328–2375. https://doi.org/10.3390/molecules18022328
75. Pignitter M, Somoza V (2012) Critical evaluation of methods for the measurement of oxidative rancidity in vegetable oils. J Food Drug Anal 20. https://doi.org/10.38212/2224-6614.2024
76. Carrasco-Pancorbo A, Cerretani L, Bendini A, Segura-Carretero A, Gallina-Toschi T,

Fernández-Gutiérrez A (2005) Analytical determination of polyphenols in olive oils. J Sep Sci 28:837–858. https://doi.org/10.1002/jssc.200500032

77. Ma S, Kim C, Neilson AP, Griffin LE, Peck GM, O'Keefe SF, Stewart AC (2019) Comparison of common analytical methods for the quantification of total polyphenols and flavanols in fruit juices and ciders. J Food Sci 84:2147–2158. https://doi.org/10.1111/1750-3841.14713
78. Sik B, Székelyhidi R, Lakatos E, Kapcsándi V, Ajtony Z (2022) Analytical procedures for determination of phenolics active herbal ingredients in fortified functional foods: an overview. Eur Food Res Technol 248:329–344
79. Naczk M, Shahidi F (2004) Extraction and analysis of phenolics in food. J Chromatogr A 1054:95–111. https://doi.org/10.1016/j.chroma.2004.08.059
80. Vidal-Casanella O, Núñez O, Granados M, Saurina J, Sentellas S (2021) Analytical methods for exploring nutraceuticals based on phenolic acids and polyphenols. Appl Sci 11. https://doi.org/10.3390/app11188276
81. Osman MF, Hassan NM, Khatib A, Tolos SM (2018) Antioxidant activities of Dialiumindum L. fruit and gas chromatography-mass spectrometry (GC-MS) of the active fractions. Antioxidants 7. https://doi.org/10.3390/antiox7110154
82. Ifeanacho MO, Ikewuchi CC, Ikewuchi JC (2017) Investigation of the profile of phenolic compounds in the leaves and stems of Pandiakaheudelotii using gas chromatography coupled with flame ionization detector. Food Sci Nutr 5:646–652. https://doi.org/10.1002/fsn3.443
83. Motilva MJ, Serra A, Macià A (2013) Analysis of food polyphenols by ultra high-performance liquid chromatography coupled to mass spectrometry: an overview. J Chromatogr A 1292:66–82. https://doi.org/10.1016/j.chroma.2013.01.012
84. Žuvela P, Skoczylas M, Jay Liu J, Baczek T, Kaliszan R, Wong MW, Buszewski B (2019) Column characterization and selection systems in reversed-phase high-performance liquid chromatography. Chem Rev 119:3674–3729. https://doi.org/10.1021/acs.chemrev.8b00246
85. López-fernández O, Domínguez R, Pateiro M, Munekata PES, Rocchetti G, Lorenzo JM (2020) Determination of polyphenols using liquid chromatography–tandem mass spectrometry technique (LC–MS/MS): a review. Antioxidants 9. https://doi.org/10.3390/antiox9060479
86. Arribas AS, Martínez-Fernández M, Chicharro M (2012) The role of electroanalytical techniques in analysis of polyphenols in wine. Trends Anal Chem 34:78–96. https://doi.org/10.1016/j.trac.2011.10.015
87. Sun MF, Jiang CL, Kong YS, Luo JL, Yin P, Guo GY (2022) Recent advances in analytical methods for determination of polyphenols in tea: a comprehensive review. Foods 11. https://doi.org/10.3390/foods11101425
88. Arráez-Román D, Fu S, Sawalha SMS, Segura-Carretero A, Fernández-Gutiérrez A (2010) HPLC/CE-ESI-TOF-MS methods for the characterization of polyphenols in almond-skin extracts. Electrophoresis 31:2289–2296. https://doi.org/10.1002/elps.200900679
89. Weingerl V (2012) A comparative study of analytical methods for determination of polyphenols in wine by HPLC/UV-vis, spectrophotometry and chemiluminometry. Macro Nano Spectrosc 448:357–368
90. Savic IM, Savic Gajic IM (2021) Optimization study on extraction of antioxidants from plum seeds (Prunus domestica L.). Optim Eng 22:141–158. https://doi.org/10.1007/s11081-020-09565-0
91. Tomaz I, Huzanić N, Preiner D, Stupić D, Andabaka Ž, Maletić E, Kontić JK, Ašperger D (2019) Extraction methods of polyphenol from grapes: extractions of grape polyphenols. In: Polyphenols in plants, pp 151–167. https://doi.org/10.1016/B978-0-12-813768-0.00010-4
92. León-González ME, Rosales-Conrado N (2017) Determination of ibuprofen enantiomers in breast milk using vortex-assisted matrix solid-phase dispersion and direct chiral liquid chromatography. J Chromatogr A 1514:88–94. https://doi.org/10.1016/J.CHROMA.2017.07.072
93. Lee J, Durst RW, Wrolstad RE, Barnes KW, Eisele T, Giusti MM, Haché J, Hofsommer H, Koswig S, Krueger DA, Kupina S, Martin SK, Martinsen BK, Miller TC, Paquette F, Ryabkova A, Skrede G, Trenn U, Wightman JD (2001) Determination of total monomeric anthocyanin pigment content of fruit juices, beverages, natural colorants, and wines by the pH differential method: collaborative study. J AOAC Int 88:1269–1278
94. Mónica Giusti M, Wrolstad RE (2001) Characterization and measurement of anthocyanins by UV-visible spectroscopy. Curr Protoc Food Anal Chem 1:F1–F2

95. Kuskoski EM, Rios JJ, Bueno JM, Fett R, Troncoso AM, Asuero AG (2012) Capillary gas chromatography-mass spectrometry (CGC-MS) analysis and antioxidant activities of phenolic and components of guarana and derivatives. Open Anal Chem 6:1–8
96. Xi J, Shouqin Z (2007) Antioxidant activity of ethanolic extracts of propolis by high hydrostatic pressure extraction. Int J Food Sci Technol 42:1350–1356. https://doi.org/10.1111/j.1365-2621.2006.01339.x
97. Saitta M, Lo CS, Salvo F, Di Bella G, Dugo G (2002) Gas chromatographic-tandem mass spectrometric identification of phenolic compounds in Sicilian olive oils. Anal Chim Acta 466:335–344
98. Gómez-Mejía E, Sacristán I, Rosales-Conrado N, León-González ME, Madrid Y (2023) Effect of storage and drying treatments on antioxidant activity and phenolic composition of lemon and clementine peel extracts. Molecules 28. https://doi.org/10.3390/molecules28041624
99. Sommella E, Pepe G, Pagano F, Ostacolo C, Tenore GC, Russo MT, Novellino E, Manfra M, Campiglia P (2015) Detailed polyphenolic profiling of Annurca apple (M. pumila Miller cv Annurca) by a combination of RP-UHPLC and HILIC, both hyphenated to IT-TOF mass spectrometry. Food Res Int 76:466–477. https://doi.org/10.1016/j.foodres.2015.05.044
100. Liu Y, Benohoud M, Galani Yamdeu JH, Gong YY, Orfila C (2021) Green extraction of polyphenols from citrus peel by-products and their antifungal activity against Aspergillus flavus. Food Chem X 12. https://doi.org/10.1016/j.fochx.2021.100144
101. Fanali C, Rocco A, Aturki Z, Mondello L, Fanali S (2012) Analysis of polyphenols and methylxantines in tea samples by means of nano-liquid chromatography utilizing capillary columns packed with core-shell particles. J Chromatogr A 1234:38–44. https://doi.org/10.1016/j.chroma.2011.12.103
102. Brighenti V, Groothuis SF, Prencipe FP, Amir R, Benvenuti S, Pellati F (2017) Metabolite fingerprinting of Punica granatum L. (pomegranate) polyphenols by means of high-performance liquid chromatography with diode array and electrospray ionization-mass spectrometry detection. J Chromatogr A 1480:20–31. https://doi.org/10.1016/j.chroma.2016.12.017
103. Giordano M, Pinela J, Dias MI, Calhelha RC, Stojković D, Soković M, Tavares D, Cánepa AL, Ferreira ICFR, Caleja C, Barros L (2021) Ultrasound-assisted extraction of flavonoids from kiwi peel: process optimization and bioactivity assessment. Appl Sci 11. https://doi.org/10.3390/app11146416
104. Gómez-Mejía E, Mikkelsen LH, Rosales-Conrado N, León-González ME, Madrid Y (2021) A combined approach based on matrix solid-phase dispersion extraction assisted by titanium dioxide nanoparticles and liquid chromatography to determine polyphenols from grape residues. J Chromatogr A 1644:462128

Chapter 2

Extraction and Quantification of Plant-Based Carotenoids

Yogesh Gat, Bhavna Negi, Kirtan Kotak, and Arya Kadam

Abstract

Carotenoids are an essential part of the human diet and are found in a lot of fruits and vegetables. Maintaining human health requires consuming a diet high in several key vitamins and minerals.

A type of naturally occurring fat-soluble pigment known as carotenoid, or tetraterpenoid, is made up of eight molecules of five-carbon isoprenoid. A hydrocarbon with 40 carbon atoms and two terminal rings that can give rise to yellow, orange, or red hue makes up a carotenoid. Nowadays, more than 600 carotenoids have been identified in nature, and the average human diet contains roughly 40 of them. Two categories of carotenoids exist: the oxygen-containing xanthophylls (which include lutein, astaxanthin, and zeaxanthin) and the hydrocarbon carotenoids, or carotenes (which include lycopene, β-carotene, and α-carotene). This chapter explains different techniques used for the extraction of carotenoids: conventional methods, such as Soxhlet extraction, maceration extraction, and accelerated solvent extraction, and novel methods, which include ultrasound extraction (UAE), microwave-assisted extraction (MWE), and enzyme-assisted extraction (EAE).

Key words Carotenoids, α-carotene, β-carotene, Extraction, Xanthophyll, Quantification

1 Introduction

Color is a key factor in food and beverage acceptability, enhancing their appearance and quality. Natural pigments from plants contribute to the vibrant and attractive colors of fruits and vegetables. The main plant pigments used in the food industry are carotenoids, chlorophyll, betalains, and anthocyanins. Carotenoids are hydrophobic metabolites with minimal water solubility but exhibit high solubility in oil. These compounds possess a chromophore, characterized by a central polyene chain of conjugated double bonds, which maximally absorbs light at wavelengths between 400 and 500 nm. This absorption imparts the characteristic yellow-orange and reddish hues to carotenoids.

The most abundant class of pigments in nature, carotenoids include over 600 structurally defined members and an estimated annual production of 100 million tons. They are found in all

Ashwani Kumar, Vidisha Tomer and Piyoosh Kumar Babele (eds.), *Phytochemicals: Extraction, Quantification, Isolation, and Characterization*, Methods and Protocols in Food Science, https://doi.org/10.1007/978-1-0716-4840-7_2,

photosynthetic organisms and are in charge of the majority of the yellow-toned colors found in fruits and flowers. Carotenoids are utilized commercially as food coloring and nutritional supplements [1], with an anticipated $935 million worldwide market by 2025. Dietary carotenoids draw great attention because of their antioxidant qualities and potential to treat technological diseases. The substantial progress made in carotenoid synthesis and deposition in plants have coincided with the substantial quantity of studies on this subject.

Plant compounds are consumed as part of a diet or as a nutraceutical and are good sources of carotenoids. A class of tetraterpenoids are called carotenoids. The fundamental framework of carotenoid structures consists of isoprenoid units. Carotenoids mainly belong to two classes: xanthophylls and carotenes. The pigments lycopene and carotenoids are mainly contributing to the color of different fruits and vegetables. Carotenoids aid in photosynthesis and give plant products their pigment. Carotenoids cannot be synthesized by animals. Therefore, the sole source of carotenoids is plant-based goods.

The primary dietary sources of carotenoids are yellow-orange fruits and vegetables. Only the red stigmata of crocus blooms contain the saffron apocarotenoids [2]. Conversely, lutein, violaxanthin, β-carotene, and neoxanthin are abundant in green leafy vegetables, and β-cryptoxanthin is mostly present in papaya, peaches, and citrus fruits like oranges and tangerines [3]. α-carotene is abundant in sweet potatoes, carrots, pumpkin, apricots, cantaloupe, and dark green vegetables like broccoli and spinach [4]. Since the majority of non-plant species, including humans, are unable to manufacture carotenoids, they must consume them through their regular diet for their bodies to function normally. The most common carotenoids identified in human plasma are lutein, zeaxanthin, β-cryptoxanthin, α-carotene, and β-carotene [5].

In the photosynthetic bacteria, algae, and plants, carotenoids and chlorophylls are necessary molecules that participate in photosynthesis and photo-protection. Through the singlet–singlet excitation transfer mechanism, carotid endosomes capture light energy and transmit it to chlorophylls. During photosynthesis, a lower energy state transition known as singlet–singlet transfer is employed.

During photosynthesis, oxygen and light combine to form reactive O_2 species, such as singlet oxygen, hydroxy radicals, and superoxide anion radicals. Singlet oxygen quenching is a notable property of carotenoids containing more than 11 conjugated double bonds. Carotenoids absorb singlet oxygen's thermal energy and release it through the vibration of polyenes.

In higher plants, lutein and lutein epoxide are components of the xanthophyll cycles. Diatoxanthin and diadinoxanthin make up

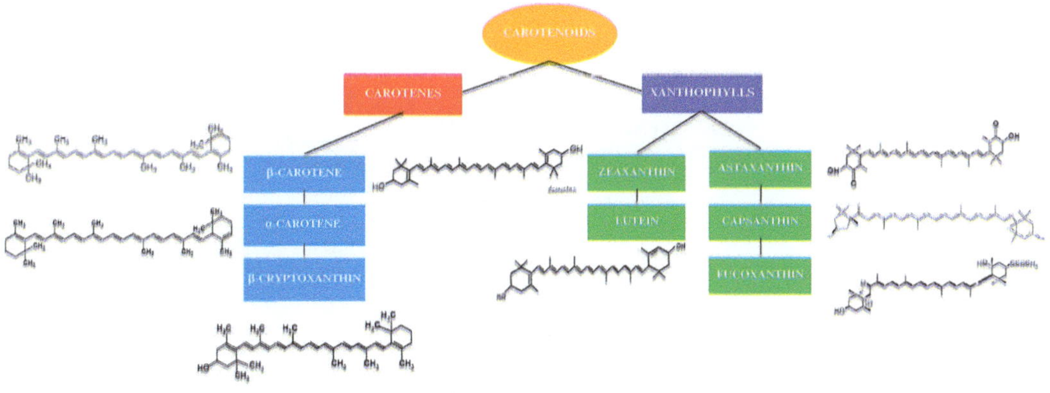

Fig. 1 Classification of carotenoids

the xanthophyll cycle in diatoms [6]. Less than 10% of the 600 identified carotenoids from natural sources function as vitamin A precursors. Human blood and tissues contain several dietary carotenoids, both with and without provitamin A activity. Beta-carotene, which makes up about 15–30% of all blood carotenoids, is the most nutritionally active kind of carotenoid. The main process that produces vitamin A is the oxygen-dependent central cleavage of provitamin α-carotenoids, and beta-carotene. A number of carotenoids have been shown to improve immunological response, prevent mutagenesis, lessen induced nuclear damage, and protect against different types of cancer in tissues, cells, and entire animals.

Recently, the extraction, identification, and purification of natural colorants, including carotenoids have drawn a lot of interest from food technologists and processors. Carotenoids can be generically categorized into two classes: xanthophylls (zeaxanthin, lutein, astaxanthin, etc.) and carotenes (α-carotene, β-carotene, etc.) as shown in Fig. 1.

Low temperature, shielding from light, and oxygen exclusion (e.g., by vacuum or hot filling, oxygen-impermeable packing, inert environment) reduce carotenoid degradation during storage. Carotenoids are extremely unsaturated and, therefore, susceptible to oxidation and isomerization as shown in Fig. 2. The conversion of trans-carotenoids, which are found in nature in their typical configuration, to cis-isomers is facilitated by heat treatment, light exposure, and acid exposure.

The methods of carotenoid extraction are divided into two major categories: conventional methods and novel methods. Conventional methods for carotenoid extraction include maceration, Soxhlet extraction, and accelerated solvent extraction technique as shown in Fig. 3. These techniques involve usage of organic solvents for the extraction. These techniques are usually associated with high extraction yields without utilizing sophisticated instruments and ensure the highest recovery of carotenoids.

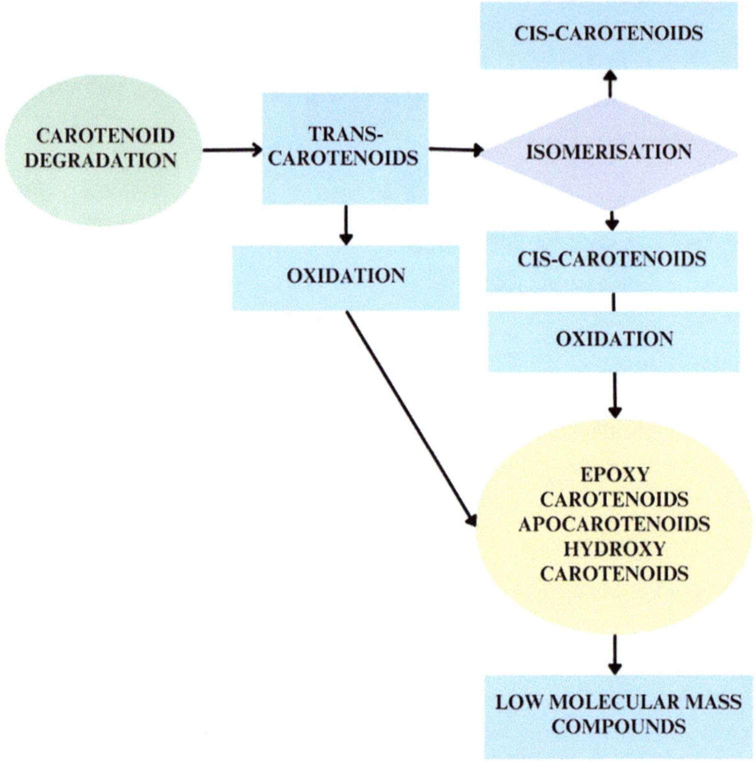

Fig. 2 Compounds produced on degradation of carotenoids

Maceration involves soaking plant material in a solvent to extract molecules like carotenoids. The solvent penetrates the plant material, dissolving the desired compounds over time. This method is simple, cost-effective, and suitable for heat-sensitive compounds. However, it can be time-consuming and may not extract all compounds efficiently, requiring longer extraction times or multiple extractions for optimal results [7].

Soxhlet extraction relies on the thorough extraction of organic compounds, or analytes, using an organic solvent in a Soxhlet system. The sample is kept inside a porous thimble and the organic solvent is continually refluxed through it as per Fig. 4. Soxhlet extraction, developed by Franz Von Soxhlet, involves placing plant materials in a thimble connected to a flask containing solvent. Vapor rises, condenses on the material, and falls back into the flask, repeating until extraction is complete [8].

The accelerated solvent extraction technique involves the extraction of carotenoids from dehydrated solid, semisolid samples with a liquid solvent. Extraction is carried out using different organic solvents, which initiate saponification. Saponification hydrolyzes esters, which leads to the release of pigments.

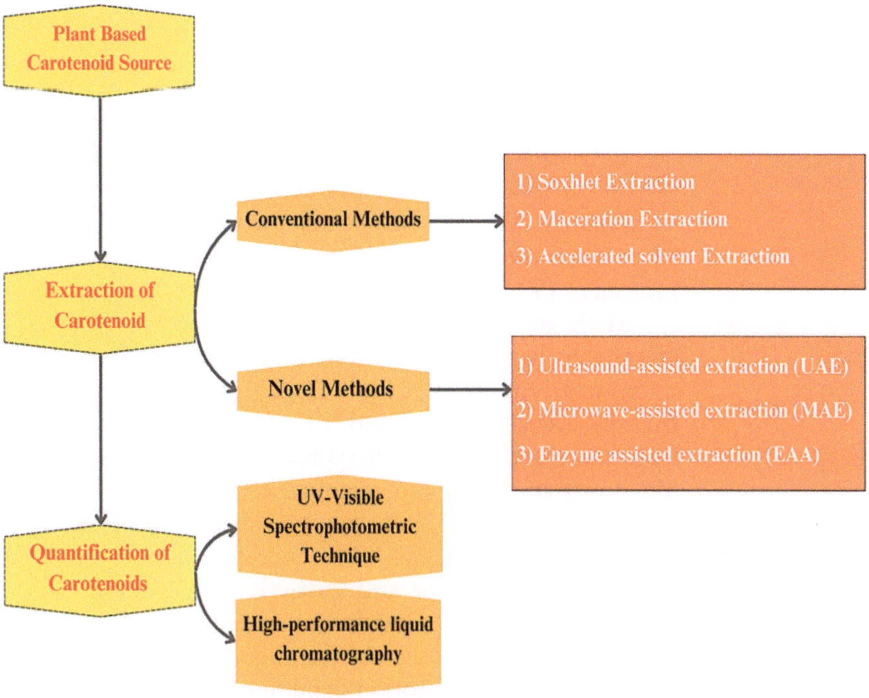

Fig. 3 Different steps involved in the extraction and analysis of carotenoids from natural sources

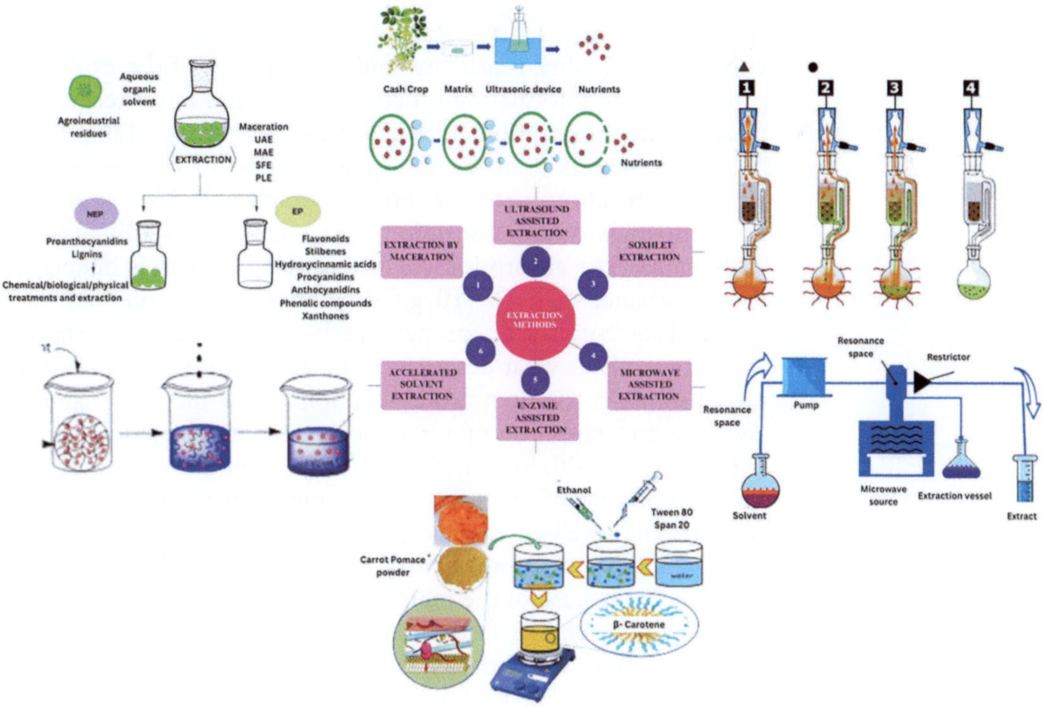

Fig. 4 Pictorial representation of different conventional and novel methods of extraction of carotenoids

Nowadays, environment-friendly extraction methods like ultrasound-assisted extraction (UAE), microwave-assisted extraction (MAE), and enzyme-assisted extraction (EAE) are used to avoid the negative aspects of conventional extraction methods.

UAE uses high-frequency sound waves with a minimal amount of solvent to extract bioactive molecules like carotenoids from a solid matrix. The high-frequency waves produce locally high pressures that lead to cavitation and the extraction of bioactive molecules [9].

MAE separates analytes from the sample's matrix into the solvent by heating the solvent in contact with the sample using microwave energy. EAE is a highly efficient and environment-friendly extraction method that separates bioactive molecules from raw materials using enzymes. Enzymes act as catalysts, breaking down cell walls and membranes to release the desired compounds. Commonly used enzymes include proteases, pectinases, cellulases, tannases, and carbohydrates. The novel methods of extraction are efficient and cost-effective. They require minimal solvent and have short extraction times. EAE offers several advantages over other extraction techniques, including higher yields, lower energy consumption, and increased selectivity for target compounds [10]. The use of enzymes and control of pH and temperature also made this process expensive [11]. Novel methods not only reduce energy or time consumption but also increase solvent recovery, reduce the amount of solvent needed, and enhance the quality, sensitivity, and effectiveness of the extraction processes. The information on conventional and novel techniques used and their extraction efficiency is also provided in Table 1.

The methods used for the quantification of carotenoids are either spectrophotometric or chromatographic. The major portion of carotenoids shows absorption in the spectrum's 400–500 nm visible area. The absorption coefficient ($A^{1\%}$), which is defined as the absorbance of a 1% (10 g/L) solution of a carotenoid, in a specified solvent, in a 1-centimeter path-length cuvette, at a specific wavelength (l), is what determines the absorbance (A) of carotenoid in solution. Carotenoids in solution follow the Beer-Lambert law. The concentration of a purified (standard) carotenoid can be measured using this information. Molarity is another way to express the absorption coefficient. The values might have some differences, considerable levels of error, or uncertainty. It is noteworthy that certain published numbers may have a considerable degree of uncertainty because of the inherent challenges in the processes for precisely determining absorption coefficients. The spectrum of a substance under study should be directly compared to an original or pure standard. The traditional method of separating carotenoids for quantitative measurement is called open-column

Table 1
Comparative assessment of different conventional and novel extraction techniques used for extraction of carotenoids

Method of extraction	Yield of carotenoids	Remark	References
Solvent extraction	Apricots peel: 17.3 to 222.09 μg/g FW Apricot flesh: 2.53 to 98.17 μg/g FW	Extraction using a hexane, acetone, ethanol, and butylhydroxytoluene mixture revealed that apricot peels, among 41 varieties studied, contain higher levels of carotenoids. Additionally, 13 different types of carotenoids were identified	[1]
	Carrot: 18.3 mg/100 g of sample Spinach: 5.6 mg/100 g of sample Capsicum: 2.4 mg/100 g of sample Beetroot: 1.9 mg/100 g of sample Broccoli: 1.3 mg/100 g of sample	Extraction was carried out with the help of 1:1 hexane-acetone mixture. Broccoli showed lower β-carotene levels after TLC purification	[2]
	Tomato peels: 608.94 ± 10.05 mg/kg (Hexane) 320.35 ± 3.36 mg/kg (ethyl acetate) 284.53 ± 7.03 mg/kg (ethanol)	Study indicates that the extraction method greatly affects lycopene recovery, with a 99.98% certainty. The SC-CO2 method produced more lycopene than solvent extraction, even though lycopene doesn't dissolve well in supercritical fluid	[3]
	Goji berries: 5.5 mg/g of dry weight	As compared to other organic solvents, ethanolic extraction indicates higher amount of Zeaxanthin	[4]
Ultrasonic assisted extraction	Carrot pomace: 21.67 ± 0.40 μg/g of sample	Mixing carrots and flaxseed oil in a 1:1 ratio produced the most carotenoids when using a high shear disperser at 20,000 rpm for 12 min	[5]
	Orange waste powder: 131.05 to 206.68 ppm/5 g of sample	Study indicates that usage of ultrasound in combination with pectinase enzyme gives maximum yield as compared to usage of ultrasonication alone. Increasing the ultrasound time from 10 to 120 min resulted in higher β-carotene extraction due to the cavitation effect	[6]
	Pumpkin peel: 31.067 ± 2.45 μg/g of sample	Various extraction methods were studied. The extraction efficiency was UAE >MAE>CE	[7]

(continued)

Table 1 (continued)

Method of extraction	Yield of carotenoids	Remark	References
Microwave-assisted extraction	Passion fruit: 1207.93 μg/100 g (OO—extracting medium) 1185.45 μg/100 g (SO—extracting medium)	Two solvents were used and effect of temperature & time on carotenoids extraction were studied for MAE. The result showed that OO is the best solvent for extraction of pigment from PFP	[8]
	Gac fruits: 156 to 236 mg/100 g DW	The study showed that higher microwave power (360W) quickly increased carotenoid extraction to 236 mg/100 g DW in 4 min. Lower powers (240W and 120W) were also effective, yielding carotenoids that increased steadily with longer extraction times. The best extraction (262 mg/100 g DW) was achieved at 120W over 25 min	[9]
	Tomato peel:13.362 mg/100 g of sample	Study optimized microwave parameters, finding that using ethyl acetate at 400W for 1 min yielded the highest all-trans-lycopene extraction at approximately 13 mg/100 g	[10]
Enzyme assisted extraction	Pumpkin: 0.38 mg/100 g of sample	Two enzymes (cellulose and pectinase) were used for extraction of carotenoids. Cellulose shows good results for extraction	[11]
	Red capsicum: 41.72 to 279.83 mg/100 g of sample	Four different enzymes (Viscozyme, pectinase, cellulase, amylase) were used to study the extraction yield. Viscozyme showed higher yield as compared to other enzymes	[12]
	Tomato processing waste: 342 ± 18 mg/100 g of sample	This study demonstrates that combining enzymes with cellulolytic and pectinolytic activities significantly boosts lycopene recovery from tomato peel waste during processing	[13]
	Sunflower: 28–147 ppm	Viscozyme with the help of various commercial green solvents were used for extraction of carotenoids	[14]

CE conventional extraction, *DW* distilled water, *PFP* passion fruit pulp, *MAE* microwave-assisted extraction, *OO* olive oil, *SO* sunflower oil, *TLC* thin layer chromatography, *SC-CO2* supercritical fluid CO_2 extraction

chromatography (OCC), which involves chromatography in descending, gravity-flow columns (often with a small amount of pressure provided by a water aspirator). It is also helpful in the isolation and purification of carotenoids for use as HPLC standards. Optical isolation of the carotenoid pigments is monitored. Another option is to apply low pressure at the top of the column (for example, using N2 gas); this method is known as flash column chromatography.

2 Materials

2.1 Maceration

- Conical flasks (volume as per the sample), Water bath.
- Solvents like hexane/acetone/ethanol or a mixture of more than one solvent can be selected based on the suitability of the sample.

2.2 Soxhlet Extraction

- Soxhlet extraction apparatus [flat-bottomed flask, condenser, and Soxhlet extraction thimble (glass or cellulose)].
- Solvents like hexane, acetone, ethanol, methanol, hexane, ethyl acetate can be used alone or in combination with water.

2.3 Accelerated Solvent Extraction

- Weighing balance, Extraction vessel, Explosion proof shaft mixer, Separating funnel.
- Acetone, n-Hexane, or hexane/acetone 1:1 (v/v).
- 40% (w/v) KOH in methanol (*see* **Note 1**).
- 10% (w/v) sodium sulfate (Na_2SO_4) (*see* **Note 1**).
- Anhydrous sodium sulfate (Na_2SO_4).
- Whatman no. 42 filter paper.

2.4 Ultrasound-Assisted Extraction

- Weighing balance, Ultrasonicator, Extraction flask, Hot plate.
- Organic solvents of analytical grade.
- Whatman No. 41 paper.

2.5 Microwave-Assisted Extraction

- Weighing balance, Microwave, Extraction flask, Ice bath, Centrifuge (up to 5000 rpm), Deep freezer.
- Organic solvents of analytical grade (e.g., Ethyl ether).
- Whatman No. 41 paper.

2.6 Enzyme-Assisted Extraction

- Weighing balance, Screw top conical flask, Water bath with shaker, Magnetic stirrer.
- Organic solvents of analytical grade (e.g., hexane).
- Required enzymes (Pectinases, cellulases, proteases, etc.).
- 0.45 m nylon filter.

2.7 Quantification of Carotenoids by Using UV-Visible Spectrophotometric Techniques

- Weighing balance, UV-Spectrophotometer, Refrigerator, 0.45 mm filter.
- Carotenoid standards.
- Solvents for dissolving carotenoid standards.

2.8 Quantification of Carotenoids by Using Chromatographic Techniques

- Weighing balance, UV-Spectrophotometer, Refrigerator, 0.45 mm filter.
- Carotenoid standards.
- Solvents for dissolving carotenoid standards.

3 Methods

3.1 Maceration

1. Grind the plant material to a fine powder using a mortar and pestle or a mechanical grinder.
2. Place the powdered sample in a clean, dry container. Add enough solvent to cover the sample completely.
3. Seal the container and shake it gently to ensure good contact between the solvent and the sample.
4. Allow the mixture to stand at room temperature for a specified period, usually 24 to 48 h, with occasional shaking or stirring.
5. After maceration, filter the mixture using a filter paper and a funnel to separate the liquid extract from the solid residue.
6. Concentrate the extract using a rotary evaporator or by gentle evaporation under reduced pressure at a low temperature to obtain the carotenoid extract.

3.2 Soxhlet Extraction

1. Grind the sample (e.g., plant material) to a fine powder. Weigh an appropriate amount of the sample (around 2–5 g) accurately.
2. Place the sample in a Soxhlet extraction thimble. Insert the thimble into the Soxhlet extractor.
3. Add 150–200 mL of hexane or hexane-acetone mixture to the round-bottomed flask.
4. Heat the flask gently to reflux the solvent.
5. Allow the extraction to continue for 6–8 h, ensuring that the solvent completely cycles through the sample.
6. After extraction, remove the solvent from the flask using a rotary evaporator under reduced pressure at a low temperature (around 40 °C) to obtain the carotenoid extract.

3.3 Accelerated Solvent Extraction

1. Take a 500 mL extraction vessel and put dehydrated plant material along with a filter paper into it.
2. To suspend paste in solvent, add 3 vol (v/w) of solvent (1:1 (v/v) hexane/acetone or hexane alone), and mix for 15 min using an explosion-proof shaft mixer (*see* **Note 2**).
3. Further, filter the solution using Whatman no. 42 filter paper and save the filtrate.

4. Carefully remove the filtrate from the filter paper. Then if at all feasible, extract again with two volumes of solvent. Reduce the solvent volume used for extractions until the filtrate has no discernible color. Save and combine all of the filtrates (*see* **Note 3**).
5. Concentration of extract to ~40 mL is carried out at ≤55 °C with the help of a rotary evaporator attached to a vacuum pump.
6. After adding 3 mL of saponifying solution, stir for 45 min at 56 °C.
7. Saponifying solution used is 40% (w/v) KOH in methanol.
8. Further, one volume of salting out solution (10% (w/v) Na2SO4) should be added to a 125-mL separatory funnel containing the saponified extract.
9. Mixture will be separated in the form of bottom layer and upper layer. After removing the bottom layer, wash the upper layer thrice with 10 mL water.
10. Add 3g of anhydrous Na_2SO_4 followed by filtration using Whatman no. 42 filter paper.

3.4 Ultrasound-Assisted Extraction

1. Ultrasound-assisted extraction of carotenoids can be carried out in an ultrasonic processor of 750W at amplitude level and frequency and of 70% and 20 kHz respectively (*see* **Note 4**).
2. Ultrasonic intensity for plant-based material (e.g., Carrot pomace) can be 48 W cm^{-2}.
3. Grind dried sample pomace is to be placed in a 100 mL jacketed vessel at the ratio of 1 g/50 mL organic solvent.
4. Once the desired temperature is achieved, subject the sample to sonication (*see* **Note 5**).
5. After extraction is completed, filter the mixture with the help of Whatman No. 41 filter paper, followed by centrifugation at 4000 rpm for 10 min.
6. Evaporate the solvent by heating the extract at 30 °C and store at 4 °C for further use.

3.5 Microwave-Assisted Extraction

1. Pretreatment can be applied for the sample, if needed, to enhance the carotenoid extraction (for example, blanching in water).
2. Weigh a suitable amount of the sample (e.g., orange, carrot, tomato) and place it in an extraction flask.
3. Add the solvent to the flask according to the optimized solid-to-liquid ratio (e.g., 1:20 for ethyl acetate).

4. Secure the flask in the MAE extractor chamber. Set the MAE system to the optimized parameters (e.g., 400 W power for 60 s).
5. Apply intermittent microwave radiation, if desired, to minimize thermal degradation (e.g., with an intermittency ratio of $\alpha = 1/4$).
6. Remove the flask from the chamber after every 5 min and place it in an ice bath to cool, to avoid excessive temperature rise. Collect the extracts after the extraction process is complete.
7. Centrifuge the extracts to separate the oil and residue (e.g., at 4500 rpm for 45 min).
8. Store the extracts for further analysis (e.g., at −20 °C).

   Precautions:

   The optimization of microwave power, solvent volume, and radiation frequency is essential in MAE to prevent carotenoid breakdown.

3.6 Enzyme-Assisted Extraction

1. Choose suitable enzymes for the extraction.
2. Prepare a mixture of the selected enzymes in equal volumes (e.g., 50:50 Pec–Cel) for optimal efficiency.
3. Place 0.5 g of the sample material containing carotenoids in a screw-top conical flask.
4. Add the enzyme solution to the flask containing the sample. Ensure thorough mixing by magnetic stirring.
5. Place the flask in a thermostated water bath at a controlled temperature (e.g., 35 ± 1 °C).
6. After the enzyme treatment, add a known volume of a suitable solvent (e.g., 30 mL of hexane) to the flask.
7. Maintain agitation at the run temperature for the required time to facilitate extraction. Stop stirring and allow the organic and aqueous phases to separate.
8. Take a 2-mL sample of the hexane supernatant and filter it through a 0.45-μm nylon filter.
9. Analyze the filtered sample for the target compound content using suitable analytical methods.

3.7 Quantification of Carotenoids by Using UV-Visible Spectrophotometric Techniques

This step involves dissolving commercially bought carotenoid standards in an appropriate solvent and measuring the absorbance at the maximum wavelength. The steps involved in quantifying a sample of individual carotenoids using a standard are outlined below.

1. Take (1–5 mg) carotenoids and dissolve in suitable solvents to make up the volume (10–15 mL) in a volumetric flask (*see* **Note 6**).

2. Diluting the solution (e.g., 1:50) in the chosen solvent to yield 0.3–0.7AU as measured with a spectrophotometer at max wavelength (*see* **Note 7**).
3. Follow the manufacturer's instructions to warm up the spectrophotometer.
4. In a cuvette, zero the spectrophotometer using a solvent (*see* **Note 8**).
5. Place a cuvette containing the carotenoid solution into the spectrophotometer's sample cell holder (*see* **Note 9**).
6. Find the absorbance at the max wavelength. Note down the readings of each sample indicated by spectrophotometer (*see* **Note 10**).
7. Scan to enable fine structural measurement.
8. Calculate the carotenoid concentration by using the formula:

$$\text{Carotenoid concentration} = \frac{A \times V_1}{A^{1\%}} \times C^{1\%}$$

   Where,

   A = absorbance reading of diluted sample

   V_1 = dilution factor

   $A^{1\%}$ = absorbance of 1% solution

   $C^{1\%}$ = concentration of 1% solution

9. To determine the chromatographic purity of the solution, HPLC analysis should be performed after the carotenoid concentration has been measured.

In dietary, biological, or natural extracts, where a mixture of carotenoids would be expected, this approach uses calculations employed for specific carotenoids of high purity to determine the "total carotenoids" present. Higher precision can be attained as extracts are refined to contain only one component. Since the fine structure of a mixture of carotenoids can only be detected after HPLC separation and identification.

3.8 Quantification of Carotenoids by Using Chromatographic Techniques

A. **Separation and Analysis**

Both the more polar xanthophylls and the nonpolar carotenes are commonly found in natural product samples. For quantitative analysis of carotenoids, reversed-phase HPLC using C18 columns has been the recommended method for carotenoids separation. Because of its weak hydrophobic interactions with the analytes (which should make it less damaging than the polar forces in normal-phase OCC), wide commercial availability, compatibility with most carotenoid solvents, and compatibility with the polarity range of carotenoids, the C18 column is widely used. Manufacturers offer a wide range of

C18 reversed-phase materials that vary in terms of end capping, bonded phase type (monomeric or polymeric), and carbon loading intensity.

B. **Isocratic Carotenoid Identification by Using Polymeric C18 Wide Pore**

In order to separate the carotenoids, this method mainly depends on the special properties of the column and needs less sophisticated equipment. It integrates the use of a polymeric C18 column, which has been demonstrated to provide exceptional selectivity for molecules with identical structures, including geometric isomers. The simultaneous measurement of tocopherols, but not retinol, is possible with the inclusion of a second detector or the use of a diode-array detector. A solvent gradient can be used in the procedure to allow for the measurement of retinol as shown in Fig. 5.

1. Mix 900 mL of HPLC-grade suitable organic solvent (e.g., methanol), 100 mL of HPLC-grade suitable organic solvent (e.g., acetonitrile), and 1 mL of HPLC-grade triethylamine (TEA) to prepare the mobile phase (*see* **Note 11**).
2. Use an inline vacuum degasser, vacuum filtration, or ultrasonic agitation to remove the mobile phase's gas.

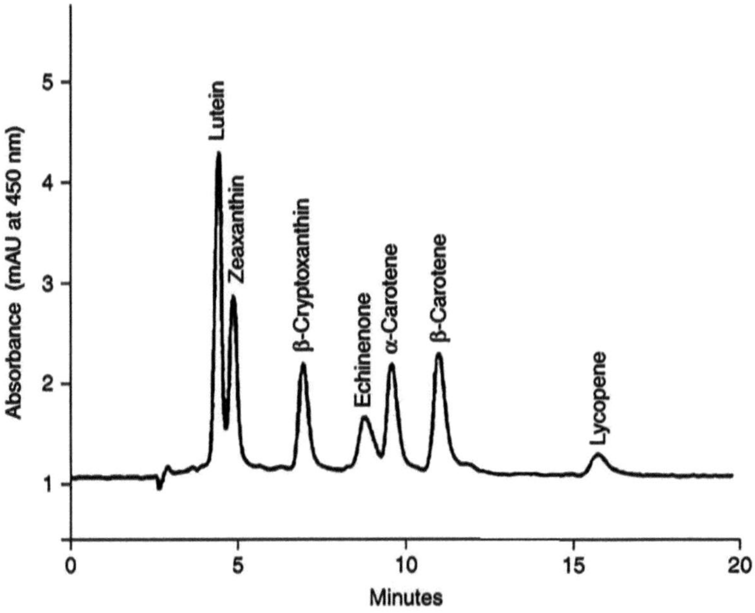

Fig. 5 Isocratic carotenoid standard analysis using HPLC. The column has a temperature of 25 °C, a visual detection at 450 nm, a 90:10 methanol: acetonitrile mobile phase mixture, and a 5-mm × 250-mm × 4.6-mm 201TP column, reproduced from [15]

3. Set a suitable flow rate (e.g., 1.0 mL/min) with the HPLC pump.
4. Set the UV-Vis detector at 450 nm (or 436 nm, if utilizing a filter photometer).
5. To create a standard curve, inject each individual standard and the standard mixes, along with any optional internal standards.
6. Add 10–50 milliliters of the sample and any optional internal standard that has been dissolved in a solvent that is miscible with the mobile phase such as acetonitrile, methanol, or ethanol (*see* **Note 12**).
7. Multiply the analytic peak areas by the calibration response factors to determine the final concentrations of carotenoids in the samples. To determine the carotenoid concentration in the original sample (i.e., initial concentration), apply sample weight and dilution factors (*see* **Note 13**).

4 Notes

1. To prepare percent solutions, measure the solute in grams equivalent to the desired percentage, dissolve it in a small quantity of distilled water, and adjust the final volume to 100 mL in a volumetric flask. For example, to make a 10% sodium sulfate solution, accurately weigh 10 g of sodium sulfate, dissolve it in a small amount of distilled water, transfer this to a volumetric flask, and ensure the final volume is precisely 100 mL. For larger volumes, the calculations can be adjusted accordingly.
2. No heating is required during mixing.
3. If the filtrate is adhered to filter paper, then both filtrate and paper should be placed in an extraction vessel.
4. Ultrasonic probe should be properly dipped into the sample.
5. Time and temperature combination varies, and depends on physicochemical properties of the plant material.
6. The material must dissolve fully. When working with crystalline samples, dissolution can be facilitated by adding a small amount of a more effective solvent such as dichloromethane at the beginning before the volume is made for spectrophotometric measurement. No more than 10% of the total volume should be needed for the majority of commonly assayed carotenoids; lycopene, however, may require up to 100%. It is convenient to dissolve carotenoids provided in sealed ampoules by adding repeated tiny aliquots of the more efficient solvent to the ampule and transferring to a volumetric flask. Since it is difficult to determine total dissolution visually, it is best to filter the

solution using a 0.45 mm filter that is appropriate for the solvent. Stock solutions that have been refrigerated or otherwise stored should be allowed to reheat to room temperature, refiltered, and have a "new" concentration calculated before making a new working solution that will be subjected to a chromatographic purity test.

7. To ensure the measurement confidence, it is advised to make at least two separate dilutions.
8. Filling the cuvettes to two-thirds full is usually sufficient; filling them to the brim is not necessary. This helps to prevent unintentional spills. Cuvettes need to be maintained incredibly clean; don't handle the cell from its surfaces.
9. If necessary, use a fine tissue to thoroughly clean the optical faces of the cells. Avoid any spilling before inserting them into the cell holder.
10. To prevent solvent evaporation or deterioration, spectrophotometric measurements should be taken as soon as the solution has been placed in the cuvette.
11. TEA acts as a modifier to stop oxidation and nonspecific adsorption.
12. This approach shows how to separate carotenoid standards. It takes 20 min to separate lutein from lycopene. In contrast to many other approaches, this one uses a different elution order: lutein, zeaxanthin, b-cryptoxanthin, echinenone, a-carotene, b-carotene, and lycopene. Column temperature affects carotenoid retention and separation; lutein and zeaxanthin may not separate efficiently at temperatures over 20–25 °C. A diode array detector, a second UV detector set at 280–300 nm, or a fluorescence detector set at 296 nm excitation and 336 nm emission can all be used to test tocopherols simultaneously.
13. The ratio of the concentration of analyte versus the peak area or height generated during calibration under the specified set of parameters such as wavelength, solvent composition, and column is known as the response factor. Response factors for every analyte are typically produced by data systems using calibration data.

References

1. Fraser PD, Bramley PM (2004) The biosynthesis and nutritional uses of carotenoids. Prog Lipid Res. 43:228–265. https://doi.org/10.1016/j.plipres.2003.10.002
2. Bouvier F, Isner JC, Dogbo O et al (2005) Oxidative tailoring of carotenoids: a prospect towards novel functions in plants. Trends Plant Sci 10:187–194. https://doi.org/10.1016/j.tplants.2005.02.007
3. Priyadarshani AMB, Jansz ER (2014) A critical review on carotenoid research in Sri Lankan context and its outcomes. Crit Rev Food Sci Nutr 54:561–571. https://doi.org/10.1080/10408398.2011.595019
4. Farré G, Sanahuja G, Naqvi S et al (2010) Travel advice on the road to carotenoids in plants. Plant Sci 179:28–48. https://doi.org/10.1016/j.plantsci.2010.03.009

5. Aizawa K, Inakuma T (2007) Quantitation of carotenoids in commonly consumed vegetables in Japan. Food Sci Technol Res 13:247–252. https://doi.org/10.3136/fstr.13.247
6. Maoka T (2020) Carotenoids as natural functional pigments. Nat Med 74:1–16. https://doi.org/10.1007/s11418-019-01364-x
7. Fajriati I, Ikhsani AY, Monitasari A et al (2022) The effect of extraction method on the extract yield in the carotenoid pigment encapsulation for halal natural pigment. Indones J Halal Res. 4:97–106. https://doi.org/10.15575/ijhar.v4i2.17188
8. Adadi P, Barakova NV, Krivoshapkina EF (2018) Selected methods of extracting carotenoids, characterization, and health concerns: a review. J Agric Food Chem. 66:5925–5947. https://doi.org/10.1021/acs.jafc.8b01407
9. Liu Y, Liu HY, Xia Y et al (2021) Screening and process optimization of ultrasound-assisted extraction of main antioxidants from sweet tea (Lithocarpus litseifolius [Hance] Chun). Food Biosci 43:101277. https://doi.org/10.1016/j.fbio.2021.101277
10. Miękus N, Iqbal A, Marszałek K et al (2019) Green chemistry extractions of carotenoids from Daucus carota L. – supercritical carbon dioxide and enzyme-assisted methods. Molecules 24:4339. https://doi.org/10.3390/molecules24234339
11. Nath P, Kaur C, Rudra SG (2016) Enzyme-assisted extraction of carotenoid-rich extract from red capsicum (Capsicum annuum). Agric Res 5:193–204. https://doi.org/10.1007/s40003-015-0201-7
12. Zhou W, Niu Y, Ding X et al (2020) Analysis of carotenoid content and diversity in apricots (Prunus armeniaca L.) grown in China. Food Chem 330:127223. https://doi.org/10.1016/j.foodchem.2020.127223
13. Rebecca LJ, Sharmila S, Das MP et al (2014) Extraction and purification of carotenoids from vegetables. J Chem Pharm Res 6:594–598
14. Kehili M, Kammlott M, Choura S et al (2017) Supercritical CO2 extraction and antioxidant activity of lycopene and β-carotene-enriched oleoresin from tomato (Lycopersicum esculentum L.) peels by-product of a Tunisian industry. Food Bioprod Process 102:340–349. https://doi.org/10.1016/j.fbp.2017.02.002
15. Wrolstad RE (ed) (2005) Handbook of food analytical chemistry. Wiley, Hoboken
16. Patsilinakos A, Ragno R, Carradori S et al (2018) Carotenoid content of Goji berries: CIELAB, HPLC-DAD analyses and quantitative correlation. Food Chem 268:49–56. https://doi.org/10.1016/j.foodchem.2018.06.013
17. Tiwari S, Upadhyay N, Singh AK et al (2019) Organic solvent-free extraction of carotenoids from carrot bio-waste and its physico-chemical properties. J Food Sci Technol 56:4678–4687. https://doi.org/10.1007/s13197-019-03920-5
18. Shahram H, Dinani ST (2019) Optimization of ultrasonic-assisted enzymatic extraction of β-carotene from orange processing waste. J Food Process Eng 42:e13042. https://doi.org/10.1111/jfpe.13042
19. Sharma M, Bhat R (2021) Extraction of carotenoids from pumpkin peel and pulp: comparison between innovative green extraction technologies (Ultrasonic and microwave-assisted extractions using corn oil). Foods 10:787. https://doi.org/10.3390/foods10040787
20. Chutia H, Mahanta CL (2021) Green ultrasound and microwave extraction of carotenoids from passion fruit peel using vegetable oils as a solvent: optimization, comparison, kinetics, and thermodynamic studies. Innov Food Sci Emerg Technol 67:102547. https://doi.org/10.1016/j.ifset.2020.102547
21. Chuyen HV, Nguyen MH, Roach PD et al (2018) Microwave-assisted extraction and ultrasound-assisted extraction for recovering carotenoids from Gac peel and their effects on antioxidant capacity of the extracts. Food Sci Nutr 6:189–196. https://doi.org/10.1002/fsn3.546
22. Ho KK, Ferruzzi MG, Liceaga AM et al (2015) Microwave-assisted extraction of lycopene in tomato peels: effect of extraction conditions on all-trans and cis-isomer yields. LWT Food Sci Technol 62:160–168. https://doi.org/10.1016/j.lwt.2014.12.061
23. Ghosh D, Biswas PK (2016) Enzyme-aided extraction of carotenoids from pumpkin tissues. Indian Chem Eng 58:1–11. https://doi.org/10.1080/00194506.2015.1046697
24. Zuorro A, Fidaleo M, Lavecchia R (2011) Enzyme-assisted extraction of lycopene from tomato processing waste. Enzyme Microb Technol 49:567–573. https://doi.org/10.1016/j.enzmictec.2011.04.020
25. Ricarte GN, Coelho MAZ, Marrucho IM (2020) Enzyme-assisted extraction of carotenoids and phenolic compounds from sunflower wastes using green solvents. 3 Biotech 10:405. https://doi.org/10.1007/s13205-020-02393-0

Chapter 3

Extraction of Fats and Quantification of Health Promoting Fatty Acids

Shabnum Showkat, Aasima Rafiq, and Oroofa

Abstract

Fatty acids are the building blocks of fats. They are chemically composed of carboxylic acids with long aliphatic chains and can be branched, unbranched, saturated, or unsaturated. Fatty acids, especially the unsaturated, perform several biological functions and possess various health benefits. Therefore, a strong focus is observed on investigating fatty acids, specifically those that could have biomedical applications. This chapter discusses the traditional and modern extraction methods of fatty acids like microwave-assisted technique, supercritical fluid extraction, and ultrasound extraction. The methods of quantification of fatty acids like thin layer chromatography, gas chromatography, high performance liquid chromatography, supercritical fluid chromatography, mass spectrometry, and capillary electrophoresis have also been discussed.

Key words Fatty acids, Extraction methods, Gas chromatography, Thin layer chromatography, High performance liquid chromatography, Supercritical fluid chromatography, Mass spectrometry, Capillary electrophoresis

1 Introduction

Fatty acids are defined as carboxylic acids with long aliphatic chains. They have a carboxylic acid (COOH) group at one end and a methyl (CH_3) group at the other [1]. They are the building blocks of fats where they are attached by ester bonds to a trihydric alcohol, glycerol, and are referred to as triglycerides or triacylglycerols. They can be branched, unbranched, saturated and unsaturated as shown in Fig. 1. Saturated fatty acids lack carbon-carbon double bonds and have a linear, straight structure. Examples include butyric (C4:0), lauric (C12:0), myristic (C14:0), palmitic (C16:0), and stearic acid (C18:0). Unsaturated fatty acids have one or more carbon–carbon double bonds. Examples include oleic acid (C18:1), linoleic acid (C18:2), linolenic acid (C18:3), eicosapentaenoic acid (C20:5), and docosahexaenoic acid. The unsaturated fatty acids are further categorized as omega-3 (ω-3), omega-6 (ω-6) and omega-9

Ashwani Kumar, Vidisha Tomer and Piyoosh Kumar Babele (eds.), *Phytochemicals: Extraction, Quantification, Isolation, and Characterization*, Methods and Protocols in Food Science, https://doi.org/10.1007/978-1-0716-4840-7_3,

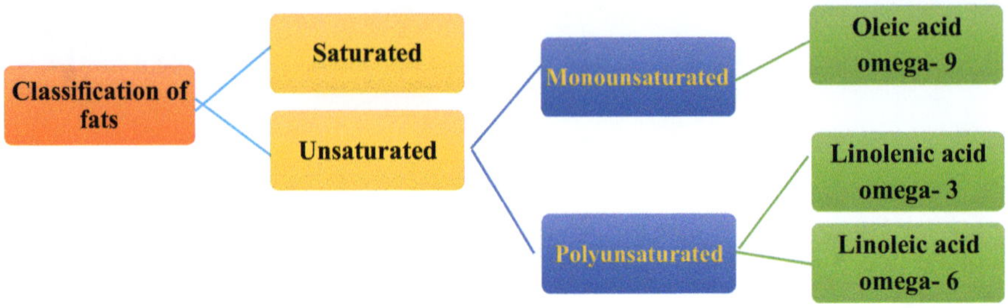

Fig. 1 Classification of fats

(ω-9) fatty acids, where, omega (ω) numbering indicates the position of the first double bond from methyl end. Major ω-3 fatty acids found in food are alpha-linoleic acid (ALA), eicosapentaenoic acid (EPA), and docosahexaenoic acid (DHA). Among these, ALA is an essential fatty acid that cannot be synthesized by human body, although, ALA can be converted in the body to EPA and DHA to some extent. It is primarily present in plant oils like flaxseed, soybean, and canola oils, while DHA and EPA are abundant in fish and other seafood. ω-3 fatty acids act as reservoirs of bioactive molecules within cell membranes and maintains the fluidity and flexibility of cell membranes. They are also essential for proper cell signaling and protein functioning. The enrichment of cells with ω-3 fatty acids modulates the expression of adhesion proteins [2]. Additionally, omega-3 provides energy to the body and serves various functions in the heart, blood vessels, lungs, immune system, and endocrine system. The benefits of omega-3 fatty acids extend to both the body and the brain. Many studies have claimed the efficacy of fatty acid supplementation in treating educational behavioral problems in children with developmental coordination disorder. The recommended daily intake of DHA and EPA in many countries is 300–650 mg/day (International Fish Meal and Fish Oil Organization). This can be achieved by consuming two servings of fatty fish per week. The adequate intake for ALA is 1600 mg for males and 1100 mg for females.

The ω-6 PUFAs are characterized by having the first double bond between the sixth and seventh carbon atom from the methyl terminal of fatty acid. The major fatty acid of this group is linoleic acid (18:2) that accounts for 80–90% of dietary ω-6 fatty acids. The other major members of this group are gamma-linolenic acid (18:3), eicosadienoic acid (20:2), Arachidonic acid (20:4), and Adrenic acid (22:4). These fatty acids exhibit both anti-inflammatory and pro-inflammatory effects. These also help to withstand cellular damage and boost cell repair through their conversion to ω-6 eicosanoids. This group of fatty acids are also necessary constituents of cell membrane [3]. Omega-9 fatty acids are a group of unsaturated fatty acids with a double bond in the ninth position from the methyl end. They can be either monounsaturated or polyunsaturated and are classified as "non-essential" fatty acids. The ω-9 group includes hypogeic acid, oleic acid, elaidic acid, gondoic acid, mead acid, erucic acid, and nervonic acid. Among these, oleic acid is the most abundant and is extensively found in various vegetable oils such as sesame oil, flaxseed oil, olive oil, rapeseed oil, and pumpkin seed oil. The other foods abundant in ω-9 fatty acids are safflower, sunflower, macadamia nut, hazelnut, olive oil, soybean oil, almond butter, avocado oil, and canola oil. This group of fatty acids have recently gained attention due to new studies and findings on their biological effects and potential risks. Various ω-9 fatty acids exhibit a range of pharmacological effects, such as regulating inflammation, lipids, cardiovascular health, and cancer-related conditions. Many studies have established the disease preventing and health promoting benefits of omega fatty acids. Nowadays, they are fortified into functional foods to give additional health benefits. Their high concentration capsules are also available in the market and contribute a large share in the functional foods market.

1.1 Extraction of Oils

The first step in the isolation of fatty acids is the extraction of oils from their respective sources. The preliminary unit operations for the oil expulsion are cleaning, dehulling, drying, and grinding. The cleaning operation is important to remove the foreign seeds, crop residue, and dirt. The presence of these particles reduces the extraction efficiency and yield of oil as they absorb the oil. Dehulling is an optional step for selective oil seeds. Drying and grinding are two major steps as the extraction efficiency of oil depends mainly on moisture content and particle size along with the temperature and time of extraction [4]. The water removal makes the samples more suitable for lipid extraction. In the traditional methods of oil extraction, the oil seeds are sun dried or roasted and the oil is extracted by pressing using screw expeller or human or animal operated press. Hydraulic presses are also used these days. However, this method is not suitable for the non-conventional sources of oils. Therefore, the other methods of oil extraction such as solvent extraction and supercritical fluid extraction are used. In the solvent extraction, especially Soxhlet extraction, samples must be dried as many organic solvents used in extraction are immiscible with water, making it difficult to efficiently extract lipids from samples having high water content. Microwave drying, vacuum dying, and freeze drying are also used as per the requirement of the sample. Microwave heating leads to rapid heating and moisture evaporation, which in turn breaks the seed structure, enhances the penetration of the solvent and increases the overall oil yield. However, this has been also reported to induce degradation, increase peroxide value and decrease sensory and physiochemical properties [5]. Vacuum and freeze drying are the most preferred methods for the drying of aromatic plants. Grinding serves two major purposes, that is, homogenization, and increase in surface area. Homogenization means uniform distribution of lipids in the sample. The increased surface area enhances the lipid extraction efficiency by increasing the sample-solvent interaction. The grinding should be performed at low temperatures to minimize the risk of lipid oxidation as the oxidation can adversely affect the quality of extracted lipids. In the presence of water, a molecule can undergo two different types of chemical processes called acid hydrolysis and alkaline hydrolysis, where the former takes place in acidic environments and the latter in basic ones.

In the samples having lipids attached to proteins and carbohydrates, acid hydrolysis technique is used to release bound lipids. This technique increased the extracted fat percentage in dried egg, flour, noodles, and semolina by 1–6%. Acid hydrolysis is not suitable for dairy products and hence alkaline hydrolysis is done in that case. The various methods for lipid extraction from food matrices include traditional methods like Soxhlet, Folch, Bligh and Dyer, and the novel methods like Microwave-Assisted Extraction (MAE), Ultrasound-Assisted Extraction (UAE), and Supercritical Fluid Extraction (SFE).

The Soxhlet extraction method is a conventional and extensively employed technique for the extraction of lipids from food samples. The Soxhlet extraction is a standardized method with a long-standing history and dependable outcomes. It is frequently employed to ascertain the total fat content in various food matrices and serves as a standard for comparative analysis in lipid research. It provides comprehensive and thorough lipid extraction and is capable of extracting both polar and nonpolar lipids, depending on the solvent utilized. The limitations of this method are long durations of extraction, extensive solvent use, and the need for adequate monitoring and ventilation due to the volatility of samples. The Folch method, established by Jordi Folch, is a prominent and dependable technique for the quantitative extraction of lipids [6, 7]. It is particularly effective for extracting lipids from lipid-dense foods and biological specimens, such as avocados, eggs, muscles, skin, and mayonnaise. It is especially esteemed for its high lipid recovery rates and reproducibility [8, 9]. This method is reliable across a diverse array of lipid-rich matrices, is highly effective, and yields a clean lipid extract suitable for fatty acid profiling. The limitations are the use of a large volume of solvents and the health and environmental hazards posed by solvents. This method yields 95–99% recovery of lipids.

The Bligh and Dyer method, established in 1959, is designed for lipid extraction from tissue samples through the use of a chloroform, methanol, and water mixture. This technique is particularly advantageous for tissues with high water content, making it applicable to a variety of biological materials, notably fish muscle [10]. Following the mixing of solvents with the sample, the resulting homogenate creates a monophasic system that, upon the addition of more water, separates into two distinct phases: lipids in the lower chloroform phase and non-lipid constituents in the upper methanol-water phase. This method has the ability to recover up to 95% of total lipids, despite utilizing a relatively modest volume of solvent. This characteristic renders it a highly efficient protocol regarding both solvent consumption and lipid yield. Although the Bligh and Dyer method is perceived as somewhat less effective than the Folch method in recovering all lipid classes, it is favored for its reduced solvent requirements. Nonetheless, the use of chloroform, a toxic and environmentally detrimental solvent, has led to initiatives aimed at identifying alternative solvent systems, which often prove to be less effective and more sensitive to the moisture content of samples [11]. The limitations of these methods are safety and environmental challenges of the solvents, and reduced effectiveness with the samples having low moisture content.

The MAE technique involves combining the sample with an appropriate solvent and exposing it to microwave radiation in short intervals (e.g., 30 s per cycle). This exposure to microwave energy facilitates the release of lipids by rapidly heating the solvent-sample matrix, thereby enhancing solvent penetration and extraction efficiency [12–14]. In contrast to conventional methods, MAE does not necessitate completely moisture-free samples, which reduces the time required for sample preparation. Additionally, it minimizes solvent consumption while achieving comparable lipid extraction yields. This method is also rapid and has high extraction repeatability. Costa and Bragagnolo [15] conducted a comparison of MAE with the Folch method for lipid extraction from fish samples and found no significant differences in fatty acid profiles. Similarly, Ramalhosa et al. [8] assessed MAE against various traditional methods, including Soxhlet, Bligh and Dyer, Modified Bligh and Dyer, and Folch techniques. Their findings indicated that MAE, Folch, and Modified Folch methods yielded the highest lipid extraction efficiency, with MAE demonstrating superior repeatability. Conversely, the Soxhlet and Modified Bligh and Dyer methods exhibited lower reproducibility. The potential drawbacks of MAE include the risk of lipid oxidation and changes in fatty acid composition due to intense localized heating [16, 17]. Therefore, meticulous control of extraction time and temperature is essential to mitigate these risks. The SFE method involves placing the sample in a pressurized (20–40 MPa) extraction chamber, where it is subjected to mild heating. Supercritical CO_2 is subsequently introduced to dissolve and extract lipid components [18, 19]. This gas selectively solubilizes nonpolar lipids, effectively separating them from polar constituents. Upon completion of the extraction, the pressure is decreased, allowing CO_2 to transition back to its gaseous state, thereby leaving behind solvent-free lipids. While SC-CO_2 (31.1 °C and 73.8 bars) is particularly adept at recovering nonpolar lipids, the addition of co-solvents such as methanol, ethanol, or water in minimal quantities can enhance polarity and facilitate the extraction of polar lipids [20]. This technique has proven effective in the analysis of various dairy products [21]. Extraction efficiency is notably affected by factors such as particle size (250–1000 μm), temperature (40–60 °C) and moisture content. High moisture levels can impede CO_2 diffusion, making freeze-drying a recommended preparatory step [6]. The notable advantages of this method are elimination of organic solvents, high selectivity and recovery rates, reduced extraction times, reusability of solvents, and suitability for thermally sensitive compounds due to operations at low temperatures. According to Berg et al. [22], SFE can be used to determine total fat and analyze lipid classes in meat products with precision comparable to traditional solvent extraction methods. When ethanol-modified CO_2 is used, combined with a hydro matrix and a small amount of cyclohexane, the extraction time is significantly reduced compared to methods like Bligh and Dyer. Ultrasonication is the application of high-frequency sound waves (usually above 20 kHz) to a liquid medium which creates tiny bubbles by a process known as cavitation. These bubbles can break down cell walls, emulsify materials, and improve mass transfer because they produce strong localized heat and shear forces when they collapse. Applications such as cell disruption, particle dispersion, and extraction procedures make extensive use of this idea.

1.2 Quantification of Fatty Acids

There are various analytical techniques that can be utilized for lipid analysis and fatty acid determination. Initially, lipid analysts played a significant role in advancing chromatographic techniques, particularly in thin layer chromatography (TLC) and gas chromatography (GC). Additionally, other methods such as high-performance liquid chromatography (HPLC), supercritical fluid chromatography (SFC), and capillary electrophoresis (CE) have been employed in lipid analysis. TLC, being a "low-tech" and straightforward method, is often preferred due to its rapidity and ability to provide on-site identification. Consequently, it finds frequent application in analytical preparations and organic chemistry [23–26]. TLC is an efficient and economical chromatographic method employed for the separation and semi-quantitative analysis of fatty acids. TLC is a classical and effective technique for separating and quantifying fatty acids and lipid classes based on their polarity [27, 28]. Always run a test gradient or TLC first to optimize solvent composition and avoid wasting time and reagents on poor separations. Before the chromatographic quantification of fatty acids, they are first converted into fatty acids methyl esters via transesterification. Gas Chromatography is a widely used method for quantifying fatty acids after their conversion to fatty acid methyl esters (FAMEs). The process involves lipid extraction, methylation (transesterification), and analysis using GC coupled with a Flame ionization detector (FID) or Mass Spectrometry (MS) [28–31]. HPLC is often used as an alternative to GC when analyzing thermolabile, non-volatile, or polar fatty acids without requiring derivatization (in some cases). Common detectors include UV, fluorescence, and Evaporative light scattering detector (ELSD) [32].

SFC represents a hybrid analytical technique that employs supercritical CO_2 as the primary mobile phase. This method is particularly adept at separating non-polar and moderately polar compounds, including various fatty acids. During the analytical process, a polar modifier, such as methanol, is generally introduced to enhance the separation of more polar fatty acids. The prepared sample is then injected into the SFC system, where it traverses a column under supercritical conditions. The separation of components occurs based on their polarity and their interactions with both the stationary phase of the column and the mobile phase [33]. This chapter discusses in detail various methods of fat extraction and provides an overview of the methods of fatty acid quantification. The analysis of fatty acids via mass spectrometry commences with the extraction of lipids, followed by optional derivatization to improve the volatility and ionization of the analytes [34]. The mass-to-charge ratios (m/z) of the ions are then measured, yielding mass spectra that provide both qualitative and quantitative information regarding individual fatty acids [31, 35].

In capillary electrophoresis methods, fatty acids are extracted from the sample matrix and may undergo derivatization to improve ionization and detection. The prepared sample is introduced into the capillary, which contains an appropriate electrolyte buffer. An electric field is applied, leading to the migration of charged fatty acid derivatives at varying velocities based on their charge-to-size ratios, which resultis in effective separation [36].

This chapter provides in depth detail of the methods of the extraction of the fat and the methods of quantification of fatty acids.

2 Materials

2.1 Extraction Methods

2.1.1 Soxhlet Extraction

(a) *Apparatus*:

- Soxhlet extractor.
- Round-bottom flask (for collecting the solvent).
- Condenser (typically water-cooled).
- Heating mantle or water/oil bath.
- Analytical balance (for weighing before and after extraction).
- Permeable thimble (commonly made of cellulose).

(b) *Reagents*:
- Organic solvent (frequently n-hexane or petroleum ether).

2.1.2 Folch Method

(a) *Apparatus*:
- Homogenizer or mechanical stirrer for optimal mixing of the sample and solvents.
- Glass beakers or centrifuge tubes for solvent amalgamation.
- Centrifuge or separatory funnel for phase separation.
- Pipettes for transferring solvent layers.
- Fume hood and personal protective equipment (due to the use of chloroform).

(b) *Reagents*:
- Purified and distilled chloroform (*see* **Note 1**).
- Purified and distilled methanol (*see* **Note 2**).
- Water or saline solution (for facilitating phase separation).
- Solvent ratio: 2 parts chloroform to 1 part methanol (v/v).
- 1% BHA-BHT antioxidant solution: Dissolve 1 g of butylated hydroxyanisole (BHA) and 1 g of butylated hydroxytoluene (BHT) in 100 mL C-M mixture (optional).
- Anhydrous sodium sulphate.

2.1.3 Bligh and Dyer Method

(a) *Apparatus*:
- Glass beakers or centrifuge tubes for the amalgamation of solvents and tissue.
- Homogenizer or mechanical stirrer for comprehensive blending of samples.
- Centrifuge (optional, for phase separation).
- Pipettes or separatory funnels for the extraction of solvent layers.
- Protective laboratory gear (gloves, goggles, fume hood) due to chloroform's toxicity.

(b) *Reagents*:
- Chloroform.
- Methanol.
- Distilled water.
- Solvent ratio: 2 parts chloroform, 2 parts methanol, 1.8 parts water.

2.1.4 Microwave-Assisted Extraction

(a) *Apparatus*:
- Microwave extraction apparatus or a laboratory-grade microwave oven with adjustable power and timing controls.
- Extraction vessels that are transparent to microwaves.
- Optional cooling system to prevent overheating and oxidation.
- Filtration apparatus (e.g., Whatman filter paper or equivalent for extract separation).

(b) *Reagents*:
- Suitable organic solvent(s) (e.g., mixtures that are compatible with microwave heating and the solubility of target lipids).

2.1.5 Supercritical Fluid Extractor

(a) *Apparatus*:
- Supercritical fluid extractor (fitted with a pressurized chamber and temperature control system).
- CO_2 gas supply with pressure regulation.
- Separator vessel for the collection of lipid fractions.
- Heating system (to gently warm the sample within the chamber).
- Pressure release mechanism (to revert CO_2 to its gaseous form and facilitate lipid separation).

(b) *Reagents*:
- Supercritical carbon dioxide (SC-CO_2) [18].
- Co-solvents (optional): Methanol, ethanol, or water (to enhance polarity).
- Water adsorbent: Hydro matrix (optional).
- Cyclohexane (optional, utilized in certain modified protocols to improve lipid recovery).

2.1.6 Ultrasonic Extraction

(a) *Apparatus*:
- 100 mL Erlenmeyer flask.
- Ultrasonic bath (frequency: 20 kHz, power: 150 W).
- Vacuum filtration setup.
- Temperature control system (for comparative studies, up to 70 °C).

(b) *Reagents*:

- Extraction solvent: Isopropanol/n-hexane mixture (40 mL) or Food-grade ethanol or ethanol-water mixture (70–90%) or Hexane or chloroform-methanol (2:1).
- Anhydrous sodium sulphate (for dehydration).

(c) *Others*:

- Grade 1 Whatman filter paper.

2.2 Quantification Techniques

2.2.1 Thin Layer Chromatography (TLC)

(a) *Sample*:

- A Sample containing lipids or fatty acids.

(b) *Apparatus*:

- TLC plates (made of glass, aluminum, or plastic, coated with silica gel or cellulose).
- Capillary tubes or micropipettes (for sample application).
- TLC developing chamber (a sealed glass tank with a lid).
- Hot air oven (for activating and drying the plates).
- UV lamp (for visualizing spots when using UV-sensitive reagents).
- Densitometer or image analysis software (for measuring spot intensity and quantification).
- Ruler or TLC scanner (to assess migration distances and Rf values).

(c) *Reagents*:

- Solvents for lipid extraction (e.g., chloroform, methanol).
- Optional derivatizing agents (to enhance spot visibility).
- Standard solutions with known concentrations of fatty acids.
- TLC developing solvent system (e.g., hexane: diethyl ether: acetic acid, or other specific mixtures tailored to target lipids).
- Visualization reagents (e.g., iodine vapor, UV-reactive dyes, or stains such as ninhydrin and sulfuric acid spray).

2.2.2 Gas Chromatography (GC)

(a) *Sample*:

- A sample containing lipids (such as food products, biological specimens, or dietary supplements).

(b) *Apparatus*:

- Gas Chromatograph system consisting of:
  - *Injector* (manual or auto sampler).
  - *Capillary column* (polar stationary phase).
  - *Flame ionization detector (FID)*.

- Heating block or water bath (for methylation reactions).
- Vortex mixer and centrifuge (optional, for homogenization and clarification).
- Analytical balance (for accurate sample measurement).
- GC vials and syringes.

(c) *Reagents*:

- A solvent mixture for lipid extraction, typically comprising chloroform and methanol.
- Reagents for the methylation process (for instance, methanolic HCl or BF_3-methanol) to transform fatty acids into FAMEs.
- Standard solutions with known concentrations of fatty acids for calibration purposes.
- A carrier gas, usually helium or nitrogen.
- Hydrogen and air, necessary for the operation of the Flame Ionization Detector.

(d) Data acquisition and analysis software.

2.2.3 High Performance Liquid Chromatography (HPLC)

(a) *Sample*:

- A sample containing lipids or fatty acids.

(b) *Apparatus*:

- HPLC system comprising:
  - Injector (manual or autosampler).
  - Column (typically reverse-phase C18 silica-based).
  - Pump system (for accurate delivery of the mobile phase).
  - Detector (UV-Vis absorbance detector is standard; fluorescence detectors may also be employed).
- Ultrasonic bath or vortex mixer (for homogenizing samples).
- Centrifuge (for phase separation following extraction).
- Evaporator or nitrogen dryer (if concentration of the extract is necessary).
- Microsyringes or autosampler vials (for accurate sample injection).

(c) *Reagents*:

- Solvents for extraction (e.g., chloroform, methanol).
- Derivatizing agents (e.g., DNPH, phenacyl bromide—if necessary for improved detection).

- Standard solutions with known concentrations of individual fatty acids.
- HPLC-grade mobile phase solvents (e.g., methanol, acetonitrile, water, acetic acid, or other acidic modifiers).
- Filtration units or syringe filters (for filtering samples and mobile phases).

(d) Data acquisition software (for chromatogram analysis and quantification).

2.2.4 Supercritical Fluid Chromatography (SFC)

(a) *Sample*:

- A Sample containing fatty acids or lipids.

(b) *Apparatus*:

- SFC system, comprising:
  - CO_2 pump and modifier pump.
  - Injector (manual or autosampler).
  - SFC-compatible column (typically filled with polar-modified silica-based stationary phases).
  - Back-pressure regulator (to sustain supercritical conditions).
  - Detector (UV-Vis detector, mass spectrometer, or fluorescence detector).
  - Data acquisition and processing software.
- Temperature-controlled oven (to regulate column temperature).
- Gas cylinder containing CO_2 (with regulator).
- Centrifuge or separator (utilized during sample preparation).
- Ultrasonic bath or vortex mixer (for sample mixing or homogenization).

(c) *Reagents*:

- Solvents for extraction (e.g., chloroform, methanol).
- Derivatizing agents (if required to improve detection).
- Calibration standards with known concentrations of fatty acids.
- Supercritical fluid: Carbon dioxide (CO_2).
- Mobile phase modifier: Methanol, ethanol, or other polar solvents (to facilitate the separation of polar analytes).
- Filtration units (e.g., syringe filters).
- Vials and sample syringes.

2.2.5 Mass Spectrometry (MS)

(a) *Sample*:

- A sample containing fatty acids, which may originate from dietary sources, biological fluids, or dietary supplements.

(b) *Apparatus*:

- A mass spectrometer, which may consist of:
  - An ionization source, such as electrospray ionization (ESI), atmospheric pressure chemical ionization (APCI), electron ionization (EI), or chemical ionization (CI).
  - Mass analyzers, including quadrupole, time-of-flight (TOF), ion trap, or orbitrap.
  - A detector that quantifies ion signal intensity.
  - A chromatographic system (optional but frequently employed).
  - GC-MS for analyzing volatile fatty acid derivatives (which necessitates derivatization).
  - LC-MS for non-volatile or polar fatty acids, suitable for intact lipids (FAMEs).
  - Data acquisition and analysis software for peak integration and quantification.
  - A nitrogen gas supply for drying or desolvation during ESI.
  - A vacuum system and ion source housing integrated within the mass spectrometer.

(c) *Reagents*:

- Solvents for extraction, such as chloroform and methanol.
- Derivatization agents utilized for esterification or transesterification processes, including methanol combined with sulfuric acid or boron trifluoride.
- Calibration standards that possess known concentrations of fatty acids.
- Isotopically labeled internal standards, such as D-labeled fatty acids, to ensure precise quantification.
- Solvents suitable for mass spectrometry, including acetonitrile, water, and formic acid for electrospray ionization (ESI) compatible solutions.
- Sample vials and filtration devices.
- Microcentrifuge tubes, pipettes, and a centrifuge.

2.2.6 Capillary Electrophoresis

(a) *Sample*:

Sample types (e.g., food matrices, biological specimens, dietary supplements).

(b) *Apparatus*:

- *Capillary electrophoresis system:*
  - Fused silica capillary with a small internal diameter (typically 25–75 μm).
  - High-voltage power supply to generate the electric field (usually 10–30 kV).
  - Sample injection mechanism (either hydrodynamic or electrokinetic).
  - Electrolyte reservoirs for holding buffer solutions at both ends of the capillary.
  - Capillary conditioning unit for pre-run rinsing with buffer or cleaning solutions.
- *Detection system:*
  - UV-Vis detector (typically operating at 200–214 nm for fatty acids).
  - Fluorescence detector, which provides enhanced sensitivity when fluorescent tags are utilized in derivatization.
  - Optional mass spectrometer (CE-MS) for advanced structural and quantitative analysis.

(c) *Reagents*:

- Solvents for extraction (e.g., chloroform, methanol, hexane) for lipid extraction.
- Derivatization agents, if necessary, to enhance solubility and improve the detectability of fatty acids.
- Calibration standards with known fatty acid concentrations.
- Electrolyte buffer solutions (e.g., phosphate buffer, borate buffer, Tris buffer).
- Internal standards, when applicable, to enhance the accuracy of quantification.
- Sample vials, micro-pipettes, and solutions for capillary rinsing.

(d) *Data acquisition and processing software*:

- Utilized for peak integration, interpretation of electropherograms, quantification, and construction of calibration curves.

3 Methods

3.1 Extraction Methods

3.1.1 Soxhlet Extraction

1. Dry the samples before extraction for efficient fat extraction. A moisture content of less than 8% is preferred.
2. Grind the sample and place in a porous thimble made up of glass or cellulose.
3. Position the thimble containing sample within the extraction chamber of the Soxhlet apparatus.
4. Pour the solvent onto the thimble and the extraction flask.
5. Heat the solvent in the flask at 70 °C using a heating mantle. The solvent evaporates, ascends into the condenser, cools, condenses, and drips back into liquid form on the thimble/chamber containing the sample, facilitating lipid extraction as it returns to the flask.
6. This cycle is repeated continuously for a duration ranging from 6 to 24 h [6].
7. After extraction, use a rotary evaporator or gently heat the flask under a fume hood to remove solvent. Avoid open flames near flammable solvents. Store extracted fat properly. If further analysis is required (e.g., FAME analysis), store lipids in airtight containers at low temperatures (e.g., −20 °C) to prevent oxidation.
8. Calculate the fat content by using the formula:

$$\%\text{Fat} = \frac{\text{Weight of extracted fat}}{\text{Weight of sample}} \times 100$$

3.1.2 Folch Method

1. Weigh (10–50 g) the sample and homogenize it.
2. Add chloroform-methanol mixture to the sample. The volume of solvent to be added is about 20 times the mass of the sample (e.g., 20 mL for every 1 g of sample).
3. To this add 2–3 drops of antioxidant solution.
4. Homogenize for 1 min and filter through Whatman No. 1 filter paper using a Buchner funnel and vacuum pump.
5. Rehomogenize the residue and repeat the extraction 2–3 times.
6. Transfer the combined filtrate into a separating flask.
7. Add distilled water (approximately a quarter of the extract) to the separating funnel.
8. Shake gently 2–3 times, and let it stand overnight. If not separated well, then centrifuge at 8000 rpm for 10 min.
9. Drain off the chloroform phase through a Whatman No. 1 filter paper into an Erlenmeyer flask containing about 2–5 g anhydrous sodium sulphate. Wash the filter paper with solvent.

10. Shake well and allow it to rest for 5 min.
11. Decant into an evaporating flask and concentrate the extract with a rotary evaporator under reduced pressure at 40 °C (water-bath temperature). Take care not to evaporate to dryness.
12. Dissolve the concentrated extract with C-M Mixture and transfer to a 50 mL volumetric flask using a pipette and make the volume to the mark with C-M Mixture.
13. Flush with nitrogen gas and store at −20 °C.

3.1.3 Bligh and Dyer Method

1. Weigh the sample (usually between 1 and 5 grams, depending on the size of the sample).
2. Homogenize the sample using a suitable homogenizer or mortar and pestle. This helps break down the tissue and releases lipids into the solvent.
3. To the homogenized sample, add methanol in a 2:1 ratio (methanol to sample). For example, if you have 1 gram of tissue, add 2 mL of methanol.
4. Mix or vortex the solution to ensure thorough contact between the sample and methanol. Methanol helps to solubilize the polar lipids, including phospholipids and glycolipids.
5. After the sample is mixed with methanol, add chloroform in a 1:1 ratio with methanol. For example, add 2 mL of chloroform if 2 mL of methanol is used.
6. Mix the solution thoroughly by vortexing or shaking. Chloroform is used to solubilize non-polar lipids like triglycerides and sterols.
7. To facilitate phase separation, add water to the mixture, typically in a 0.8:1 ratio (water to chloroform-methanol mixture). If 4 mL of chloroform and methanol is used, add about 3.2 mL of water.
8. Mix the solution thoroughly. The addition of water helps in phase separation by allowing the organic and aqueous layers to form. The lipids will mostly partition into the chloroform layer.
9. After mixing, allow the sample to settle and separate into two distinct phases:
   (a) The lower chloroform phase (organic phase) will contain the lipids.
   (b) The upper aqueous-methanol phase will contain the aqueous-soluble components.
10. If using a separating funnel, allow the phases to separate and drain the lower chloroform phase into a clean container (e.g., a beaker or round-bottom flask).

11. The lower organic phase will contain the extracted lipids, including fats, phospholipids, and neutral lipids.
12. The extraction can be optionally repeated by adding another portion of chloroform and water to the remaining aqueous phase. This ensures maximum recovery of lipid from the sample.
13. Use a rotary evaporator to remove the chloroform and methanol. If a rotary evaporator is not available, a gentle stream of nitrogen gas can be used to evaporate the solvents. After evaporation, the lipid extract will be in the round-bottom flask.
14. To remove any remaining traces of solvent, dry the lipid extract by placing it under a stream of nitrogen gas or in a vacuum oven at a low temperature (around 40–50 °C).
15. After the solvent has been removed, weigh the lipid extract to determine the lipid content.
16. To analyze fatty acids specifically, perform transesterification to convert the fatty acids into FAMEs. These can then be analyzed using GC.

3.1.4 Microwave-Assisted Extraction

1. Weigh around 5–20 grams of the sample, depending on the nature of the material (e.g., seeds, nuts, algae, or fish). If the sample is solid, grind or crush it to increase the surface area, which facilitates better extraction. A mortar and pestle, blender, or grinder can be used.
2. A common solvent mixture used for fatty acid extraction is hexane combined with ethanol or methanol. For example, a 2:1 (v/v) ratio of hexane to ethanol is frequently used.
3. The solvent volume depends on the sample size, but usually, a ratio of 10–20 mL of solvent per gram of sample is used.
4. Transfer the homogenized sample and solvent mixture into a microwave-safe extraction vessel. Make sure the vessel is designed for microwave extraction (e.g., Teflon or glass vessels).
5. For microwave extraction process: Set the microwave system parameters:
   (a) Temperature: Typically set the temperature between 40 ° C and 90 °C, depending on the sample and fatty acids being extracted.
   (b) Time: Extraction time can range from 5 to 30 min, depending on the efficiency of the microwave system and sample type.
   (c) Power: Carefully control microwave power and time to prevent thermal degradation of fats. Excessive power or exposure time can cause localized overheating, leading to

oxidation, hydrolysis, or decomposition of delicate lipids. Using non-polar solvents like hexane or petroleum ether helps selectively extract fats while minimizing unwanted reactions. Use pulsed microwave settings or intermittent heating and combine with closed-vessel systems to better control temperature and pressure, improving extraction efficiency and preserving fat integrity.

6. After the extraction is complete, allow the vessel to cool down. If using a solvent mixture like hexane and ethanol, there will be phase separation (organic and aqueous phases). The fatty acids and lipids will primarily be in the organic phase (upper layer).
7. Use filter paper to separate any solid remnants or particulate matter from the liquid extract.
8. Transfer the organic phase containing the extracted fatty acids to a round-bottom flask, and use a rotary evaporator or vacuum system to remove the solvent.
9. Once the solvent is removed, place the lipid extract under a nitrogen stream or in a vacuum oven at a low temperature (40–50 °C) to remove any residual solvent traces.
10. After drying, weigh the lipid extract to calculate the yield of fatty acids or lipids.
11. If the lipid extract is not being analyzed immediately, store it in air-tight containers under cold conditions (e.g., in a freezer) to prevent degradation or oxidation of sensitive fatty acids.

3.1.5 Supercritical Fluid Extraction

1. Weigh 10–50 grams of fine powdered sample.
2. Load the sample into the SFE extraction vessel.
3. Set the parameters for extraction:
   (a) Set the extraction pressure: The CO_2 is typically used at pressures between 100 and 500 bar (1450–7250 psi), depending on the system and type of material being extracted.
   (b) Set the extraction temperature: The temperature usually ranges between 30 °C and 80 °C. A temperature of around 40 °C is commonly used for extracting fatty acids while keeping CO_2 in a supercritical state.
   (c) Optimize the CO_2 flow rate: Optimize pressure and temperature to selectively extract target compounds in SFE. Higher pressures generally increase solubility and extraction efficiency. Temperature influences both solubility and selectivity—moderate increases can enhance extraction, but too much may degrade sensitive compounds (like unsaturated fats or bioactive). Use co-solvents (like ethanol) in small amounts to enhance polarity and improve extraction of more polar lipids or functional compounds if needed.

4. Start the extraction process:
    (a) Pressurize the system: Gradually pressurize the system to the desired extraction pressure. This process will transform the CO_2 into a supercritical state, where it exhibits liquid-like density but gas-like diffusivity.
    (b) Flow the supercritical CO_2 through the sample: The CO_2 will pass through the sample, extracting fatty acids and other lipids. The supercritical CO_2 solvating power is enhanced due to its density and ability to diffuse through the solid material.
5. Collection of extracted fatty acids:
    (a) Separate the extract: The fatty acids and lipids extracted by the supercritical CO_2 will be carried through the system to a collection vessel. This vessel can be at a lower pressure and temperature than the extraction vessel to help precipitate the lipids.
    (b) Fractionation (optional): Some systems allow fractionation, where different lipids (e.g., triglycerides, phospholipids, or omega-3 fatty acids) are selectively extracted by adjusting pressure, temperature, or CO_2 flow rate during the process.
6. Release pressure and collect the extract:
    (a) Release the pressure: After the extraction cycle, slowly depressurize the system. This allows the CO_2 to return to its gaseous state, separating from the extracted lipids.
    (b) Collect the lipid extract: The extracted fatty acids will be collected in the receiving vessel. This extract typically contains both fatty acids and other bioactive compounds, depending on the solvent power of CO_2.
7. Post-extraction solvent removal (if required):
    (a) If any residual CO_2 remains in the extract, use a vacuum or rotary evaporator to remove it. This step ensures that the final extract is free of CO_2 or other solvents.
    (b) To ensure no solvent traces remain, dry the extract under a nitrogen stream or in a vacuum oven (at low temperatures).
8. Quantification and analysis of fatty acids:
    (a) Weigh the extract: Once the solvent is completely removed, weigh the extracted lipids to determine the yield.
9. Storage of extract (if required):
    (a) If the extracted fatty acids are not being analyzed immediately, store them in an airtight container in a dark, cool place (or in a freezer) to prevent degradation due to light or oxygen exposure.

3.1.6 Ultrasonic Extraction

1. Weigh 5-20 g of dried and ground sample.
2. Prepare the solvent/sample ratio: 10:1 to 30:1 (mL/g).
3. Place the sample-solvent mixture in a beaker or extraction vessel.
4. Insert the ultrasonic probe or place it in an ultrasonic bath.

Parameter	Typical value
Frequency	20-40khz
Power	100–500 W
Time	10–60 min
Temperature	25–50 °C
Pulsing power	5 s on/5 s off

The ultrasonic waves create cavitation, thereby facilitating the release of lipids from the sample matrix [22]. Control the ultrasound intensity and exposure time to avoid fat degradation. Excessive power or duration can degrade sensitive fats (especially unsaturated ones), cause oxidation, and increase sample temperature, which can alter the fat's chemical structure. Combine ultrasound with a mild solvent (like hexane or ethanol) and maintain temperature control (e.g., with an ice bath or water-jacketed system) to maximize yield and preserve fat quality.

5. Filtration and separation: After sonication, the extract is collected and subjected to vacuum filtration (0.8–0.95 bar) using Grade 1 Whatman filter paper or centrifuge (e.g., 5000 rpm for 10 min). Collect the supernatant (contains extracted lipids). The resulting filtrate is diluted to a final volume of 50 mL with the same organic solvent and dehydrated with anhydrous sodium sulphate. This method has been shown to be as effective as traditional liquid-liquid extraction while providing environmental advantages through reduced solvent usage. Da Porto et al. [37] conducted a comparison between UAE and the Soxhlet method for oil extraction from grape seeds, revealing that UAE produced much higher oil yields.
6. The UAE method offers a high lipid yield with significantly diminished solvent consumption and reduced processing time relative to conventional extraction techniques. Additionally, research indicates that the application of ultrasonography does not adversely impact the fatty acid profile of the extracted oil [38–40].

3.2 Quantification Methods

3.2.1 Quantification of Fatty Acids by TLC

1. *Sample preparation*
   - Extract total lipids using any of the above-mentioned extraction methods.
   - Homogenize sample in chloroform: methanol (2:1 v/v).
   - Add aqueous KCl or NaCl to separate phases.
   - Collect the lower organic (chloroform) layer containing lipids.
2. *Spotting and development*
   - Spot 5–10 μL of lipid extract and standards onto a silica gel TLC plate using a micropipette.
   - Develop the plate in a chamber saturated with hexane: diethyl ether: acetic acid (80:20:1).
   - Allow the solvent to migrate until 1–2 cm from the top (approx. 10–15 cm distance).
   - Remove and air-dry the plate.
3. *Visualization and detection*
   - Option A (charring): Spray plate with 10% $CuSO_2$ in 8% H_2PO_2 and heat at 180 °C for 10–15 min. Spots will appear as dark brown/black bands.
   - Option B (iodine vapor): Expose the plate to iodine vapor in a closed chamber for a few minutes. Lipid bands will turn yellow-brown.
4. *Quantification*
   - Compare the intensity/area of unknown bands with standards using:
   - Densitometry (TLC scanner).
   - Scraping + elution of bands followed by spectrophotometric quantification.
   - Gravimetric method (for large samples: weigh scraped material).
   - Prepare calibration curves using known concentrations of standard fatty acids for accurate quantification.

3.2.2 Quantification of Fatty Acids by GC

1. *Sample preparation*
   - Extract total lipids using any of the above-mentioned extraction methods.
   - Mix 1 part of the sample with 2 parts of chloroform: methanol (2:1, v/v).
   - Homogenize and filter the mixture.
   - Add 0.88% KCl or NaCl to separate layers.
   - Collect the lower chloroform phase (contains lipids).

2. *Transesterification to FAMEs*
   - Dry the lipid extract under nitrogen.
   - Add 1 mL of 0.5 M NaOH in methanol and heat at 60 °C for 10 min (saponification).
   - Add 2 mL of 14% BF_2 in methanol and heat at 60 °C for 30 min (methylation).
   - Cool, then add 1–2 mL of distilled water and 2 mL of hexane.
   - Vortex and allow phase separation.
   - Collect the hexane layer containing FAMEs.
   - Alternative: Acid-catalyzed methylation using H_2SO_4 in methanol.
3. *GC analysis*

   Instrumentation setup
   - Column: Fused silica capillary column (e.g., DB-23, 30 m × 0.25 mm × 0.25 μm).
   - Carrier gas: Helium or nitrogen.
   - Injector temp: 250 °C.
   - Detector temp (FID): 280 °C.
   - Split ratio: 1:20.
   - Oven program:
   - Start at 120 °C, hold 1 min.
   - Ramp to 220 °C at 4 °C/min, hold 10 min.
   - Injection Volume: 1 μL of hexane extract.
4. *Quantification*
   - Identify peaks by comparing retention times with FAME standards.
   - Use internal standard (e.g., C17:0) for quantification.
   - Calculate concentration based on peak area relative to the internal standard.
   - *Calculations*:

$$\text{FA concentration (mg/g)} = \frac{A_{\text{FA}}}{A_{\text{IS}}} \times \frac{C_{\text{IS}}}{W} \times R$$

Where,

A_{FA} = area of fatty acid peak

A_{IS} = area of internal standard

C_{IS} = concentration of internal standard (mg)

W = sample weight (g)

R = response factor (from calibration curve)

3.2.3 Quantification of Fatty Acids by HPLC

1. *Sample preparation*
   - Extract total lipids using any of the above-mentioned extraction methods.
   - Homogenize sample in chloroform: methanol (2:1 v/v).
   - Add water/saline to induce phase separation.
   - Collect lower chloroform phase containing lipids.
2. *Hydrolysis (if needed for free fatty acids)*
   - Saponify lipids by heating with 0.5 M KOH in methanol at 60 °C for 30 min.
   - Acidify with HCl to pH ~2 and extract FFAs with hexane.
3. *Derivatization (optional but recommended for UV/fluorescence detection)*
   - For UV detection: React fatty acids with 2, 4-DNPH in acidic conditions to form hydrazones.
   - For fluorescence detection: React with OPA in the presence of a thiol compound (e.g., mercaptoethanol).
4. *HPLC conditions*

   For FAMEs or FFAs

   - Column: C18 reverse-phase column (e.g., 250 mm × 4.6 mm, 5 μm).
   - Mobile phase: Gradient of acetonitrile: water or methanol: water (95:5 to 100:0).
   - Flow rate: 1.0 mL/min.
   - Injection volume: 10–20 μL.
   - Detector: UV (at 254 nm for DNPH derivatives) or ELSD.
   - For non-derivatized FFAs (ELSD or MS):
   - Use isocratic elution with acetonitrile: water (95:5) or acetonitrile: isopropanol.
5. *Quantification*
   - Identify fatty acids by comparing retention times with known standards.
   - Quantify using external or internal calibration curves.
   - Use software to integrate peak areas and calculate concentration:

$$\text{FA concentration (mg/g)} = \frac{A_{\text{sample}}}{A_{\text{std}}} + \frac{C_{\text{std}}}{w}$$

Where,

A_{sample}: Peak area of sample

A_{std}: Peak area of standard

C_{std}: Concentration of standard

W: Sample weight

6. *Observation*: Note the migration or elution times and signal intensities (peak areas) for each detected fatty acid.
7. *Documentation*: Record all analytical parameters used, such as buffer or solvent composition, flow rate, temperature settings, and instrumental conditions.

3.2.4 Quantification of Fatty Acids SFC

1. *Sample preparation*
   - Extract total lipids using any of the above-mentioned extraction methods.
   - Mix sample with chloroform: methanol (2:1, v/v).
   - Homogenize and centrifuge.
   - Separate and collect chloroform (lipid) layer.
   - Dry under nitrogen.
2. *Transesterification (Methylation of fatty acids) (Method: Morrison & Smith 1964)*
   - Add 1 mL of 0.5 M NaOH in methanol and heat at 60 °C for 10 min.
   - Add 2 mL of 14% BF_3 in methanol and heat at 60 °C for 30 min.
   - Add 1–2 mL of water and 2 mL of hexane.
   - Vortex and collect the hexane layer (contains FAMEs).
   - Dry under nitrogen and redissolve in 1 mL of isopropanol or hexane.
3. *SFC Instrument Setup*
   - Column: Diol or C18 silica column.
   - Mobile Phase: A: Supercritical CO_2, B: Methanol (modifier) with 0.1% formic acid (optional).
   - Backpressure: 150–200 bar.
   - Column Temperature: 35–40 °C.
   - Flow Rate: 1.0–2.0 mL/min.
   - Detector: UV at 205–220 nm, ELSD, or MS.
   - Injection Volume: 5–10 μL.
4. *Calibration and quantification*
   - Prepare a calibration curve using standard FAMEs at known concentrations.
   - Inject standards and samples under identical conditions.
   - Plot peak area vs. concentration for each standard.

- Identify fatty acids by comparing retention times with standards.
- Calculations:

$$\text{Concentration (mg/mL)} = \frac{\text{Peak Area}}{\text{slope of calibration curve}}$$

3.2.5 Quantification of Fatty Acids by MS

1. *Sample preparation*
   - Extract total lipids using any of the above-mentioned extraction methods.
   - Mix sample with chloroform: methanol (2:1, v/v).
   - Homogenize and centrifuge to separate phases.
   - Collect the lower organic phase containing lipids.
   - Dry under nitrogen.
   - Methylation/Transesterification (Optional, for GC-MS, *see* Subheading 3.2.2).
2. *Instrument setup*

   A. *LC-MS/MS method*
   - Column: C18 reverse-phase (e.g., 100 mm × 2.1 mm, 1.7 μm).
   - Mobile phase: A = Water + 0.1% formic acid; B = Acetonitrile or methanol + 0.1% formic acid.
   - Gradient: Start at 30% B, ramp to 95% B over 15 min.
   - Injection volume: 5–10 μL.
   - Ionization: ESI (−) mode for free fatty acids.
   - Scan mode: MRM or full-scan.
   - Detection: Monitor $[M–H]^-$ ions of fatty acids.

   B. *GC-MS method (for FAMEs)*
   - Column: DB-23 or HP-88 (30 m × 0.25 mm, 0.25 μm).
   - Carrier gas: Helium.
   - Injector temp: 250 °C.
   - Oven program: 120 °C to 230 °C at 4 °C/min.
   - MS mode: EI (electron ionization), scan range 50–550 m/z.
   - Detect and quantify: Based on characteristic fragments (e.g., m/z 74 for FAMEs).
3. *Quantification*
   - Prepare a calibration curve using a series of standard fatty acids or FAMEs at known concentrations.
   - Spike all samples with an internal standard (e.g., C17:0 or D31-palmitic acid).

- Inject sample and standards under identical conditions.
- Calculations:

$$\text{FA concentration (mg/g)} = \frac{\text{AFA}}{\text{AIS}} + \frac{\text{CIS}}{w} \times R$$

Where,

AFA: Peak area of fatty acid

AIS: Peak area of internal standard

CIS: Concentration of internal standard

W: Sample weight

R: Response factor from calibration curve

4. *Precautions*:
   - Run blanks and replicate injections.
   - Check mass spectra for diagnostic ions.
   - Use deuterated or odd-chain fatty acid as internal standard for better accuracy.

3.2.6 Quantification of Fatty Acids by Capillary Electrophoresis

1. *Sample preparation*
   - Extract total lipids using any of the above-mentioned extraction methods.
   - Dry the extract under nitrogen.
   - Dissolve the dried lipid extract in 1 mL ethanol or methanol.
   - Perform saponification: add 1 mL of 0.5 M NaOH, heat at 60 °C for 30 min to hydrolyze lipids into free fatty acids.
   - Neutralize with 1 M HCl.
   - Filter through a 0.45 μm membrane before CE injection.
2. *Capillary electrophoresis setup*
   - Equipment: Capillary Electrophoresis System with UV or DAD detector.
   - Buffer: 20–50 mM sodium tetraborate or phosphate buffer, pH ~9.0 to ionize fatty acids and enable migration.
   - Capillary conditions:
     - Fused silica capillary: 50 μm I.D., 50–75 cm length.
     - Temperature: 25 °C.
     - Voltage: 20–30 kV.
     - Injection: 10–30 s, hydrodynamic or electrokinetic.
     - Detection: 200–220 nm (UV detection), or indirect UV (e.g., with chromophoric background electrolyte).

3. *Quantification*
   - Prepare calibration standards with known concentrations of free fatty acids (FFAs), including the internal standard.
   - Run the standards to obtain migration times and peak areas.
   - Inject and run the sample under identical conditions.
   - Identify each fatty acid by its migration time compared to standards.
4. *Observations*
   - Record migration times and peak areas.
   - Ensure peak resolution (especially between similar chain-length fatty acids).
   - Check for any peak tailing or baseline instability.
5. *Documentation*
   - Buffer composition, pH, and ionic strength.
   - Capillary length, voltage, and temperature.
   - Sample preparation steps and dilutions.
   - Calibration curve data and linearity ($r^2 > 0.99$ desirable),
   - Any deviations, capillary blockages, or poor reproducibility.

4 Notes

1. To purify and distil chloroform, wash 1 liter of chloroform with 10 mL concentrated sulfuric acid in a separating funnel. Following this, wash it 2–3 times with distilled water again using a separating funnel. Add anhydrous calcium chloride to the washed chloroform. Let it stand overnight, and transfer to the distillation flask on the next day. Distill and collect fraction that distills over at 60.5 °C. To this add 1% v/v purified and distilled methyl alcohol as stabilizer. Keep in the dark. This should be used within a month.
2. To remove impurities, such as moisture, acids, and aldehydes, from methyl alcohol add granular potassium hydroxide to it. Distil the mixture and collect a fraction that distills over at 64.5 °C. Store in the dark and use within a month.
3. After extraction, use a rotary evaporator or gently heat the flask under a fume hood to remove the solvent. Avoid open flames near flammable solvents. Store extracted fat properly. If further analysis is required (e.g., FAME analysis), store lipids in airtight containers at low temperatures (e.g., −20 °C) to prevent oxidation.
4. Carefully control microwave power and time to prevent thermal degradation of fats. Excessive power or exposure time can

cause localized overheating, leading to oxidation, hydrolysis, or decomposition of delicate lipids. Using non-polar solvents like hexane or petroleum ether helps selectively extract fats while minimizing unwanted reactions. Use pulsed microwave settings or intermittent heating and combine with closed-vessel systems to better control temperature and pressure, improving extraction efficiency and preserving fat integrity.

5. Optimize pressure and temperature to selectively extract target compounds in SFE. Higher pressures generally increase solubility and extraction efficiency. Temperature influences both solubility and selectivity—moderate increases can enhance extraction, but too much may degrade sensitive compounds (like unsaturated fats or bioactives). Use co-solvents (like ethanol) in small amounts to enhance polarity and improve extraction of more polar lipids or functional compounds, if needed.
6. Control the ultrasound intensity and exposure time to avoid fat degradation. Excessive power or duration can degrade sensitive fats (especially unsaturated ones), cause oxidation, and increase sample temperature, which can alter the fat's chemical structure. Combine ultrasound with a mild solvent (like hexane or ethanol) and maintain temperature control (e.g., with an ice bath or water-jacketed system) to maximize yield and preserve fat quality.
7. Chromatography relies on differential interactions between compounds, the stationary phase, and the mobile phase. To achieve good separation, use polar stationary phases (like silica gel) for normal-phase chromatography. Use non-polar stationary phases (like C18) for reverse-phase chromatography to separate non-polar compounds. Always run a test gradient or TLC (thin-layer chromatography) first to optimize solvent composition and avoid wasting time and reagents on poor separations.

References

1. Lund J, Rustan AC (2020) Fatty acid structures and properties. In: Encyclopedia of Life Sciences, pp 283–292. https://doi.org/10.1002/9780470015902.a0029198
2. Surette ME (2008) The science behind dietary omega-3 fatty acids. CMAJ 178:177–180. https://doi.org/10.1503/cmaj.071356
3. Mazidi M, Shekoohi N, Katsiki N et al (2021) Omega-6 fatty acids and the risk of cardiovascular disease: insights from a systematic review and meta-analysis of randomized controlled trials and a Mendelian randomization study. Arch Med Sci 18:466. https://doi.org/10.5114/aoms/136070
4. Gutiérrez LF, Ratti C, Belkacemi K (2008) Effects of drying method on the extraction yields and quality of oils from Quebec Sea buckthorn (Hippophae rhamnoides L.) seeds and pulp. Food Chem 106:896–904. https://doi.org/10.1016/j.foodchem.2007.06.058
5. Lukešová D, Dostálová J, Mahmoud EAEM et al (2009) Oxidation changes of vegetable oils during microwave heating. Czech J Food Sci 27:178–181. https://doi.org/10.17221/929-CJFS
6. Senorans FJ, Luna P (2012) Sample preparation techniques for the determination of fats in food. In: Pawliszyn J, Bayona JM, Dugo P, Le XC, Lee HK, Li XF, Lord H (eds)

Comprehensive sampling and sample preparation, 1st edn. Academic Press, Cambridge

7. Fernández AL, El-Diasty MM, Martínez A et al (2011) A simple technique to rule out occlusion of right coronary artery after aortic valve surgery. Ann Thorac Surg 92:2281–2282. https://doi.org/10.1016/j.athoracsur.2011.06.016
8. Ramalhosa MJ, Paiga P, Morais S et al (2012) Lipid content of frozen fish: comparison of different extraction methods and variability during freezing storage. Food Chem 131: 328–336. https://doi.org/10.1016/j.foodchem.2011.07.123
9. Shinn SE, Proctor A (2013) Rapid lipid extraction from egg yolks. J Am Oil Chem Soc 90: 315–316. https://doi.org/10.1007/s11746-012-2155-2
10. Breil C, Abert Vian M, Zemb et al (2017) Bligh and Dyer and Folch methods for solid liquid liquid extraction of lipids from microorganisms. Int J Mol Sci 18:708. https://doi.org/10.3390/ijms18040708
11. Iverson SJ, Lang SL, Cooper MH (2001) Comparison of the Bligh and Dyer and Folch methods for total lipid determination in a broad range of marine tissue. Lipids 36:1283–1287. https://doi.org/10.1007/s11745-001-0843-0
12. Byrne FP, Jin S, Paggiola G et al (2016) Tools and techniques for solvent selection: green solvent selection guides. Sustain Chem Process 4: 7. https://doi.org/10.1186/s40508-016-0051-z
13. Batista A, Vetter W, Luckas B (2001) Use of focused open vessel microwave-assisted extraction as prelude for the determination of the fatty acid profile of fish–a comparison with results obtained after liquid-liquid extraction according to Bligh and Dyer. Eur Food Res Technol 212:377–384. https://doi.org/10.1007/s002170000240
14. Delazar A, Nahar L, Hamedeyazdan S et al (2012) Microwave-assisted extraction in natural products isolation. In: Sarker SD, Nahar L (eds) Natural products isolation, 3rd edn. Humana Press
15. Costa DDSV, Bragagnolo N (2017) Development and validation of a novel microwave-assisted extraction method for fish lipids. Eur J Lipid Sci Technol 119:1600108. https://doi.org/10.1002/ejlt.201600108
16. Hu Q, Sommerfeld M, Jarvis E et al (2008) Microalgal triacylglycerols as feedstocks for biofuel production: perspectives and advances. Plant J 54:621–639. https://doi.org/10.1111/j.1365-313X.2008.03492.x
17. Akanda MJH, Sarker MZI, Ferdosh S et al (2012) Applications of supercritical fluid extraction (SFE) of palm oil and oil from natural sources. Molecules 17:1764–1794. https://doi.org/10.3390/molecules17021764
18. Sairam P, Ghosh S, Jena S et al (2012) Supercritical fluid extraction (SFE)-an overview. Asian J Res Pharm Sci 2:112–120
19. Sahena F, Zaidul IS, Jinap et al (2009) Application of supercritical CO_2 in lipid extraction–a review. J Food Eng 95:240–253. https://doi.org/10.1016/j.jfoodeng.2009.06.026
20. Patil PD, Dandamudi KPR, Wang J et al (2018) Extraction of bio-oils from algae with supercritical carbon dioxide and co-solvents. J Supercrit Fluids 135. https://doi.org/10.1016/j.supflu.2017.12.019
21. Lebovka N, Vorobiev E, Chemat F (2012) Enhancing extraction processes in the food industry, 1st edn. CRC Press, Boca Raton
22. Berg H, Mågård M, Johansson G, Mathiasson L (1997) Development of a supercritical fluid extraction method for determination of lipid classes and total fat in meats and its comparison with conventional methods. J Chromatogr A 785(1–2):345–352. https://doi.org/10.1016/S0021-9673(97)00686-9
23. Teng H, Chen L, Huang Q et al (2016) Ultrasonic-assisted extraction of raspberry seed oil and evaluation of its physicochemical properties fatty acid compositions and antioxidant activities. PLoS One 11:e0153457. https://doi.org/10.1371/journal.pone.0153457
24. Szterk A, Ofiara K, Strus B et al (2022) Content of health-promoting fatty acids in commercial sheep, cow and goat cheeses. Foods 11:1116. https://doi.org/10.3390/foods11081116
25. Calvigioni M, Bertolini A, Codini S et al (2023) HPLC-MS-MS quantification of short-chain fatty acids actively secreted by probiotic strains. Front Microbiol 14:1124144. https://doi.org/10.3389/fmicb.2023.1124144
26. Akinyede KA, Hughes GD, Ekpo OE (2022) Comparative study of the antioxidant constituents, activities and the GC-MS quantification and identification of fatty acids of four selected Helichrysum species. Plants 11:998. https://doi.org/10.3390/plants11080998
27. Moreau RA, Whitaker BD, Hicks KB (2002) Phytosterols, phytostanols, and their conjugates in foods: structural diversity, quantitative analysis, and health-promoting uses. Prog Lipid Res 41:457–500. https://doi.org/10.1016/S0163-7827(02)00006-1

28. Bligh EG, Dyer WJ (1959) A rapid method of total lipid extraction and purification. Can J Biochem Physiol 37(8):911–917. https://doi.org/10.1139/o59-099
29. Nardelli V et al (2010) Multi-residue method for the determination of organochlorine pesticides in fish feed based on a cleanup approach followed by gas chromatography triple quadrupole tandem mass spectrometry. J Chromatogr A 1217:4996–5003. https://doi.org/10.1016/j.chroma.2010.05.052
30. Morrison WR, Smith LM (1964) Preparation of fatty acid methyl esters and dimethylacetals from lipids with boron fluoride–methanol. J Lipid Res 5(4):600–608
31. Ichihara KI, Fukubayashi Y (2010) Preparation of fatty acid methyl esters for gas-liquid chromatography. J Lipid Res 51(3):635–640. https://doi.org/10.1194/jlr.D001065
32. AOCS Official Method Ce 1-62 (1997) Fatty acid composition by gas chromatography. In Book: Sampling and analysis of commercial fats and oils
33. Kuksis A (2010) Lipidomics and metabolomics of dietary lipid peroxidation. Mass Spectrom Nutr Res 9:102
34. West C, Lesellier E (2013) Effects of mobile phase composition on retention and selectivity in achiral supercritical fluid chromatography. J Chromatogr A 1302:152–162. https://doi.org/10.1016/j.chroma.2013.06.003
35. Han X, Gross RW (2005) Shotgun lipidomics: multidimensional MS analysis of cellular lipidomes. Expert Rev Proteomics 2:253–264. https://doi.org/10.1586/14789450.2.2.253
36. Christie WW (2014) Lipid analysis: isolation, separation, identification and structural analysis of lipids. Elsevier
37. Cifuentes A (2006) Recent advances in the application of capillary electromigration methods for food analysis. Electrophoresis 27(1): 283–303. https://doi.org/10.1002/elps.200500474
38. Da Porto C, Porretto E, Decorti D (2013) Comparison of ultrasound-assisted extraction with conventional extraction methods of oil and polyphenols from grape (Vitis vinifera L.) seeds. Ultrason Sonochem 20:1076–1080. https://doi.org/10.1016/j.ultsonch.2012.12.002
39. Chemat F, Fabiano-Tixier AS, Vian MA et al (2015) Solvent-free extraction of food and natural products. Trends Anal Chem 71:157–168. https://doi.org/10.1016/j.trac.2015.02.021
40. Pico Y (2013) Ultrasound-assisted extraction for food and environmental samples. Trends Anal Chem 43:84–99. https://doi.org/10.1016/j.trac.2012.12.005

Chapter 4

Extraction and Quantification of Health-Promoting Phytosterols

Ankur Tripathi, Prashant Sahni, Shalini Pandey, and Ashwani Kumar

Abstract

Phytosterols, plant-derived sterols, have attracted significant attention for their potential health benefits. They exert antioxidant, anti-inflammatory, and anticancer activities for which they have been considered a promising candidate to be used as functional food. A multitude of phytosterols like β-sitosterol, campesterol, and stigmasterol have been investigated for their therapeutic potential. This chapter aims to describe the methodologies used for isolation and characterization of phytosterols. The isolation process typically involves the extraction of plant materials using organic solvents, followed by purification using chromatographic methods. Advanced chromatographic methods, such as high-performance liquid chromatography (HPLC) and gas chromatography-mass spectrometry (GC-MS), have been employed to ensure the purity and structural integrity of phyosterols. Additionally, spectroscopic techniques, including nuclear magnetic resonance (NMR) and infrared (IR) spectroscopy, have also been used for detailed structural characterization. This chapter consolidates methodologies for isolation and characterization of phytosterols in order to fully harness their therapeutic potential.

Key words Phytosterols, Sample preparation, Chromatographic methods, Purification

1 Introduction

The potential therapeutic benefits and nutritional significance of phytosterols has gained the attention of researchers for their isolation and characterization. Phytosterols are structurally similar to cholesterol and play a crucial role in promoting cardiovascular health. Beyond their cholesterol-lowering properties, phytosterols exert antioxidant, anti-inflammatory, and anticancer activities, which makes them a promising candidate to be used as functional food. A multitude of phytosterols have been isolated and characterized from various plant species of which β-sitosterol, campesterol, and stigmasterol are more abundant in nature [1]. Other phytosterols, such as brassicasterol, Δ^5 -avenasterol, sitostanol, and campestanol, are present in smaller amounts and have been reported from almost all plants. Isolation and characterization of

Ashwani Kumar, Vidisha Tomer and Piyoosh Kumar Babele (eds.), *Phytochemicals: Extraction, Quantification, Isolation, and Characterization*, Methods and Protocols in Food Science, https://doi.org/10.1007/978-1-0716-4840-7_4,

phytosterols is crucial for advancing their applications in health-promoting products and dietary supplements. Phytosterols isolation involves extraction, followed by purification, whereas identification and quantification of specific phytosterol, elucidation of chemical structures, and determination of purity involve various techniques such as spectroscopy and chromatography. Qualitative and quantitative studies of bioactive compounds including from plant materials mainly rely on the selection of proper methods. Quantitative analysis of phytosterols typically involves several key steps to accurately determine their concentration in samples. Here is a general outline of how this process might proceed:

(a) Sample preparation

- Extraction: Extract phytosterols from sample using an appropriate solvent. Common solvents include hexane, ethanol, or mixtures thereof (*see* Subheading 2).
- Clean up: Purify the extract to remove interfering compounds that could affect the analysis.

(b) Instrumental analysis

- High Performance Liquid Chromatography (HPLC) or Gas Chromatography (GC) is commonly used techniques.
  - GC analysis: Phytosterols are often derivatized before analysis (e.g., trimethylsilylation) to improve their volatility and separation. Most widely used detector for the quantification of phytosterols is flame ionization detection (FID).
  - HPLC analysis: It is another method widely used for phytosterols analysis due to the short analysis time, requires a small amount of sample, and has no need for derivatization.

(c) Quantification

- Internal standards: Use internal standards to accurately quantify phytosterols. Common internal standards include β-sitosterol, campesterol, or stigmasterol.
- Calibration curves: Prepare calibration standards with known concentrations of phytosterols to create a calibration curve.
- Integration: Integrate peak areas (GC) or peak heights (HPLC) of phytosterol peaks and compare to the calibration curve.

(d) Data analysis

- Calculations: Calculate the concentration of phytosterols in the original sample based on the calibration curve.

- Quality control: Evaluate the quality of the analysis through parameters such as recovery rates, precision, and limits of detection/quantification.

(e) Additional considerations:

- Matrix effects: Account for matrix effects if the sample matrix (e.g., food, plant material) affects the analysis.
- Sample size: Ensure the sample size is appropriate for the chosen analytical method and detection limits.

The aim of this chapter is not to produce a comprehensive text, but a general and, wherever necessary, simplified method for isolation and characterization of health-promoting phytosterols.

2 Materials

Before proceeding for extraction and characterization of health-promoting phytosterols, prepare all required solutions using Milli Q water and analytical grade chemicals with high purity. Preparation and storage of all reagents should be carried out at 25 °C. Dispose of all the waste material properly. Sample collection and preparation for phytosterols isolation involves grinding, sieving, and extraction of lipids. Before grinding, wash samples with Milli Q water to remove any extraneous matter and dry at room temperature. After drying, samples containing high lipid content such as nuts and seeds use mill grinder, sieve through 60 mesh, and vacuum-seal to reduce oxidation. In case of leaves and stem, dry in an oven at 60 °C for 2 h, then grind and sieve through 60 mesh. For liquid samples, store at −80 °C for 6 h, then place in a refrigerated vacuum dryer for solidification, and grind and sieve through 60 mesh. Store the ground samples at −20 °C until analysis.

2.1 Extraction of Phytosterols Using Acid Hydrolysis

- Analytical balance, Screw cap glass tube, Shaking water bath, Vortex mixture

  Absolute ethanol, 6 M hydrochloric acid: Take 50 mL of the concentrated HCl [(Density: 1.18 and 37% purity (by mass)] and dilute with Milli Q water up to 100 mL (*see* **Notes 1** and **2**), hexane and diethyl ether solution (1:1, *v/v*) (*see* **Note 3**).

2.2 Extraction of Phytosterols Using Soxhlet Extraction

- Soxhlet extraction apparatus, Cellulose or glass thimble, Rotary evaporator, n-hexane, Anhydrous magnesium sulphate.

2.3 Extraction of Phytosterols Using Maceration

- Maceration flask, Separating funnel, Shaker, Nitrogen gas, ethanol/chloroform solution (2:1, *v/v*): (*see* **Note 4**), 1M KCl solution: (*see* **Note 5**).

2.4 Extraction of Phytosterols Using Supercritical Carbon Dioxide Extraction (SFE)

- Supercritical CO_2 Extraction System (including extraction vessel, separator, and back-pressure valve); Pressure regulator (to maintain 350 bars in the extraction vessel), Temperature control system (to maintain 60 °C in the extractor and 40 °C in the separator), CO_2 Flowmeter (to regulate CO_2 at 9 kg/h), Amber vials (for light protection), Depressurization system (for controlled release of CO_2), Analytical balance (for weighing the extract), Storage freezer (−20 °C), Carbon Dioxide (CO_2).

2.5 Quantitative Analysis of Phytosterols Using Gas Chromatography

- Chloroform, Acetone, Methanol, 1 M Potassium hydroxide (KOH) in methanol (*see* **Note 6**), Dichloromethane, MSHFBA (n-methyl-n-trimethylsilylheptafluorobutyramide), 1-MIM (1-Methylimidazole), Gas Chromatograph equipped with FID detector and DB-35MS capillary column (25 m × 0.20 mm × 0.33 μm), Hydrogen as a carrier gas at a flow rate of 1.5 mL/min, Internal Standards:5α-cholestane or 5β-cholestan-3α-ol (epicoprostanol) (50 μg/mL in acetone), Standard solutions of Stigmasterol, Campesterol, Brassicasterol and β-Sitosterol (*see* **Note** 7).

2.6 Quantitative Analysis of Phytosterols Using High-Performance Liquid Chromatography (HPLC)

- HPLC Chromatograph, Column: Reversed-phase C-18 column (5 μm, 250 × 4.6 mm), Detector: UV-VIS, 0.45 μm membrane filters, 6-Ketocholestanol (120 g), Anhydrous ethanol (10 mL), HPLC-grade Methanol, Acetonitrile (HPLC-grade), Methanol: Acetonitrile solution (80:20, *v/v*) (Mobile Phase) (*see* **Note 8**), Potassium hydroxide (KOH), Nitrogen gas (for evaporation), Standard solutions of Stigmasterol, Campesterol, Brassicasterol, and β-Sitosterol (*see* **Note 9**).

2.7 Quantitative Analysis of Phytosterols Using High-Performance Liquid Chromatography-Mass Spectroscopy (HPLC-MS)

- HPLC Chromatograph, Column: Reversed-phase C-18 column (5 μm, 250 × 4.6 mm), Detector: UV-VIS, Photodiode Array (PDA) detector, Acetonitrile (HPLC-grade), Acetone (HPLC-grade) for mobile phase preparation, Nitrogen gas (for drying), Methanol (HPLC-grade), Standard solutions of Stigmasterol, Campesterol, Brassicasterol, and β-Sitosterol (*see* **Note 9**).

2.8 Quantitative Analysis of Phytosterols Using Gas Chromatography-Mass Spectrometry (GC-MS)

- GC Column: 30 m × 0.25 mm × 0.25 μm (for separation), Helium (99.999%) Carrier gas at 1.0 mL/min flow rate, Mass Spectrometry Detector (MSD): For detecting phytosterols and oxidation products. Internal Standards: 5α-cholestane or 5-β-cholestan-3α-ol (epicoprostanol) (50 μg/mL in acetone), Standard solutions of Stigmasterol, Campesterol, Brassicasterol, and β-Sitosterol (*see* **Note 10**), Dichloromethane (for dissolving and extraction), 1M Potassium hydroxide (KOH) in 95% ethanol (for saponification) (*see* **Note 11**), Acetone (for preparing standards and internal standards), Nitrogen stream (for evaporation), Gas phase injection vials (2 mL, for transferring the evaporated sample).

2.9 Separation and Extraction of Phytosterol Oxidation Products (POPs) by Solid-Phase Extraction (SPE)

- Silica Solid-Phase Extraction (SPE) cartridges (for purifying the extracted fraction), GC Column: 30 m × 0.25 mm × 0.25 μm (for separation), Helium (99.999%)—Carrier gas at 1.0 mL/min flow rate, Mass Spectrometry Detector (MSD): For detecting phytosterols and oxidation products. 19-Hydroxycholesterol (50 μg/mL in acetone) used as the internal standard, Dichloromethane (for extracting unsaponified fractions), Hexane (for activating the SPE column, 10 mL), Diethyl ether/n-hexane solution (9:1, *v/v*) (*see* **Note 12**), Primary eluent for extracting triglycerides (10 mL), Ether/hexane solution (3:7, *v/v*) (*see* **Note 13**), Secondary eluent (10 mL), Acetone (final elution solvent, 10 mL), Nitrogen gas (for evaporation under nitrogen stream), N-methyl-N-(trimethylsilyl) heptafluorobutyramide (MSHFBA) (for derivatization), 1-methyl imidazole (1-MIM) (as a catalyst for derivatization), n-Hexane (for dissolving derivatized sample).

3 Methods

Various methods have been described below to extract phytosterols from plant samples.

3.1 Acid Hydrolysis [2]

1. Weigh 30 g ground and dried sample accurately in a screw cap glass tube and add 30 mL of absolute ethanol and 150 mL of 6 M hydrochloric acid (*see* **Notes 1** and **2**) to the samples.
2. Flush nitrogen gas into the tubes, tighten a Teflon tape over them to prevent lipid oxidation, and place the tubes in a shaking water bath (100–200 rpm) at 50–60 °C.
3. Heat the sample tubes at 80 °C for 60 min, cool under running water, then fill with 250 mL of absolute ethanol and vortex for 30 s then add 360 mL deionized water.
4. Add 210 mL solution of hexane and diethyl ether (1:1, *v/v*) into the tubes and shake vigorously for 2 min and then leave undisturbed for 10 min to facilitate phase separation.
5. After collecting the upper layer, repeat the extraction step twice.
6. Dry under nitrogen steam.

3.2 Soxhlet Extraction [3]

1. Extract 30 g of ground and dried sample with n-hexane (300 mL) using a Soxhlet extractor for 8 h at 70 °C.
2. Pass the hexane-oil mixture through anhydrous magnesium sulfate layered on filter paper in a funnel, followed by rotary evaporators for the removal of the solvent at 40 °C.
3. Weigh the oil samples, and store at −20 °C until further analysis.

3.3 Maceration [4]

1. Extract 30 g of ground and dried samples by shaking with 300 mL of ethanol/chloroform solution (2:1, *v/v*) at room temperature for 2 h.
2. After filtering the mixture with paper filters, add 70 mL of 1 M KCl solution to the separatory funnel and shake gently overnight to separate into two phases (*see* **Note 14**).
3. Collect the lower phase and evaporate under gentle nitrogen flow (*see* **Note 15**).
4. For further analysis, weigh the extracted oil phase, flush with nitrogen, and store at −20 °C.

3.4 Supercritical Carbon Dioxide Extraction [5]

1. Load the dried sample into the supercritical carbon dioxide extraction vessel and flush it with CO_2.
2. Fill the extraction vessel with CO_2 to a pressure of 350 bars and 60 °C (*see* **Notes 16** and **17**).
3. Maintain the conditions at the separator at 50 bars and 40 °C by regulating the back-pressure valve between extractor and separator.
4. After 120 min of extraction process, maintain a CO_2 flow at the rate of 9 kg/h, depressurize the system, and collect the extract in an amber vial to protect it from light. Finally, weigh and store the samples at −20 °C until analysis.

3.5 Gas Chromatography Technique for the Quantification of Phytosterols [6, 7]

To analyze phytosterols, the GC system equipped with FID detector and DB-35MS column (25 m × 0.20 mm × 0.33 μm) is the most widely used. The GC-FID procedure includes addition of an internal standard (5α-cholestane), saponification with ethanolic KOH solution, washing with Milli Q water phase to remove saponifiedmaterial, and derivatization of phytosetrols with n-methyl-n-trimethylsilylheptafluorobutyramide (MSHFBA) and 1-methyl imidazole (MIM), or hexamethyldisilazane (HMDS) and trimethylchlorosilane (TMCS) in anhydrous pyridine for GC analysis.

A. **Preparation of phytosetrols solutions**

   1. Use 5α-cholestane or 5β-cholestan-3α-ol (epicoprostanol) as internal standards for quantification of phytosterols (50 μg/mL in acetone).

2. Preparation of standard stock solutions of stigmasterol, campesterol, brassicasterol, and β-sitosterol: Accurately weigh 10 mg stigmasterol, campesterol, brassicasterol, and β-sitosterol standard, dilute to a 10 mL flask with methanol/chloroform/acetone.
3. **Preparation of working concentrations from stock solution**: To obtain working concentrations of 10, 50, 100, 150, and 200 μg/mL, accurately pipette out 0.10, 0.50, 1.0, 1.5, and 2.0 mL of the standard stock solution into a 10 mL flask, and makeup to the mark with methanol.

B. **Saponification of phytosterols**

1. Saponify the oil sample (50 mg) with 1 M KOH in methanol for 18 h at room temperature (*see* **Note 6**).
2. Add Milli Q water to move saponified material.
3. Extract the unsaponifiable phytosterols with dichloromethane and evaporate under a stream of nitrogen.

C. **Derivatization of phytosterols**

Extract unsaponified fraction with dichloromethane, and evaporate under a gentle nitrogen stream, and incubate in 100 μL MSHFBA: 1-MIM (95:5, *v/v*) at 75 °C for 20 min.

D. **Instrument conditions**

1. Inject 1.0 L of sample to GC–MS column in a splitless mode.
2. For 5 min in the beginning, maintain the column temperature at 100 °C, and then increase the temperature by 25 °C/min to 250 °C, then increase by 3 °C/min to 290 °C.
3. After reaching 290 °C, maintain the temperature for 20 min.
4. Use carrier gas at a flow rate of 1.5 mL/min and set the detector temperature at 300 °C.
5. Use 5a-cholestane as an internal standard and compare the retention times (RT) (relative to 5a-cholestane) with those of commercially available standards.
6. Express the results as μg/g of oil in the samples.
7. Quantification of phytosterols: For accurate quantification of phyosterols, use peak area, retention times, and internal standard. Inject phytosterolsstandards (campesterol, stigmasterol, brassicasterol, and β-sitosterol) and internal standards (5α-cholestane) separately to get standard calibration curves for GC analysis.

3.6 HPLC Technique for the Quantification of Phytosterols [4]

A. **Preparation of solutions**

1. Weigh 50 g potassium hydroxide (KOH) and dissolve in 100 mL of Milli Q water.
2. Preparation of standard stock solutions of stigmasterol, campesterol, brassicasterol, and β-sitosterol: Accurately weigh 10 mg stigmasterol, campesterol, brassicasterol, and β-sitosterol standard, dilute to a 10 mL flask with HPLC-grade methanol and store in the dark at −18 °C.
3. Preparation of working concentrations from stock solution: To obtain working concentrations of 10, 50, 100, 150, and 200 μg/mL, accurately pipette out 0.10, 0.50, 1.0, 1.5, and 2.0 mL of the standard stock solution into a 10 mL flask, and makeup to the volume with HPLC-grade methanol and store in the dark at −18 °C.

B. **Extraction of phytosterols**

1. Mix 120 g of 6-ketocholestanol with 1 g of dried plant sample in a conical flask and soak with 10 mL of anhydrous ethanol.
2. Place the conical flak on a vibrating plate at 130 rpm at 30 ° C, and extract for 4 h.
3. Filter the ethanol extract and evaporate under a nitrogen flow then redissolve the evaporated materials in HPLC-grade methanol and filter with 0.45 μm membrane filters before chromatography analysis.

C. **Instrument conditions**

1. Use a reversed-phase, isocratic column with an ultraviolet-visible detector.
2. Use methanol; acetonitrile solution (80:20 ratio) as mobile phase @ 1.0 mL/min in the flow.
3. Use a reverse 5 μm C-18 column (250 × 4.6 mm) for separation.
4. Set the column temperature at 30 °C and run HPLC system under isocratic conditions for 30 min.
5. Set the injection volume (20 μL) and wavelength at 210 nm.

D. **Quantification of phytosterols**

1. Develop a linear calibration curve based on standard solutions of different phytosterols at different concentrations in the range of 50–500 μg/mL.
2. Carry out quantification based on internal calibration,
3. Express the final concentrations as μg/ml (*see* **Note 18**).

3.7 HPLC–MS Technique for the Quantification of Phytosterols [4]

A. **Preparation of solutions**: As in GC assays

B. **Instrument conditions**

1. Use HPLC system equipped with mass spectrometry for phytosterol analysis.
2. Equip LC–MS with a column (250 × 4.6 mm ID, 5 μm, Y) at a temperature of 25 °C for separation.
3. Set the flow rate of mobile phase [Mobile phase A (acetonitrile) and mobile phase B (acetone)] at 1.0 mL/min.
4. Generate gradient by dropping the initial concentration of phase A from 100% to 30% at 10 min, to 20% at 30 min, and to 0% at 40 min, and the initial concentration of phase B (100%) 50 min.
5. Inject 50 μL volume and set the detection wavelength at 205 nm using Photodiode Array (PDA) detector.
6. During the experimentation, maintain the following conditions: 4000 volt capillary voltage; 10 μA current; 5 mL/min drying gas flow; 350 °C drying gas temperature; 200 V fragmentor voltage; 60 psi nebulizer pressure; and 500 °C vaporizer temperature.
7. Split the quarter of the flow before it reaches the interface sprayer.
8. Detect the phytosterols and phytosterols esters using the selected ion monitoring (SIM) mode.
9. To determine the purity of standard phytosterols and phytosterols esters, apply full scan mode.

C. **Quantification of phytosterols**

1. Compare retentions times (RT) and MS data with standards.
2. Carry out quantification based on internal calibration,
3. Express the final concentrations as μg/mL.

3.8 GC–MS Technique for the Quantification of Phytosterols

A. **Preparation of solutions**

- Prepare internal standard of 19-hydroxycholesterol (50 μg/mL in acetone) and 5α-cholestane (50 μg/mL in acetone).
- Preparation of standard solutions of cholesterol, 7a-hydroxycholesterol, 7b-hydroxycholesterol, 7-ketocholesterol, 5a,6a-epoxy cholesterol, 5b,6b-epoxy cholesterol, cholestanetrio, 19-hydroxycholesterol, and 5a-cholestane: As in GC assays.

B. **Saponification of phytosterols**

1. Transfer about 50 mg of lipid samples to tubes with 10 μL 5α-cholestane (50 μg/mL in acetone).

2. Dissolve the mixture in 0.5 mL dichloromethane, and then saponify with 4 mL of 1M KOH in 95% ethanol) and leave for shaking at 25 °C in the dark for 18 h.
3. Extract the unsaponified fraction with dichloromethane.
4. Add 5 mL of dichloromethane and 3 mL of water, mix well, centrifuge at 3000 rpm for 5 min.
5. Discard the clear water phase, wash the dichloromethane phase with 5 mL of Milli Q water, and repeat three times until the organic phase is clarified.
6. Evaporate the organic phase to about 1 mL under a nitrogen stream, and then transfer to 2 mL gas phase injection vial.
7. Continue evaporation to dryness under a nitrogen stream [8].

C. **Derivatization of phytosterols**

1. Incubate the samples in 100 μL mixture of N-methyl-N-(trimethylsilyl)hep- tafluoro-butyramide (MSHFBA) and 1-methyl imidazole (1-MIM) (95:5, *v/v*) at 75 °C for 20 min, after that remove the solvent from dry samples under nitrogen stream.
2. Fill the derivatized sample with n-hexane to 1 mL.

D. **Instrument conditions**

1. Carry out separation and analysis of phytosterol and phytosterol oxidation products (POPs) with 30 m × 0.25 mm × 0.25 m GC column coupled to an MSD detector.
2. Use helium (99.999%) as the carrier gas at the flow speed of 1.0 mL/min.
3. Carry out separation of phytosterol and phytosterol oxidation products under SIM mode with a solvent delay of 13.0 min.
4. Maintain following GC–MS conditions: Helium (99.999%) as carrier gas at 1.0 mL/min; Oven temperature at 100 °C for 1 min, gradually increase up to 250 °C at a rate of 40 ° C/min, then increase to 300 °C at 2.5 °C/min, and hold for 10 min the Injection temperature at 300 °C with a splitless mode. Transfer line temperature at 290 °C and temperature of ion source at 250 °C.

E. **Quantification of phytosterols**

1. Compare retentions times (RT) and MS data with standards.
2. Carry out quantification based on internal calibration,
3. Express the final concentrations as μg/mL.

3.9 Separation and Extraction of Phytosterol Oxidation Products (POPs) by Solid-Phase Extraction (SPE) [4]

1. Carry out cold saponification (consistent with PS analysis) of about 50 mg lipid samples with 19-hydroxycholesterol (50 μg/mL in acetone) as the internal standards.
2. Extract the unsaponified fraction with dichloromethane and purify by single-step silica SPE cartridges.
3. Activate the SPE column with 10 mL of hexane.
4. Carry out elution of triglycerides from the sample using 10 mL of diethyl ether/n-hexane (9:1, *v/v*) as the primary eluent to remove the majority of less polar compounds.
5. Use 10 mL of ether/hexane (3:7, *v/v*) as secondary eluent.
6. Elute and the final volume of acetone (10 mL) and collect the filtrate.
7. Evaporate organic phase to about 1 mL under a nitrogen stream and then transfer to 2 mL gas phase injection vial.
8. Continue to evaporate to dryness under a nitrogen stream.

A. **Derivatization of phytosterols:** As in GC-MS assay

B. **Instrument conditions**: As in GC-MS assay

C. **Quantification of phytosterol oxidation products**: As in GC-MS assay

4 Notes

1. Acid hydrolysis cannot be used to determine $^{\Delta}7$ -phytosterols as they are decomposed or isomerized after short periods of hydrolysis.
2. Preparation of 6M hydrochloric acid solution: Take 50 mL of the concentrated HCl [(Density: 1.18 and 37% purity (by mass)] and dilute with Milli Q water up to 100 mL.
3. Hexane/diethyl ether solution (1:1, *v/v*): Mix equal volumes of hexane and diethyl ether.
4. Ethanol/chloroform solution (2:1, *v/v*): Mix two volume of Ethanol and one volume of chloroform.
5. 1 M KCl solution: Dissolve 74.6 g of potassium chloride (KCl) in distilled water and dilute to a final volume of 1 liter.
6. Preparation of 1 M Potassium hydroxide (KOH) in methanol (Dissolve 56 g of KOH in methanol to a final volume of 1 liter).
7. Standard solutions of Stigmasterol, Campesterol, Brassicasterol, and β-Sitosterol: Accurately weigh 10 mg of each standard separately and dilute to 10 mL with methanol. For preparation of standard curve, accurately pipette out 0.10, 0.50, 1.0, 1.5, and 2.0 mL of each standard stock solution

into a 10 mL flask separately, and makeup the final volume with methanol.

8. Methanol: Acetonitrile solution (80:20, *v/v*): Mix 80 mL of methanol with 20 mL of acetonitrile.
9. Standard solutions of Stigmasterol, Campesterol, Brassicasterol, and β-Sitosterol: Accurately weigh 10 mg of each standard separately and dilute to 10 mL with methanol. For preparation of standard curve, accurately pipette out 0.10, 0.50, 1.0, 1.5, and 2.0 mL of each standard stock solution into a 10 mL flask separately, and makeup the final volume with methanol (HPLC-grade).
10. Standard solutions of Stigmasterol, Campesterol, Brassicasterol, and β-Sitosterol: Accurately weigh 10 mg of each standard separately and dilute to 10 mL with acetone. For preparation of standard curve, accurately pipette out 0.10, 0.50, 1.0, 1.5, and 2.0 mL of each standard stock solution into a 10 mL flask separately, and makeup the final volume with acetone.
11. Preparation of 1 M Potassium hydroxide (KOH) in 95% ethanol (for saponification): For 95% ethanol, mix 5 mL Milli Q water with 95 mL of absolute ethanol, then dissolve 56 g of KOH in it.
12. Diethyl ether/n-hexane solution (9:1, *v/v*): Mix nine volume of Diethyl ether and one volume of n-hexane.
13. Ether/hexane solution (3:7, *v/v*): Mix three volume of Diethyl ether and one volume of n-hexane.
14. During maceration, centrifugation is required when plant material is too small to be filtered.
15. During maceration, solvents should be under a gentle flow of nitrogen gas to concentrate the phytosterol extract. Nitrogen flow helps remove the solvent without overheating the extract and potentially damaging the sensitive phytosterols. The nitrogen flow rate should be optimized to ensure efficient solvent removal without causing excessive loss of the extracted phytosterols.
16. Supercritical carbon dioxide extraction requires optimization of pressure, temperature, flow rate of CO_2, extraction time, the solvent to feed ratio, particle size, and moisture content of the sample.
17. Supercritical fluid extraction using carbon dioxide (CO_2) is an ideal solvent for lipid and lipidsoluble substances, but not for most pharmaceuticals and drug compounds.
18. HPLC is a highly suitable quantification method for thermally sensitive phytosterols.

References

1. Salehi B, Quispe C, Sharifi-Rad J, Cruz-Martins-N, Nigam M, Mishra AP, Konovalov DA, Orobinskaya V, Abu-Reidah IM, Zam W, Sharopov F, Venneri T, Capasso R, Kukula-Koch W, Wawruszak A, Koch W (2021) Phytosterols: from preclinical evidence to potential clinical applications. Front Pharmacol 11: 599959. https://doi.org/10.3389/fphar.2020.599959
2. Islam MA, Jeong BG, Jung J, Shin EC, Choi SG, Chun J (2017) Phytosterol determination and method validation for selected nuts and seeds. Food Anal Methods 10:3225–3234
3. Kozłowska M, Gruczyńska E, Ścibisz I, Rudzińska M (2016) Fatty acids and sterols composition, and antioxidant activity of oils extracted from plant seeds. Food chem 15(213):450–456
4. Yang B, Zhao T, Liu Y, Lu B (2023) Guidelines for extraction and quantitative analysis of phytosterols and oxidation products. eFood 4(4): e108. https://doi.org/10.1002/efd2.108
5. Mustapa AN, Martin Á, Mato RB, Cocero MJ (2015) Extraction of phytocompounds from the medicinal plant ClinacanthusnutansLindau by microwave-assisted extraction and supercritical carbon dioxide extraction. Ind Crops Prod 74: 83–94
6. Zhao Y, Yang B, Xu T, Wang M, Lu B (2019) Photooxidation of phytosterols in oil matrix: effects of the light, photosensitizers and unsaturation degree of the lipids. Food Chem. 288: 162–169
7. Alvarez-Sala A, Garcia-Llatas G, Cilla A, Barbera R, Sánchez-Siles LM, Lagarda MJ (2016) Impact of lipid components and emulsifiers on plant sterols bioaccessibility from milk-based fruit beverages. J Agric Food Chem 64(28):5686–5691
8. Wang M, Zhang L, Wu X, Zhao Y, Wu L, Lu B (2019) Quantitative determination of free and esterified phytosterol profile in nuts and seeds commonly consumed in China by SPE/GC–MS. LWT 100:355–361

Chapter 5

Extraction and Quantification of Alkaloids

Prashant Sahni, Ankur Tripathi, Shalini Pandey, Simardeep Kaur, Prerana Shere, and Ashwani Kumar

Abstract

Alkaloids, a diverse group of naturally occurring compounds primarily found in plants, possess significant pharmacological properties and are of considerable interest in the fields of medicine, nutraceuticals, and biochemistry. This chapter focuses on the extraction and quantification of these alkaloid classes, emphasizing the methods tailored for their efficient extraction from plant matrices. Various extraction techniques, ranging from traditional methods like maceration, percolation, and Soxhlet extraction to modern approaches such as microwave-assisted extraction (MAE) and supercritical fluid extraction (SFE) were discussed. These methods are compared in terms of efficiency, solvent consumption, extraction time, and their applicability to different alkaloid classes. Different methods of quantification including UV-VIS spectroscopy, HPLC and iodometric back titration analysis was discussed with their principle and detailed procedure.

Key words Bioactive, Supercritical, Spectroscopy, Nutraceutical, Medicine

1 Introduction

Alkaloids are a diverse group of naturally occurring organic compounds that predominantly contain basic nitrogen atoms. They are typically derived from plant sources, though some are also found in animals and microorganisms. Alkaloids have long been of interest due to their wide range of pharmacological activities and therapeutic potential. From morphine and quinine to nicotine and caffeine, these compounds are known for their significant impacts on human physiology and their historical use in medicine and traditional practices. Their ability to interact with biological systems often makes them potent, either as therapeutics or, in some cases, as toxins. Alkaloids are categorized into various classes based on their chemical structure, such as tropane alkaloids, indole alkaloids, and isoquinoline alkaloids, each with unique properties and applications [1].

Ashwani Kumar, Vidisha Tomer and Piyoosh Kumar Babele (eds.), *Phytochemicals: Extraction, Quantification, Isolation, and Characterization*, Methods and Protocols in Food Science, https://doi.org/10.1007/978-1-0716-4840-7_5,

The process of extracting and quantifying alkaloids from natural sources is critical in both research and industrial applications. Extraction methods vary widely, depending on the nature of the alkaloids and the matrix from which they are being isolated. Traditional methods include solvent extraction, maceration, and percolation, while modern approaches often employ techniques such as supercritical fluid extraction and microwave-assisted extraction to enhance efficiency and yield [2]. Each method has its advantages and limitations concerning the selectivity, efficiency, and purity of the extracted alkaloids. Quantification of alkaloids, on the other hand, is essential for determining their concentration and ensuring the consistency and efficacy of pharmaceutical preparations. Analytical techniques such as high-performance liquid chromatography (HPLC), gas chromatography (GC), and mass spectrometry (MS) are commonly employed to achieve accurate measurement and characterization of alkaloid compounds [3].

Understanding the intricacies of alkaloid extraction and quantification not only advances pharmacognosy but also enhances the development of new drugs and agricultural products. The ability to efficiently extract and accurately measure these compounds enables researchers and manufacturers to ensure high-quality products and to explore new therapeutic potentials. With ongoing advancements in analytical technologies and extraction methods, the study of alkaloids continues to evolve, providing deeper insights into their roles and applications. This chapter will delve into the methodologies used for extracting and quantifying alkaloids, examining the techniques, their effectiveness, and the challenges involved in this complex field of study [1–3].

1.1 Classes of Alkaloids

Alkaloids are a large and diverse group of nitrogenous organic compounds that are primarily derived from plants, although some are also found in animals and microorganisms. They are classified into various classes based on their chemical structures, which often include complex ring systems. Principal classes of alkaloids include:

1.1.1 Tropane Alkaloids

Tropane alkaloids are characterized by a bicyclic structure that includes a tropane ring—a seven-membered ring fused to a six-membered ring. This class includes several notable alkaloids with significant pharmacological effects. These alkaloids generally exhibit moderate to high polarity due to the presence of nitrogen atoms in their rings, which can form hydrogen bonds.

I. **Atropine**: Found in *Atropa belladonna* (deadly nightshade), atropine is well known for its anticholinergic properties. It is used to treat bradycardia (slow heart rate) and to induce pupil dilation during ophthalmic procedures [1].

II. **Scopolamine**: Derived from Datura *stramonium* (Jimson weed), scopolamine is utilized primarily to prevent motion sickness and postoperative nausea due to its anticholinergic and antiemetic effects [1].

1.1.2 Indole Alkaloids

Indole alkaloids are characterized by the presence of an indole ring system, which consists of a fused benzene and pyrrole ring. This class of alkaloids is notable for its diverse biological activities. This class generally exhibits a range of polarities based on additional functional groups attached to the indole structure. The indole ring contributes to moderate polarity, but additional substituents can influence overall polarity.

I. **Reserpine**: Extracted from *Rauvolfia serpentina* (Indian snakeroot), reserpine was historically used as an antihypertensive agent and is also employed in the treatment of certain psychiatric disorders due to its ability to deplete monoamine neurotransmitters [2].

II. **Vincristine**: Obtained from *Catharanthus roseus* (periwinkle), vincristine is an important chemotherapeutic agent used in the treatment of various cancers, including leukemia and lymphoma, due to its role in inhibiting cell division [2].

1.1.3 Isoquinoline Alkaloids

Isoquinoline alkaloids feature an isoquinoline ring, which consists of a benzene ring fused to a pyrrole-like ring. They are known for their potent physiological effects. The isoquinoline ring contributes to moderate polarity, with additional functional groups affecting overall solubility.

I. **Morphine**: Derived from *Papaver somniferum* (opium poppy), morphine is a powerful analgesic used in the management of severe pain. It acts primarily on opioid receptors in the central nervous system [3].

II. **Codeine**: Also derived from the opium poppy, codeine is used for its analgesic and antitussive (cough-suppressant) properties. It is often prescribed for moderate pain relief and to alleviate cough [3].

1.1.4 Pyrrolidine Alkaloids

Pyrrolidine alkaloids contain a pyrrolidine ring, and they are known for their stimulant and psychoactive properties. These alkaloids generally have lower polarity compared to other classes. The pyrrolidine ring itself is relatively non-polar, but the presence of additional polar functional groups can increase polarity.

I. **Cocaine**: Derived from *Erythroxylum coca* (coca plant), cocaine is both a stimulant and a local anesthetic. It has significant effects on the central nervous system, including euphoria and increased energy, but it is also highly addictive and has a potential for abuse [1].

II. **Nicotine**: Extracted from *Nicotiana tabacum* (tobacco plant), nicotine is a potent stimulant that affects the central nervous system and is responsible for the addictive properties of tobacco products [1].

1.1.5 Quinoline Alkaloids

Quinoline alkaloids are characterized by the presence of a quinoline ring system—a fused benzene and pyridine ring. The quinoline ring contributes to moderate polarity, and additional substituents can modify this polarity.

I. **Quinine**: Derived from Cinchona species, quinine has been historically used to treat malaria. It works by interfering with the growth and replication of the Plasmodium parasites that cause the disease [3].

1.2 Solvents Used in the Extraction of Alkaloids

In the extraction of alkaloids, selecting the appropriate solvent is crucial for optimizing yield and purity. The selection of an appropriate solvent for the extraction of alkaloids is a critical step in the analytical processing of these compounds. Alkaloids, a diverse group of naturally occurring organic compounds, exhibit varying degrees of polarity, solubility, and chemical reactivity. This diversity necessitates careful consideration of solvent properties to achieve efficient and selective extraction. The choice of solvent impacts the yield, purity, and overall effectiveness of the extraction process, influencing the quality of the final alkaloid product. Solvents used in alkaloid extraction must be capable of dissolving the target alkaloids from complex plant matrices while minimizing the co-extraction of unwanted substances. Factors such as solvent polarity, boiling point, miscibility, and environmental and safety considerations play a pivotal role in the selection process. Following solvents can be employed for extraction of different alkaloids.

1.2.1 Ethanol

Ethanol is a commonly used solvent due to its effectiveness in dissolving a wide range of alkaloids and its relative safety and environmental friendliness. Ethanol's moderate polarity makes it suitable for extracting both polar and moderately non-polar alkaloids from plant materials. Ethanol is typically used in concentrations ranging from 50% to 95% (v/v) for alkaloid extraction. Higher concentrations (90–95%) are used when a more efficient extraction is desired, as ethanol's polarity can help in solubilizing alkaloids effectively [4].

1.2.2 Methanol

Methanol is another widely used solvent known for its high polarity, which enhances its ability to dissolve a broad spectrum of alkaloids. Its low boiling point and high miscibility with water make it a versatile choice for laboratory-scale extractions. Methanol is often used in concentrations of 70% to 100% (v/v). A concentration of 100% methanol is typically used for extracting alkaloids from plant materials where a high degree of extraction efficiency is required [5]. Quinine extraction from Cinchona bark often utilizes methanol to ensure efficient recovery of the alkaloid [2].

1.2.3 Water

Water is employed particularly when extracting water-soluble alkaloids or when used in conjunction with organic solvents in a biphasic extraction process. Its high polarity ensures effective extraction of polar alkaloids, although it often requires combination with other solvents to improve extraction efficiency. Extraction of caffeine from coffee beans often starts with water, as caffeine is moderately soluble in water [1].

1.2.4 Chloroform

For the extraction of alkaloids with moderate polarity, chloroform is useful. It is a moderately polar solvent that is not miscible with water but effective when used in liquid-liquid extraction systems. Typically used as a pure solvent (100% chloroform) for extracting alkaloids with non-polar characteristics. It is often used in combination with other solvents to improve selectivity in extraction. The extraction of nicotine from tobacco leaves can involve chloroform, which selectively dissolves nicotine due to its non-polar nature [1].

1.2.5 Hexane

Hexane, a non-polar solvent, is particularly useful in the initial stages of extraction to remove non-polar compounds before extracting alkaloids with other solvents. Its non-polar nature makes it suitable for isolating non-polar substances [6]. Typically used as 100% hexane for initial extraction phases to separate non-polar components from plant materials. the extraction process of certain tropane alkaloids, hexane may be used in the initial extraction stages [3].

1.2.6 Acetone

Acetone serves as another versatile solvent, especially for alkaloids that are not very polar. Its miscibility with water and many organic solvents, combined with its low boiling point, makes it effective in various extraction scenarios [7].

1.2.7 Dimethyl Sulfoxide

Dimethyl sulfoxide (DMSO), with its high polarity and ability to penetrate biological membranes, is used for extracting highly polar alkaloids. It can be particularly effective in extracting compounds from plant tissues due to its unique solvent properties [7].

1.2.8 Supercritical CO_2

Supercritical CO_2 is employed for its high selectivity and efficiency in extracting alkaloids, especially when a green solvent with minimal environmental impact is desired. Its supercritical state allows it to act as a non-polar solvent while offering the benefits of reduced solvent residues [7].

1.3 Steps for the Extraction of the Alkaloids [7]

1.3.1 Preparation of Plant Material

The plant material should be dried and ground into a coarse or fine powder, depending on the desired extraction efficiency. The surface area of the plant material should be increased for more efficient extraction. Drying can be done in an oven at 40–60 °C

1.3.2 Selection of Solvent

To choose a solvent that efficiently dissolves the target alkaloids, solvents should be selected based on the polarity of the alkaloids. Common solvents include ethanol, methanol, and water.

1.3.3 Volume Determination

Calculate the volume of solvent needed to fully immerse the plant material. Typically, a solvent-to-plant material ratio of 1:5 to 1:10 (w/v) is used.

1.3.4 Extraction Process

Alkaloid extraction involves various techniques, each suited for different types of alkaloids and extraction needs. For details see the next section (*see* Subheading 1.3.5) on extraction methods.

1.3.5 Filtration

After the maceration period, filter the mixture using filter paper or a filtration apparatus to separate the liquid extract from the plant residue.

1.3.6 Concentration of Extract

For manual evaporation, place the filtrate in an evaporating dish and heat gently using a water bath to evaporate the solvent. If using a rotary evaporator, transfer the filtrate to the evaporator flask and set the appropriate temperature (usually around 40–60 °C) and reduced pressure (20–200 mbar). This will remove the solvent efficiently while preserving the alkaloids. Avoid overheating to prevent the degradation of alkaloids. Continue evaporation until a concentrated extract is obtained. The final volume should be adjusted according to the desired concentration of the alkaloid extract.

1.3.7 Storage of Extract

Transfer the concentrated extract to clean, dry glass bottles. Amber bottles are preferred to protect the extract from light-induced degradation. Store the bottles in a cool, dark place to prevent degradation of the alkaloids. Ideally, the storage temperature should be between 4 °C and 25 °C.

1.4 Extraction of Alkaloids

The major methods used for alkaloid extraction are maceration, percolation, Soxhlet extraction, supercritical fluid extraction (SFE), and microwave-assisted extraction (MAE). Among these, maceration, percolation, and Soxhlet extraction are the conventional methods, while SFE and MAE are the novel methods. Maceration involves soaking the plant material in a solvent to dissolve the alkaloids [7]. The principle behind maceration is based on the diffusion of alkaloids from the plant matrix into the solvent over time. The solvent dissolves the alkaloids as it penetrates the plant material. This method provides moderate extraction efficiency, which is influenced by the solubility of the alkaloids and the duration of extraction. This method may not be fully effective for alkaloids with low solubility such as caffeine from *Coffea arabica* and morphine from *Papaver somniferum*. This method require basic equipment, is easy to use, uses inexpensive solvents and materials, and can be performed with a range of solvents and plant materials. These attributes make this method simple, cost effective, and flexible. The limitations of this method are long extraction time, need of large quantities of solvent, and incomplete extraction.

Percolation is a more dynamic extraction method compared to maceration, involving the continuous flow of solvent through a column of plant material [2]. The principle of percolation is based on the solvent continuously passing through the plant material, which ensures fresh solvent contact and improves the efficiency of alkaloid extraction. Percolation generally yields higher extraction efficiency compared to maceration due to continuous solvent flow and more effective contact with the plant material. This method is suitable for scalability and large-scale extractions. It is suitable for the alkaloids such as quinine from *Cinchona* species and atropine from *Atropa belladonna*. The limitations are complex setup and requirement of more time, solvent, and labor compared to simpler methods.

Soxhlet extraction is an efficient technique used for the exhaustive extraction of alkaloids. It involves continuous refluxing of the solvent through the plant material [1]. The principle of Soxhlet extraction involves the repeated condensation and evaporation of the solvent, which allows for continuous extraction and concentration of alkaloids in the solvent. Soxhlet extraction provides high extraction efficiency and is capable of extracting a broad range of alkaloids with high yields. This method provides near-complete extraction of alkaloids and has automation potential for large-scale extractions. This method is effective for the extraction of codeine from *Papaver somniferum*, and nicotine from *Nicotiana tabacum*. The limitations are high solvent use, high energy requirement as heating is involved, and long extraction times.

SFE uses supercritical fluids, such as CO_2, to extract alkaloids. This method is known for its efficiency and selectivity [8]. In SFE, supercritical fluids exhibit properties of both gases and liquids, allowing them to diffuse through plant material and selectively dissolve alkaloids. The extraction process is carried out under high pressure and temperature conditions. This technique is highly effective for non-polar to moderately polar alkaloids, with the degree of extraction adjustable by varying the extraction conditions. This method allows targeted extraction of specific alkaloids and gives high efficiency with minimal solvent use. This method is also non-toxic and leaves no residues. This method is well suited for the extraction of caffeine from *Coffea arabica*, and capsaicin from *Capsicum annuum*. The major limitations are higher cost of equipment and complex operation. The process is carried under high pressure conditions and requires precise control of temperature and pressure. It also possesses limited polarity range and is less effective for very polar alkaloids without modifications.

MAE uses microwave radiation to rapidly heat solvents and plant material, improving the extraction efficiency of alkaloids [9]. Microwave energy heats the solvent and plant material quickly, enhancing the solubility and diffusion of alkaloids through dielectric heating. This method accelerates the extraction process compared to conventional techniques [9]. This method often achieves higher yields and reduced extraction times compared to traditional methods. The solvent consumption is also low and is energy efficient due to reduced extraction times. This method provides good extraction yields for the alkaloids such as quinine from *Cinchona* species, and reserpine from *Rauvolfia serpentina*. The demerits of this method are requirement of specialized microwave extraction equipment, degradation of sensitive alkaloids due to high temperatures, and limited applicability, as it is not suitable for all types of plant materials.

1.5 Quantification of Alkaloids

The major techniques used for the quantification of alkaloids are spectrophotometric, chromatographic, and titrimetric. UV-Vis spectrophotometry quantifies alkaloids based on their interaction with Bromocresol Green (BCG), which forms a yellow-colored complex with alkaloids [10]. The concentration of alkaloids is determined by measuring the absorbance of this complex at a specific wavelength using Beer's Law.

$$\text{Alkaloid} + \text{BCG} \rightarrow \text{Alkaloid–BCG Complex (Yellow)}$$

The chromatographic technique used for the quantification of alkaloids is high-performance liquid chromatography (HPLC). HPLC separates alkaloids based on their interactions with a stationary phase and a mobile phase. The alkaloids are detected and quantified based on the area under the chromatographic peaks. The quantification relies on comparing the peak areas of sample components with those of known standards [11]. In the iodometric back titration method, the alkaloids react with iodine to form a complex. The excess iodine that does not react with the alkaloids is then titrated with sodium thiosulfate. The amount of sodium thiosulfate used is proportional to the amount of excess iodine, which in turn helps in determining the amount of iodine that reacted with the alkaloids. This allows for the quantification of the alkaloids in the sample [11, 12]. The series of reactions that take place in iodometric back titration are as follows:

1. Reaction of alkaloids with iodine:

$$\text{Alkaloid} + I_2 \rightarrow \text{Alkaloid} - \text{Iodine Complex}$$

2. Titration of excess iodine:

$$I_2 + 2\ Na_2S_2O_3 \rightarrow 2\ I- \ + 2\ Na_2S_4O_6$$

2 Materials

2.1 Maceration

(a) **Apparatus**:
- Sample grinder.
- Maceration apparatus.
- Stirrer.

(b) **Reagents**:
- Solvents (e.g., ethanol, water).

2.2 Percolation

(a) **Apparatus**:
- Sample grinder.
- Percolator.

(b) **Reagents**:
- Solvents (e.g., ethanol, water).

2.3 Soxhlet Extraction

(a) **Apparatus**:
- Sample grinder.
- Soxhlet apparatus.
- Round bottom flask.
- Thimble (to contain plant material).
- Heater.

(b) **Reagents**:
- Organic solvents (e.g., Petroleum ether).

2.4 SFE

(a) **Apparatus**:
- Supercritical fluid extractor.

(b) **Reagents**:
- CO_2 or other supercritical fluid.

2.5 MAE

(a) **Apparatus**:
- Microwave extraction unit.

(b) **Reagents**:
- Ethanol/methanol/water, etc.

2.6 Quantification by UV-Vis Spectrophotometric Method

(a) **Apparatus**:
- UV-Vis Spectrophotometer.
- Analytical Balance.
- Magnetic Stirrer (optional).

(b) **Reagents**:
- Bromocresol Green (BCG) solution (0.1% w/v) (*see* **Note 1**).
- Standard alkaloid solutions (*see* **Note 2**).
- **Sample solution**: Alkaloid extract or plant material dissolved in a suitable solvent (e.g., methanol).

(c) **Others**:
- Volumetric flasks.
- Pipettes.
- Cuvettes.

2.7 Quantification by HPLC

(a) **Apparatus**:
- HPLC Chromatograph.
- Column: C18 reverse-phase column (e.g., 250 mm × 4.6 mm, 5 μm particle size).
- HPLC vials for sample injection.
- Detector: UV.
- Analytical balance.
- Centrifuge (if necessary for sample preparation).

(b) **Reagents**:
- Mobile phases
  - Water (HPLC Grade).
  - Organic solvent (e.g., methanol or acetonitrile, HPLC-grade).
- Standards: Alkaloid standards of known concentrations (*see* **Note 2**).
- Methanol or any other solvent for sample preparation.

2.8 Quantification by Iodometric Back Titration Analysis

(a) **Apparatus**:
- Analytical balance.
- Burette.
- Pipettes.
- Conical flasks.
- Beakers.
- Volumetric flasks.
- Stirring rod.
- Graduated cylinders.
- Funnel.
- pH meter.

(b) **Reagents**:
- Iodine solution (0.01 M) (*see* **Note 3**).
- Sodium thiosulphate solution (0.1 M) (*see* **Note 4**).
- Starch indicator solution (*see* **Note 5**).
- Standard alkaloid solution (Prepare a solution of known concentration, i.e., 0.1 mg/mL) (*see* **Note 6**).

3 Methods

3.1 Maceration

1. **Mixing**: Place the powdered plant material into a glass container. Add the chosen solvent to the container, ensuring the plant material is completely submerged.
2. **Stirring**: Stir the mixture occasionally to facilitate the dissolution of alkaloids. This can be done manually with a stirring rod or using a magnetic stirrer.
3. **Maceration time**: Allow the mixture to stand at room temperature (usually 20–25 °C) for an extended period, typically 24 to 72 h. The time can be adjusted based on the plant material and solvent used.
4. **Separation**: Separate the liquid extract from the plant residue by filtration.

3.2 Percolation

1. **Saturate plant material**: Initially, saturate the plant material with solvent to initiate the extraction process. This is done by pouring a sufficient amount of solvent over the plant material and allowing it to drain through.
2. **Start percolation**: After saturation, start the percolation process by allowing the solvent to flow through the plant material at a controlled rate. The flow rate should be slow enough to maximize extraction, typically a few drops per minute.
3. **Collect extract**: Collect the eluate (the liquid that has passed through the plant material) in a clean container. Continue the percolation until the solvent coming through is no longer extracting significant amounts of alkaloids.
4. **Repeat extraction**: Depending on the desired concentration, you may need to repeat the percolation process with fresh solvent to ensure complete extraction of alkaloids.

3.3 Soxhlet Extraction

1. **Assemble apparatus**: Set up the Soxhlet apparatus by connecting the Soxhlet extractor to a round-bottom flask with the solvent and a condenser.

2. **Prepare plant material**: Place the powdered plant material into a thimble (typically made of paper or metal) and insert it into the Soxhlet extractor.
3. **Add solvent**: Pour the selected solvent into the round-bottom flask. The volume should be sufficient to ensure that the solvent circulates through the plant material.
4. **Connect components**: Ensure that all components (Soxhlet extractor, round-bottom flask, and condenser) are securely connected to prevent any leakage or loss of solvent.
5. **Heating**: Heat the solvent in the round-bottom flask using a heating mantle or hot plate. The solvent should be heated to its boiling point, typically around 40–60 °C, depending on the solvent used.
6. **Refluxing**: Allow the solvent to evaporate and rise into the Soxhlet extractor, where it condenses and drips down through the plant material. This process is repeated continuously, ensuring thorough extraction.
7. **Extraction time**: Continue the extraction process for a period usually ranging from 4 to 24 h, depending on the plant material and the solvent used. The process duration can be adjusted to achieve the desired concentration of alkaloids.

3.4 Supercritical Fluid Extraction (SFE) [8]

1. **Assemble system**: Connect the extraction vessel to the high-pressure pump and the CO_2 reservoir. Ensure all connections are secure to prevent leaks.
2. **Load plant material**: Place the powdered plant material into the extraction vessel. Use appropriate loading techniques to avoid compaction.
3. **Set parameters**: Program the system with the desired temperature and pressure settings. Typical supercritical CO_2 extraction conditions are around 31–35 °C (supercritical temperature) and 70–250 bar (supercritical pressure).
4. **Pressurization**: Increase the pressure in the extraction vessel using the high-pressure pump until the CO_2 reaches its supercritical state.
5. **Heating**: Heat the CO_2 to the desired temperature to ensure it remains in a supercritical state. Typical temperatures range from 31 °C to 50 °C, depending on the alkaloids' solubility.
6. **Extraction**: Allow the supercritical CO_2 to flow through the plant material. The CO_2 will dissolve the alkaloids as it passes through the material. The process typically runs for 1–4 h, depending on the type of plant material and desired yield.

7. **Separation**: The CO_2-alkaloid mixture is then passed to a separator where the pressure is reduced, causing the alkaloids to precipitate out. The CO_2 is recycled back into the system.

3.5 Microwave-Assisted Extraction

1. **Assemble system**: Set up the MAE unit according to the manufacturer's instructions. Ensure that all components, including the microwave cavity, sample holder, and temperature control, are correctly assembled.
2. **Prepare plant material**: Place the powdered plant material into microwave-compatible extraction vessels. Use appropriate containers that can withstand microwave irradiation and pressure changes.
3. **Add solvent**: Add the chosen solvent to the plant material in the extraction vessel. Ensure that the solvent completely covers the plant material.
4. **Set parameters**: Program MAE unit with appropriate microwave power, temperature, and extraction time. Typical microwave power settings range from 300 to 1000 watts, and temperatures range from 50 °C to 120 °C.
5. **Extraction**: Start the extraction process. Microwaves will rapidly heat the solvent and plant material, causing the solvent to penetrate the plant material and dissolve the alkaloids. Extraction times typically range from 5 to 30 min, depending on the plant material and solvent used.
6. **Cooling**: After the extraction period, allow the system to cool down. Some MAE systems automatically cool down after the extraction process.

3.6 Quantification by UV-Vis Spectrophotometric Method

A. **Preparation of sample solution**:
   1. **Material**: 100 mg of dried, powdered plant material or alkaloid extract.
   2. **Preparation**:
      (i) Dissolve the sample in 10 mL of methanol or a suitable solvent.
      (ii) Filter through a 0.45 μm filter to remove any particulates.

B. **Reaction of alkaloids with BCG**:
   1. Pipette 1 mL of standard alkaloid solution or sample solution into a clean test tube.
   2. Add 1 mL of BCG solution to the test tube.
   3. Mix thoroughly using a magnetic stirrer or vortex mixer.
   4. Allow the reaction to proceed for 10 min at room temperature (approximately 25 °C).

C. **Measurement**:

1. **Equipment setup**:

    (i) Turn on the UV-Vis spectrophotometer and allow it to warm up as per the manufacturer's instructions.

    (ii) Set the wavelength to 425 nm, which is the λ max for the alkaloid-BCG complex.

2. **Blank preparation**:

    (i) Prepare a blank by mixing 1 mL of methanol with 1 mL of BCG solution.

    (ii) Zero the spectrophotometer using this blank.

3. **Sample measurement**:

    (i) Transfer the reaction mixture from the test tube to a cuvette.

    (ii) Measure the absorbance of the sample solution at 425 nm.

    (iii) Record the absorbance values for all standard solutions and the sample.

D. **Data analysis and calculations**:

1. **Calibration curve**:

    (i) Plot the absorbance values of the standard solutions against their concentrations to create a calibration curve.

2. **Quantification**:

    (i) Determine the concentration of alkaloids in the sample by comparing its absorbance with the calibration curve. Use Beer's law to calculate the concentration:

$$\text{Concentration (mg/mL)} = \text{Absorbance/Slope of Calibration Curve}$$

    (ii) For more accurate results, perform the analysis in triplicate and average the results.

E. **Observations**:

(i) **Peak absorbance**: Ensure the absorbance maximum is observed at 425 nm.

(ii) **Color intensity**: The yellow intensity should correlate with the concentration of alkaloids.

F. **Documentation**:

(i) Document all wavelengths, concentrations, and absorbance values.

(ii) Record any deviations from expected results or anomalies during the experiment.

3.7 Quantification by HPLC [11]

A. **Preparation of sample**:

1. **Material**: 100 mg of dried, powdered plant material or alkaloid extract.
2. **Preparation**:
   (i) Dissolve the sample in 10 mL of methanol or suitable solvent.
   (ii) Centrifuge if needed to remove particulates.
   (iii) Filter the solution through a 0.45 μm filter into HPLC vials.

B. **HPLC system setup**:

1. **Column**: Install an appropriate column, such as a C18 reverse-phase column (e.g., 250 mm × 4.6 mm, 5 μm particle size).
2. **Mobile phase**: Prepare the mobile phase (e.g., 80% water and 20% acetonitrile).
3. **Flow rate**: Set the flow rate to 1.0 mL/min.
4. **Temperature**: Maintain the column temperature at 25 °C.

C. **Injection and separation**:

1. **Injection volume**: Inject 20 μL of each standard and sample solution into the HPLC system.
2. **Elution conditions**: Run the samples under the following conditions:
   (i) Start with 80% water and 20% acetonitrile for 5 min, then increase the acetonitrile percentage linearly to 60% over 20 min.
   (ii) Set the UV detector to 254 nm (or a wavelength specific to the alkaloids being analyzed).

D. **Data collection and analysis**:

1. **Chromatograms**: Record chromatograms for each injection.
2. **Peak identification**: Identify peaks corresponding to the alkaloids based on retention times compared to standards.
3. **Quantification**: Measure the area under each peak using the HPLC software. Compare the peak areas of the samples to those of the standards to determine the concentration of alkaloids in the samples.

E. **Calculations**:

1. **Calibration curve**: Plot a calibration curve of peak area versus concentration for each standard.
2. **Concentration calculation**: Use the calibration curve equation (typically $y = mx + c$, where y is the peak area,

m is the slope, x is the concentration, and c is the intercept) to calculate the concentration of alkaloids in the sample:

$$\text{Concentration (mg/mL)} = \text{Peak Area} - c/m$$

3. **Sample concentration**: Convert the concentration to the final concentration in the extract or plant material.

F. **Observations**:

1. Record the retention times and peak areas of each alkaloid.
2. Ensure that all peaks are well-resolved and that there is no overlap between peaks.

G. **Documentation**:

1. Document the conditions used (flow rate, temperature, mobile phase composition) and ensure they are consistent across samples and standards.
2. Record any deviations or anomalies observed during the analysis.

3.8 Quantification by Iodometric Back Titration Analysis [11, 12]

A. **Reaction with iodine**:

1. Pipette 25 mL of the standard alkaloid solution into a conical flask.
2. Add 25 mL of 0.01 M iodine solution to the flask.
3. Mix thoroughly and allow the reaction to proceed for 15 min at room temperature (approximately 25 °C).
4. Ensure the solution turns a dark brown color, indicating the formation of the alkaloid-iodine complex.

B. **Back titration with sodium thiosulfate**:

1. After the reaction period, add a few drops of 1% starch solution to the reaction mixture. This will form a blue complex with iodine and help in detecting the endpoint.
2. Fill a burette with 0.1 M sodium thiosulfate solution.
3. Titrate the reaction mixture with sodium thiosulfate until the blue color disappears, indicating that all excess iodine has reacted.
4. Record the volume of sodium thiosulfate used.

C. **Calculations**:

1. **Determine the amount of iodine reacted**:

   (i) Calculate the amount of iodine that had not reacted based on the volume of sodium thiosulfate used:

$$\text{Amount of Iodine} = \text{Volume of } Na_2S_2O_3 \times \text{Molarity of } Na_2S_2O_3$$

D. **Quantification of alkaloids**:

1. Determine the amount of iodine that reacted with the alkaloids:

$$\text{Iodine reacted} = \text{Initial iodine added} - \text{iodine left (Titrated with } Na_2S_2O_3\text{)}$$

2. Calculate the concentration of alkaloids using the amount of iodine that reacted with the alkaloids:

$$\text{Concentration of alkaloids} = \frac{\text{Iodine reacted} \times \text{Volume of sample}}{\text{Weight of sample}}$$

E. **Observations**:

1. **Endpoint**: The endpoint of the titration is indicated by the disappearance of the blue color in the presence of starch.
2. **Consistency**: Perform the titrations in duplicate or triplicate to ensure accuracy and reliability of the results.

F. **Documentation**:

1. Record all measurements including the volumes of iodine, sodium thiosulfate, and the calculated alkaloid concentrations.
2. Document any deviations or issues encountered during the experiment.
3. Note any anomalies or deviations from expected results.

4 Notes

1. To prepare 0.1% BCG solution, accurately weigh 100 mg or 0.1 g BCG, dissolve it in small quantities of methanol or distilled water, transfer to a 100 mL volumetric flask and make a final volume of 100 mL with methanol or distilled water. Stir until fully dissolved, and filter if necessary.
2. Take alkaloid standards such as morphine and quinine. Morphine is used for opiate alkaloids, quinine for cinchona alkaloids, atropine or scopolamine for tropane alkaloids, and nicotine for nicotine estimation. Weigh 10 mg each standard using an analytical balance. Dissolve in small volume of methanol and make final volume 100 mL by adding more methanol. The concentration of alkaloid obtained will be 0.1 mg/mL stock solution. Prepare a series of standard solutions by diluting the stock solution (e.g., 0.01 mg/mL, 0.05 mg/mL, 0.1 mg/mL) in methanol.
3. To prepare iodine solution, dissolve 1.27 g of iodine crystals in 50 mL of 0.1 M potassium iodide solution. Dilute it to 250 mL with distilled water to prepare 0.01 M iodine solution.

4. To prepare sodium thiosulfate solution dissolve 24.8 g of sodium thiosulfate ($Na_2S_2O_3{\cdot}5H_2O$) in small quantities, if using distilled water, and make to 1 liter in a volumetric flask with the addition of more distilled water. Standardize the solution, if necessary, using a standard iodine solution.
5. To prepare the starch solution. Weigh 1 g of starch powder, add it slowly to about 100 mL boiling distilled water, let it dissolve completely, remove from heater, cool, and make a final volume of 100 mL with distilled water in a volumetric flask.
6. To prepare standard solution for iodometric test weigh 100 mg of alkaloid sample using an analytical balance, dissolve in 100 mL of distilled water to prepare 1 mg/mL stock solution. Further dilute to prepare working solutions (e.g., 0.1 mg/mL).
7. The procedures mentioned for quantification includes general steps. For the estimation of specific alkaloid, suitable modification in the procedure may be required depending on the sample and should be referred to the related literature for research purpose [13–22].

References

1. Wink M, Roberts MF (1998) Alkaloids: biochemistry, ecology, and medicinal applications. Plenum Press
2. Baxter H, Puri B, Harborne JB, Hall A et al (1998) In Phytochemical dictionary: a handbook of bioactive compounds from plants. CRC Press
3. Uchegbu NN, Fasuan TO, Oni K (2023) Biocompounds and amino acid in Vernonia amygdalina: impact of post-harvest treatments. J Food Meas Charact 17(1):385–407
4. Gurib-Fakim A (2006) Medicinal plants: traditions of yesterday and drugs of tomorrow. Mol Asp Med 27(1):1–93
5. Yubin JI, Miao Y, Bing W et al (2014) The extraction, separation and purification of alkaloids in the natural medicine. J Chem Pharma Res 6(1):338–345
6. Babiker F, Jamal P, Mirghani MES, at al (2017) Characterization, purification and identification of some Alkaloids in Datura stramonium. Int Food Res J 24.
7. Harborne JB (1998) Phytochemical methods: a guide to modern techniques of plant analysis. Chapman and Hall
8. Rios JL, Recio MC (2005) Medicinal plants and their alkaloids: a review. J Ethnopharmacol 100(1–2):1–12
9. Chemat F, Abert-Vian M, Fabiano-Tixier AS (2012) Green extraction of natural products: concept and principles. In: Green extraction processes for biobased products and biofuels. Springer, pp 1–31
10. Fazel S, Hamidreza M, Rouhollah G et al (2010) Spectrophotometric determination of total alkaloids in some Iranian medicinal plants. J App Hortic 12(1):69–70
11. Skoog DA, Holler FJ, Crouch SR (2017) Principles of instrumental analysis. Cengage Learning
12. Asogwa FC, Ugodi GW, Ebugosi RS, Achara IN, Atim HA (2023) UV-spectroscopic and iodometric back titration analysis of total caffeine content of energy drinks available in markets in cross river state, South-South, Nigeria. Glob J Pure Appl Sci *29*(1):1–10
13. Kamal J (2011) Quantification of alkaloids, phenols and flavonoids in sunflower (Helianthus annuus L.). African Journal of Biotechnology 10(16):3149–3151
14. Bournine L, Bensalem S, Wauters JN, Iguer-Ouada M, Maiza-Benabdesselam F, Bedjou F, Frédérich M (2013) Identification and quantification of the main active anticancer alkaloids from the root of Glaucium flavum. International journal of molecular sciences 14(12): 23533–23544

15. Wang T, Frandsen HL, Christiansson NR, Rosendal SE, Pedersen M, Smedsgaard J (2019) Pyrrolizidine alkaloids in honey: Quantification with and without standards. Food Control 98:227–237
16. Lu HY, Wang SS, Cai R, Meng Y, Xie X, Zhao WJ (2012) Rapid discrimination and quantification of alkaloids in Corydalis Tuber by near-infrared spectroscopy. Journal of pharmaceutical and biomedical analysis 59:44–49
17. Suchomelová J, Bochořáková H, Paulová H, Musil P, Táborská E (2007) HPLC quantification of seven quaternary benzo [c] phenanthridine alkaloids in six species of the family Papaveraceae. Journal of Pharmaceutical and Biomedical Analysis 44(1): 283–287
18. Ding B, Zhou T, Fan G, Hong Z, Wu Y (2007) Qualitative and quantitative determination of ten alkaloids in traditional Chinese medicine Corydalis yanhusuo WT Wang by LC–MS/MS and LC–DAD. Journal of pharmaceutical and biomedical analysis 45(2):219–226
19. Tipke I, Bücker L, Middelstaedt J, Winterhalter P, Lubienski M, Beuerle T (2019) HILIC HPLC-ESI-MS/MS identification and quantification of the alkaloids from the genus Equisetum. Phytochemical analysis 30(6):669–678
20. Sharma A, Kamble SH, León F, Chear NJY, King TI, Berthold EC, Avery BA (2019) Simultaneous quantification of ten key Kratom alkaloids in Mitragyna speciosa leaf extracts and commercial products by ultra-performance liquid chromatography– tandem mass spectrometry. Drug testing and analysis 11(8): 1162–1171
21. Liu HL, Luo R, Chen XQ, Ba YY, Zheng L, Guo WW, Wu X (2015) Identification and simultaneous quantification of five alkaloids in Piper longum L. by HPLC–ESI-MSn and UFLC–ESI-MS/MS and their application to Piper nigrum L. Food chemistry 177:191–196
22. Tabasum S, Khare S, Jain K (2016) Spectrophotometric quantification of total phenolic, flavonoid, and alkaloid contents of Abrus precatorius L. seeds. Asian J Pharm Clin Res 9(2) 371–374

Chapter 6

Extraction and Quantification of Dietary Fibers and Health-Promoting Oligosaccharides

Aditi Sachdeva, Abhay Tiwari, Khushi Keshari, Garima Chouhan, and Praveen Yadav

Abstract

Dietary fiber (DF) constitutes a significant portion of our regular diet and plays a crucial role in maintaining gastrointestinal health in the human body. It also offers many health benefits and reduces the risk of chronic diseases such as diabetes, heart disease, and certain types of cancer. This chapter provides an overview of how these fibers can be extracted from various food sources. They are mainly found in plants and involve distinct extraction and quantification methods. Extraction methods may include mechanical, chemical, enzymatic, and synergistic methods. These methods can be paired with quantification methods such as UV-VIS-spectrophotometer, Fourier-transform infrared spectroscopy (FTIR), X-ray diffraction (XRD), High-performance liquid chromatography (HPLC), Differential scanning calorimetry (DSC), Thermal gravimetric analysis (TGA), and Nuclear Magnetic Resonance (NMR). The extraction and quantification of dietary fibers are important steps in their isolation in purified forms. These purified fibers can be used to design novel health and functional foods.

Key words Dietary fibers, Extraction, Quantification, Isolation

1 Introduction

Dietary Fibers (DF), one of the most important macronutrients, are complex polysaccharides that show resistance to hydrolysis by the human body's digestive enzymes and help modify structural properties. They also comprise non-carbohydrate components. These fibers are mostly found in the edible plant's cell walls. DFs are categorized into soluble and insoluble types based on solubility and linear and nonlinear forms based on structure. Soluble fibers are primarily composed of pectin, gums, and mucilage in contrast with insoluble fibers constituted by cellulose, lignin, and hemicellulose [1]. Insoluble fibers are indigestible by the small intestine,

Aditi Sachdeva and Abhay Tiwari contributed equally to this work.

Ashwani Kumar, Vidisha Tomer and Piyoosh Kumar Babele (eds.), *Phytochemicals: Extraction, Quantification, Isolation, and Characterization*, Methods and Protocols in Food Science, https://doi.org/10.1007/978-1-0716-4840-7_6,

adding to the increase in stool bulk and preventing conditions like constipation and hernia. Conversely, the consumption of soluble fibers aids in lowering insulin, blood sugar, and cholesterol levels and reducing colon cancer risk. A balanced diet rich in dietary fibers offers various health benefits, enhances gut microbiota or microflora that facilitates efficient digestion, aids in weight management by regulating satiety—a prolonged feeling of fullness, reduces insulin and blood sugar levels, thereby enhancing metabolism, promotes bowel regularity to decrease the likelihood of diarrhea, and shows anti-inflammatory properties [2].

Additionally, various medicinal qualities are inherent in DF-containing plant sources. Common natural sources of dietary fiber include cereals, legumes, fruits, vegetables, and nuts. Inulin, pectin, guar gum, and carboxymethylcellulose are some examples of soluble fibers. On the other hand, apple skins, asparagus stalks, and broccoli stems are some rich sources of insoluble fibers that can be consumed to bulk up the fecal matter.

The extraction of these complex DF polymers requires a strenuous procedure to receive the extracted yield with efficiency. The extraction of dietary fibers can be done through distinct methods involving thermal, enzymatic, and chemical processes. Thermal treatments for the extraction of DF comprise hot water, microwave heating, and subcritical water methods. Similarly, enzymatic extraction involves commercial and non-commercial brand enzymes, while chemical extraction uses alkaline, acid, oxidant/bleach, solvent, or chelating agents for the treatment [3]. An illustration of the methods in detail is depicted below in Fig. 1 and also in Tables 1

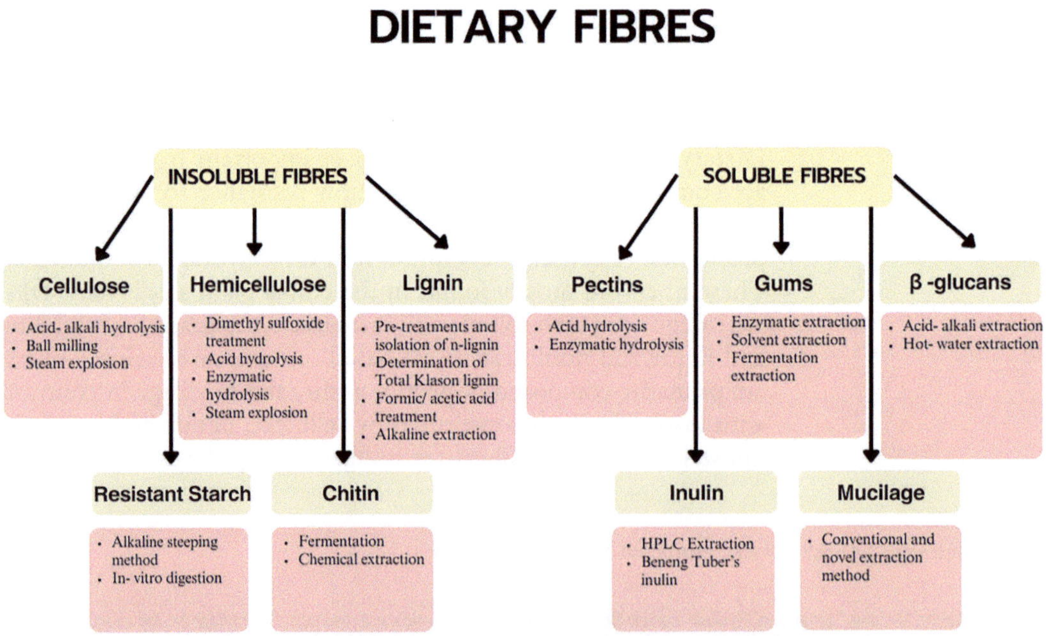

Fig. 1 Different methods for the extraction of soluble and insoluble dietary fibers

Table 1
Extraction methods of insoluble fibers

Insoluble fibers	Method of extraction	References
Cellulose	1. Acid-alkali hydrolysis 2. Ball milling 3. Steam explosion	[5] [6] [9]
Hemicellulose	4. Dimethyl sulfoxide treatment 5. Acid hydrolysis 6. Steam explosion	[10] [10] [13]
Lignin	7. Pretreatment and isolation of n-lignin 8. Determination of total klason lignin 9. Formic acid/acetic acid treatment 10. Alkaline extraction	[14] [15] [16] [17]
Chitin	11. Fermentation-based extraction 12. Chemical extraction	[18] [18]
Resistant starch	13. Alkaline steeping extraction 14. In vitro digestion	[20] [21]

Table 2
Extraction of soluble fibers

Soluble fibers	Method of extraction	References
Pectins	1. Acid hydrolysis 2. Enzymatic extraction	[22] [23, 24]
Gums	3. Enzymatic extraction 4. Solvent extraction 5. Fermentation extraction	[25] [26–28] [29]
Inulin	6. NaOCl-based extraction 7. Ethanol-based extraction	[30] [31, 32]
β-glucans	8. Acid-alkali extraction 9. Hot-water extraction	[33] [34]
Mucilage	10. Conventional extraction	[35, 36]

and 2. The extraction methods may vary according to the types of fibers present in a food product and highly depend on its constitution, viz., presence of oligosaccharides, their complexity, nature, and degree of polymerization [4].

2 Materials

Prepare and dilute all the solutions in distilled or deionized water only. Preparation and storage of solutions are to be done at room temperature (unless indicated otherwise).

2.1 Insoluble Dietary Fibers

2.1.1 Cellulose

Acid-Alkali Hydrolysis

1. Sodium hydroxide (NaOH) (5% w/v)/g.
2. Hydrogen peroxide (H_2O_2) (16% *v*/*v*).
3. Sulfuric acid (H_2SO_4) (60% v/v)/g.
4. Ultrasonicator (VCX 750).
5. Dialysis membrane.

Ball Milling

1. Planetary micro mill.
2. Cellulosic material.
3. 1410 μm mesh.
4. NaOH (5%).
5. Drying oven.
6. FTIR.
7. XRD.

Steam Explosion

1. High-temperature/high-pressure reactor (HPHT).
2. Supercritical carbon dioxide ($ScCO_2$) extraction unit.
3. H_2O_2 (30%).
4. Ethanol.
5. Filtration system.
6. FTIR.

2.1.2 Hemicellulose

Dimethyl Sulfoxide Treatment

1. Dioxane (containing 0.05 mol HCl).
2. Formic acid.
3. 90% aqueous solution.
4. Acetic acid.
5. Ethanol.
6. Dimethyl sulfoxide.
7. Potassium hydroxide (KOH) 8%.
8. Neutral dioxane, 80%.

Acid Hydrolysis

1. Sulfuric Acid (H_2SO_4).
2. Distilled water.

Steam Explosion

1. Stainless steel digester.
2. Blowing tank.
3. Electro-valves for steam admission.
4. Ball valve.
5. Sealed polyethylene bags.

2.1.3 Lignin

Pre-Treatments and Isolation of n-Lignin

1. Wiley mill.
2. 7.5 wt% aqueous ammonia.
3. H_2SO_4: 6 M solution.
4. Aqueous ammonia cooking liquor, pH 8.5.
5. Drying Oven.
6. UV-Vis spectrophotometer, FTIR spectrometer, and NMR spectrometer.

Determination of Total Klason Lignin

1. Weighted thimble tube.
2. Ethanol-benzene solvent.
3. Desiccator.
4. UV spectrometer.
5. H_2SO_4: 73% concentration.

Formic Acid/Acetic Acid Treatment

1. 85% organic acid (Formic acid/acetic acid = 70:30 v/v).
2. Buchner funnel.
3. 80% formic acid.

Alkaline Extraction

1. Sodium hydroxide (NaOH): concentration of 2% (w/v).
2. Ethanol.
3. Hydrochloric acid (HCl): 0.1 M solution.
4. Filtration setup.
5. Drying Oven.

2.1.4 Chitin

Fermentation-Based Extraction

1. *Lactobacillus acidophilus* (10^8–10^9 CFU/mL).
2. pH meter.
3. Di-nitro-salicylic acid.
4. Biofilm solutions.
5. Shaking incubator.
6. *Exiguobacterium profundum.*
7. FTIR/SEM/XRD.

Chemical Extraction

1. Hydrochloric acid (HCl), 1.5 mol/L: 1:10 (w/v).
2. Reflux device.
3. NaOH (10%).

2.1.5 Resistant Starch

Alkaline Steeping Method

1. Pancreatin.
2. Sodium acetate buffer: 0.1 M, pH 5.2.
3. Amyloglucosidase.
4. Sodium bisulfite (ACS reagent).
5. Absolute methanol.
6. Shaking water bath.
7. Absolute ethanol.
8. Mesh sieves: 80.

In Vitro Digestion Method

1. Sodium maleate buffer (pH 6.0).
2. Pancreatic alpha-amylase and amyloglucosidase.
3. Water Bath.
4. Aqueous ethanol.
5. Centrifuge.
6. KOH (2 M).
7. GOPOD (glucose oxidase peroxidase reagent).
8. Sodium acetate buffer.

2.2 Soluble Dietary Fibers

2.2.1 Pectin

Acid Hydrolysis

1. Electromagnetic induction (EMI Heater).
2. Water Bath.
3. Cheesecloth.

Enzymatic Extraction

1. Freeze-drier.
2. Sodium citrate buffer: pH 5.2.
3. Cellulase.
4. α-amylase.
5. Pronase.

2.2.2 Gums

Enzymatic Extraction

1. Thermostable α-amylase (termamyl).
2. Acetic acid: 2 M solution.
3. Water bath.
4. Anhydrous ethanol.
5. Ethanol: 96% concentration.
6. Vacuum oven.

Solvent Extraction

1. Alkaline solution.
2. Ethanol or acetone.

Fermentation Extraction

1. Fermentation broth.
2. Non-polar solvents.

2.2.3 Inulin

HPLC Extraction

1. 4% Sodium Hypochlorite (NaClO).
2. Sucrose.
3. Agar.
4. Hot Air Oven.
5. Freeze Dryer.
6. Water Bath.
7. Rotary Evaporator.
8. Vacuum Filter.
9. HPLC.

Ethanol-Based Extraction

1. Water bath.
2. Centrifuge.
3. Filter cloth.
4. Ethanol.
5. FTIR/XRD.

2.2.4 Beta-Glucans

Acid-Alkali Extraction

1. NaOH: 1 M solution.
2. HCl or Acetic Acid: 1 M solution.
3. Hot air oven.
4. Sodium hypochlorite (NaClO) (2%).

Hot-Water Extraction

1. Ethanol: 80% concentration.
2. Citric acid.
3. Digital pH meter.
4. Sodium bicarbonate ($NaHCO_3$).
5. Hot air dryer.

2.2.5 Mucilage

Conventional Extraction

1. Alcohol.
2. Extraction enzymes.

2.3 Quantification of Dietary Fiber

1. Scanning electron microscope (SEM).
2. Fourier transform infrared (FTIR) spectroscopy.
3. X-ray diffraction.

4. Dynamic light scattering (DLS) equipment.
5. Nuclear magnetic resonance.

3 Methods

Methods of extraction for dietary fibers involve distinct methods listed in Tables 1 and 2.

3.1 Insoluble Dietary Fibers

3.1.1 Cellulose

Acid-Alkali Hydrolysis

1. Immerse the fiber-containing product in deionized water at 80 °C for 1 h and place it in the oven for drying at 100 °C for 3 h.
2. Treat the sample with 20 mL of sodium hydroxide (5% w/v)/g biomass at 90 °C for 1 h and stir it continuously (*see* **Note 1**). Wash the obtained sample with deionized water and dry them again in the oven at 60 °C unless the mass becomes constant.
3. Treat the sample with 40 mL hydrogen peroxide solution (16% v/v) and sodium hydroxide (5% w/v)/g biomass at 55 °C for 90 min while stirring continuously (*see* **Note 1**). Rinse it with deionized water and dry the obtained sample in an oven at 60 °C until it shows constant mass.
4. Now, pour 20 mL of sulfuric acid (60% v/v)/g biomass and rest it for 45 min at room temperature while stirring. Halt the reaction by immersing the samples in an ice bath and adding 200 mL of cold deionized water (*see* **Note 1**). Rinse with deionized water through a centrifuge for 15 min at 9000 rpm [5].
5. Ultrasonicate the samples after neutralizing their pH with 40% amplitude and 1 s pulse. Dry the remaining content using a hot air oven at 60 °C for 5–10 min (*see* **Note 2**).

Ball Milling

1. Stir the sample in 1.5 L hot water at 80 °C for 3 h; this removes all the dirt particles and other soluble compounds from the sample. Dry the sample at 100 °C for 4 h. Mill the samples in a grinding machine and sieve through 1410 μm mesh. Remove all the soluble and non-polar compounds through Soxhlet extraction [7].
2. Isolate the cellulose using an NaOH (5%) solution at 80 °C for 5 h. Bleach the obtained sample and neutralize its pH before drying (*see* **Note 3**). Carry simultaneous runs in FTIR and XRD to obtain the results [6].

Steam Explosion

1. Prepare the sample to obtain a water-to-raw material ratio of 10:1.
2. Set the reactor at 200 °C with no pressure release for 30 min to 4 h. After this treatment, depressurize the reactor, followed by overnight cooling. Keep the pressure at 0.69–4.83 MPa.

3. Apply $ScCO_2$ as it impacts structural and physicochemical properties of cellulose. Filter and wash the solution several times to obtain the solid residue and supernatant separately.
4. Make the remaining solids undergo oxidation with the assistance of 60 mL hydrogen peroxide. Purify the fibers, followed by washing the mixture with distilled water and diluted ethanol (1:1).
5. Bleach the mixture and agitate continuously for an hour at 70 ° C.
6. FTIR or XRD can be used to detect and quantify cellulosic fibers (*see* **Note 4**) [8, 9].

3.1.2 Hemicellulose

Dimethyl Sulfoxide Treatment

1. This method extracts hemicellulose without pretreatment (*see* **Note 5**).
2. Take dimethyl sulfoxide, aqueous solutions of 90% neutral dioxane, 80% dioxane, and 8% KOH for extraction from the required sample [10].

Acid Hydrolysis

1. Weigh the sample and indirectly heat it with the help of a laboratory reactor with saturated steam to 140, 150, and 160 °C for 10 and 20 min (*see* **Note 6**).
2. The concentrations of sulfuric acid should be 70 mg_{acid}/g_{dm} and 100 mg_{acid}/g_{dm}, and the solid-liquid ratio must be taken as 1:10.
3. Cool the reactor after this process and choose the best condition for hydrolysis from (140 °C for 10 min with 100 mg_{acid}/g_{dm}, 140 °C for 20 min with 100 mg_{acid}/g_{dm} and 150 °C for 20 min with 70 mg_{acid}/g_{dm}) [11, 12].

Steam Explosion

1. Carry out steam explosion in a stainless-steel digester and set its temperature at 190 °C and pressure at 16 kg/cm^2. Treat the sample with steam for 5–10 min and discharge pressure 6 kg/cm^2. Carry out multiple cycles, repeating the same steps.
2. Put the sample in the blowing tank to calculate the severity factor (*see* **Note 7**). After the readings, remove the sample, wash it with cold water, air-dry it, and note the weight.
3. Measure the moisture loss to determine the extraction process's total yield [13].

3.1.3 Lignin

Pre-Treatments and Isolation of n-Lignin

1. Grind the sample in a Wiley mill to a size that can easily pass through a 1 mm screen. Apply pretreatment to the sample hydrothermally at 140 °C for 5 h.
2. Collect the solid portion after the treatment and wash it with water. Dry this at 40 °C and measure the weight after moisture loss.

3. Carry out the aqueous ammonia treatment with a small portion of the sample and keep the parameters likewise (Temperature—100, 130, 160 °C and ammonia %—0, 4, 12, 20%) [14].
4. Use the pressing method to extract the treated solid fraction with aqueous ammonia at a liquid-to-solid ratio of 7. Heat the reactor at 140 °C for 3 h and filter the liquid part from the solid using hermetic Nutsche equipment.
5. Evaporate the residual ammonia from the solid residue at ambient pressure by acidification at pH 5 and trap it in 6 M sulfuric acid. You may observe the lignin as a precipitate after centrifugation and freeze-drying (*see* **Note 8**) [14].

Determination of Total Klason Lignin

1. Heat the sample at 47 °C for 2 h for hydrolyzing and solubilizing the carbohydrate.
2. Transfer the sample to a weighted thimble tube and extract with the help of ethanol-benzene solvent. Continue this process for 5 h, and then dry the residue in an oven at about 100 °C. Cool the residue at room temperature using a desiccator.
3. Use the extractive-free biomass to treat with 72% H_2SO_4 for 2 h and constantly stir at room temperature. Dilute the solution to make the concentration 3% and continue reflux for 4 h at 80 °C.
4. The difference in the weight of the sample before and after oven drying gives the insoluble lignin content. For soluble lignin, use a UV spectrophotometer (*see* **Note 9**) to determine the absorbance and soluble lignin [15].

Formic Acid/Acetic Acid Treatment

1. Extract the pulp by crushing/cutting the sample. Add organic acid to the pulp at a ratio of 1:8 (fiber to liquor) and heat it till the boiling starts for almost 2 h. Use a hot plate or heating mantle.
2. Cool the mixture at room temperature and filter the fibers with a Buchner funnel. Wash the obtained fibers with formic acid and hot distilled water afterward to neutralize the pH.
3. Filter and bleach the solution. Afterward, leave the solution undisturbed to isolate lignin [16].
4. FTIR, TGA (thermogravimetric analysis), and DSC (Differential Scanning Calorimetry) can be used for the characterization of the extracted lignin sample (*see* **Note 10**).

3.1.4 Chitin

Fermentation-Based Extraction

1. Wash the samples thoroughly with distilled water and dry them before further treatment. Grind the dried sample into 2–3 mm-sized pieces. Isolate the sample [17] and incubate for 3 days. Identify the colonies.

2. Store the isolated bacterial strains in a 25% glycerol preservation tube at −80 °C. Two main colonies are observed: *E. profundum* and *L. acidophilus.* The medium for fermentation as described by Xie et al. [18].
3. In a shaking incubator, set the rotations at 140 rpm at room temperature for 120 h alongside the collected supernatant. Determine the pH and titratable acidity of the supernatant.
4. Determine the glucose consumption using the DNS method (*see* **Note 11**). By the plate counting method, the biofilms are diluted serially prior to plating to obtain the plates, such as if they contain at least one plate with 20–200 CFUs.
5. Collect the fermented supernatant after filtration and wash it with deionized water. Dry the sample at 60 °C for 48 h.
6. Use either SEM, FTIR, or XRD for quantification of chitin [18].

Chemical Extraction

1. Treat the sample with prepared HCl for 6 h at 25 °C; this removes all the minerals. Wash with pure water after treating with HCl until the solution gets neutralized.
2. Remove the excess water from the solution and transfer it to a reflux device. Add 20 mL of freshly prepared NaOH at 90 °C for almost 3 h, and wash it again till it is neutralized. Dry this for 48 h at 60 °C to obtain the results [18].

3.1.5 Resistant Starch

Alkaline Steeping Method

1. Hydrolyze the digestible starch whilst subjecting the sample to pancreatin and amyloglucosidase.
2. Add about 4 g sample to the enzyme solution made with 3.2 g of pancreatin, 320 mL of 0.1 M sodium acetate buffer (*see* **Note 12**), and 3.2 mL of amyloglucosidase. Vortex and incubate the obtained solution at 37 °C for 16 h in a shaking water bath.
3. Add 160 mL absolute ethanol and vortex again to deactivate the enzymes. A resistant starch pellet can be observed after centrifuging at 3500 rpm for 10 min.
4. Wash the solution with ethanol twice and let it dry at 40 °C followed by screening through 80 mesh sieves. This step removes the digestible starch from the pellet. Store at 4 °C and verify the purity by AACC method [19, 20].

In Vitro Digestion Method

1. Mix the sample with sodium maleate buffer (pH 6.0) (*see* **Note 13**) with the suitable enzymes.
2. Incubate the samples in the water bath for 16 h at 37 °C. Add ethanol and centrifuge the remaining mixture at 3000 rpm for 10 min.

3. Rinse 1–2 times with aqueous ethanol. Separate the supernatant and the residues by stirring continuously.
4. Dissolve the resistant starch pellet in 2 M KOH in the ice water bath.
5. Neutralize the solution using sodium acetate buffer along with resistant starch hydrolyzation.
6. Centrifuge the sample and mix it with GOPOD reagent, followed by incubation at 50 °C for 20 min.
7. Note the absorbance of each solution at 510 nm with the help of a Spectrophotometer [21].

3.2 Soluble Dietary Fibers

3.2.1 Pectin

Acid Hydrolysis

1. For pretreatment, wash the sample with water and blanch it at 90 °C for 10 min using a ratio of solid to liquid as 1:25 (w/v). Filter the mixture with the help of a multilayered cheesecloth and dry the residues using an air convection oven at around 50 °C.
2. Treat these dry residues with 85% ethanol at 70 °C in a water bath for 20 min, making the solid-to-liquid ratio 1:5 (w/v). Again, filter the mixture and dry the residues for 16 h to obtain acid-insoluble solids (AIS) (*see* **Note 14**).
3. Take 10 g of AIS and add distilled water in a ratio of 1:50 (w/v). Adjust its pH to 1.2 by adding a suitable amount of 1 M H_2SO_4 and provide heat at 80 °C for 90 min to get the extract by constant stirring.
4. Similarly, repeat the process for EMI heating. Set the extraction time and suitable power level (4, 5, 6, 7, 8 and 9) at the apparatus necessary to reach 80 °C to extract pectin. For correlation between time and power levels, refer to Zouambia et al. (2019) [22].

Enzymatic Extraction

1. Make powder of the sample by grinding and freeze-drying.
2. Suspend the prepared sample in 1.5 L sodium citrate buffer, 600 mg cellulase, 500 mg α-amylase and stir the mixture continuously for 20 h at 30 °C. Centrifuge for 30 min.
3. Collect the residue and supernatant separately. Resuspend the pellet remains in 100 mL distilled water and centrifuge again for 15–20 min. Combine the supernatants and treat them with absolute ethanol.
4. Collect the precipitate, solubilize it in distilled water, freeze-dry, and grind. This makes A fraction.
5. For fraction B, pour 0.5 g/100 mL solution of fraction A and mix with the pronase for 1 h at 50 °C for reduction of protein content present in the sample. Heat the solution vigorously at 85 °C for 30 min, cool it down at room temperature and centrifuge for another 20 min.

6. Strain the supernatant and treat it with absolute ethanol. Collect the resulting precipitate, dissolve it in distilled water, freeze-dry it, and grind it to obtain fraction B [23, 24].

3.2.2 Gums

Enzymatic Extraction

1. Defat the sample (*see* **Note 15**) and transfer it into a falcon tube; add 30 mL of water at 9.5 pH.
2. Mix the samples for 1 h using a rotator at 70 rpm. Following this, centrifuge at 8000× *g* for 10 min and collect the supernatant. Add thermostable α-amylase and incubate with enzymes at 80 °C.
3. Cool down the sample, then add acetic acid until the protein solubility is reduced. Place it in a boiling water bath and centrifuge afterward.
4. Mix the supernatant with anhydrous ethanol in a ratio of 1:3 (v/v) to obtain the precipitate. Filter the solution to collect residue after centrifugation and washing.
5. Clean gum can be observed after washing with ethanol and drying using a vacuum oven at 40 °C [25].

Solvent Extraction

1. Agitate the sample. Extract the sample with the help of a solvent [26].
2. Separate the solvent mixture using distillation or evaporation and observe the precipitate as concentrated gum extract. Filter and purify to further refine [27, 28].

Fermentation Extraction

Culture the desired microorganism in a nutrient-rich medium and incubate it to produce the gums. Process the fermentation broth to get rid of any impurities and microbial cells. Filter the precipitate using non-polar solvents (*see* **Note 16**) [29].

3.2.3 Inulin

NaOCl-Based Extraction

1. Prepare the sample by cutting the sample up to 1 cm in length and treating it with NaOCl. Rinse the sample with sterile water. Inoculate the sample on culture media with pH 5.7, which should be supplemented by 30 g/L of sucrose and 5 g/L of agar. Maintain the conditions at around 25 °C for 16 h under suitable light. Remove the moisture from the sample using a dry air oven.
2. Carry out extraction by making a decoction in hot water (80 ° C) using a water bath. Stir constantly. Keep the water-to-sample ratio at 5:1.
3. Use vacuum filtration to filter out the sample and obtain the concentrated sample in the rotary evaporator at 45 °C. Continue this process until the volume reaches 30 mL.
4. Lyophilize the solution using a freeze dryer and store at – 18 °C [30].

5. Detection and quantification of inulin can be done by HPLC (High-Performance Liquid Chromatography) (*see* **Note 17**).

Ethanol-Based Extraction

1. Clean the sample, peel it, and cut it into fine slices. Blend the sample with water in the ratio 1:2.
2. Heat at 80 °C for 30 min in the water bath and then cool down for a while. Filter the solution with a cloth and collect the filtrate.
3. Make different concentrations by adding different amounts of ethanol by volume (30%, 40%, 50% and 96%). Store these solutions in a freezer to initiate the precipitation of inulin.
4. Centrifuge the precipitate after 18 h at 1500 rpm for 10–15 min. Then add the precipitate to water in a ratio of 1: 2, followed by heating for a few minutes so that the solution dries and inulin is obtained.
5. FTIR and XRD can be used to quantify inulin [31].

3.2.4 ß-Glucans

Acid-Alkali Extraction

1. Mix the sample with fivefold diluted NaOH and incubate at 80 °C in a stirrer for 2 h. Collect the pellets by centrifuging them for 25 min at 4 °C and suspend them in distilled water.
2. Dissolve the pellets in a fivefold diluted HCl solution or acetic acid and incubate at 80 °C in a stirrer for 2 h. Repeat the centrifugation process to obtain the pellets.
3. Wash the pellets 3–4 times with distilled water and dry them in the oven at 60 °C. Alternate methods for extraction of ß-glucans have been studied by Noppawat Pengkumsri and colleagues [32].
4. Use FTIR to detect the presence of glucans.

Hot-Water Extraction

1. Reduce the particle size of the sample to less than or equal to 1 mm by grinding.
2. Mix the prepared sample with ethanol in 1:6 and heat at 80 °C for 2 h. Blend this mixture with water in a ratio 1 to 10 at a reduced temperature of 55 °C while stirring it continuously for 2 h.
3. Centrifuge the mixture for 20 min at a speed of 4025× *g* at 40 °C. Adjust the pH to 8.5 with the help of $NaHCO_3$ and stir continuously for 30 min at 55 °C. Centrifuge again following the same parameters.
4. Add citric acid to the supernatant and adjust the pH to 4. Centrifuge the mixture now at a speed of 21,913× *g* for 25 min at 4 °C.
5. Collect the ß-glucans pellets with the help of precipitation, dry them for 16 h at 43 °C.
6. SEM, FTIR, or DLS (Dynamic Light Scattering) can be used to detect the presence of ß-glucan [33].

3.2.5 Mucilage

Conventional Extraction

1. Prepare the sample by de-hulling, crushing (if needed) and homogenize properly.
2. Extract mucilage via hot or cold distilled water with continuous stirring. Filter the solution and add alcohol to the filtrate for the precipitation of mucilage.
3. Depending on the nature of the sample, use water or organic solvent such as methanol for the extraction [34, 35].
4. Select suitable enzymes for solvent-based extraction among protease, α-amylase, mannanase, galactanase, xylanase, arabinase, cellulase, pectinase, hemicellulase, and β-glucanase [32]. Novel methods are also similar to conventional methods; however, they involve some analytical techniques and instruments [36].

3.3 Quantification of Fibers

3.3.1 Insoluble Dietary fibers

Quantifying insoluble dietary fibers such as cellulose, hemicellulose, lignin, chitin, and resistant starch is primarily based on Scanning Electron Microscopy (SEM), FTIR, and X-ray Diffraction (XRD). The cellulose can be obtained via steam explosion and analyzed by SEM at 3 kv and 300× magnification [6, 9]. The extracted lignin using acetic acid treatment can be characterized by FTIR (8400s, 500–4000 cm^{-1}, 50 scans) and thermal properties are evaluated by DSC (Q2000, 5 °C/min from 30 to 400 °C under nitrogen) and TGA (Q500, 30–800 °C at 10 °C/min) [16]. Chitin derived via fermentation can be examined through SEM, FTIR (scanning 4000–400 cm^{-1} at 4 cm^{-1} resolution), and XRD (Cu Kα1 radiation at 40 kV and 40 mA). The degree of deacetylation for chitin can be determined using FTIR absorbance ratios between 1655 and 3450 cm^{-1} [18].

3.3.2 Soluble Dietary Fibers

The quantification of soluble dietary fibers, including pectins, gums, inulin, β-glucans, and mucilage, can be primarily achieved through FTIR and additional techniques like Dynamic Light Scattering (DLS) for β-glucans. Analyze the extracted pectins through acid hydrolysis using a spectrophotometer with KBr disks, scanning from 4000 to 400 cm^{-1} at a 4 cm^{-1} resolution over 128 scans [22]. Perform the enzymatic-assisted ultrasound extraction of pectin using an Elmasonic S120H at 37 kHz at either 20 °C or 55 °C for 15 min, followed by FTIR and high-resolution ^{1}H and ^{13}C NMR spectroscopy using a spectrometer operating at 25 °C [23, 24]. Inulin extracted via ethanol-based methods can be characterized using FTIR over 4000–400 cm^{-1} and further quantify by X-ray diffraction (XRD), scanning from 10° to 80° 2θ at 0.04° steps [31]. Mucilage samples can be analyzed through FTIR using KBr disks [36]. β-Glucans can be characterized using SEM for morphology, FTIR for functional group determination, and DLS for measuring particle size distribution and zeta potential.

4 Notes

1. Dissolve 5 g of NaOH pellets for every 100 mL of distilled water to make 5% NaOH solution. Sodium hydroxide is used particularly for its nature to remove non-cellulosic components, disrupt hydrogen bonds, swell cellulose fibers, degrade impurities, and form reactive intermediates. Similarly, the use of acid is done to hydrolyze the hemicellulose and disrupt the cell wall structure to improve the extraction of cellulose. Hydrogen peroxide is used as a bleaching agent in this process to enhance the final properties of cellulose.
2. Ultrasound is used here for the detection of the amount of cellulose present in the sample. It enhances efficiency and effectiveness and allows real-time monitoring and distinct applications.
3. During the bleaching and neutralization process, carefully monitor the pH levels and adjust slowly to avoid over-neutralization, which can affect the quality and properties of the isolated cellulose.
4. FTIR and XRD are generally employed to detect and quantify such fibers due to their ability to provide detailed chemical and structural information, crystallinity, and overall quality of the extracted cellulose. Handle this equipment safely or with the help of the operator.
5. To make an 8% KOH solution, dissolve 8 g of KOH pellets in 100 mL distilled water with continuous stirring. For dioxane (with 0.05 mol HCl), add 0.05 mol or approximately 1.83 mL conc. HCl to 100 mL dioxane and stir. As dimethyl sulfoxide may pose inhalation risks on heating, it is advised to use it inside a fume hood or in a well-ventilated area. Keep DMSO away from other chemicals to prevent any kind of reaction.
6. Add 60 mL of conc. sulfuric acid to 40 mL distilled water slowly, with continuous cooling and stirring to make 60% sulfuric acid solution. Steam extraction is a milder process which is efficient for the extraction of hemicellulose. Efficiency of extraction is increased as the steam helps in breaking down of lignin and cellulosic matrix that encases hemicellulose. Keep a check on the temperature and pressure of the steam.
7. Use severity factor to calculate:

$$\log R = \log(t \cdot \exp(T_2 - T_1)/14.75) \tag{1}$$

Where:

t is the reaction time in minutes.

T_2 is the reaction temperature in degrees Celsius.

T_1 is a reference temperature commonly set at 100 °C.

14.75 is an empirical constant derived from the Arrhenius equation for the rate of chemical reactions.

8. For 6 M sulfuric acid, carefully add 330 mL of concentrated sulfuric acid to distilled water and make the volume up to 1 liter. Lyophilization is another word for freeze drying, which is used to preserve perishable materials by removing water content through sublimation. Avoid overloading while using a freeze dryer, and keep a check on sealing. Consider the cryoprotectants, which are stable as per the extraction process.
9. Calibrate the spectrophotometer and take the reading of the blank accurately before moving on to the sample readings. Prepare samples at appropriate concentrations to fall within the buffer linear range of the spectrophotometer. Cuvettes need to be properly cleaned and dried carefully.
10. Calibrate the instrument with the help of the user manual or the operator and set the environmental conditions as per your requirement. Run a pre-test for TGA and DSC before running the samples.
11. The DNS method is used to quantify reducing sugars such as glucose in a given sample. It results in a color change as the DNS reagent reacts with the reducing sugars to form a reddish brown complex whose intensity is proportional to the concentration of reducing sugars present.
12. Dissolve 8.2 g sodium acetate in ~900 mL water, adjust pH to 5.2 with acetic acid, and make up to 1 liter, to obtain sodium acetate buffer (0.1 M, pH 5.2).
13. Dissolve appropriate amounts of sodium maleate in water, adjust pH to 6.0 with HCl or NaOH, and make up to desired volume to get a sodium maleate buffer.
14. Acid-insoluble solids refer to the portion of solid material that remains undissolved or insoluble in acid after a specified treatment or digestion process. Choose the acid and its concentration wisely with minimum contamination and homogeneity of the sample.
15. Defatting of the sample can be done by Soxhlet apparatus with the help of a suitable solvent such as hexane or petroleum ether.
16. Non-polar solvents can be selected among hexane, petroleum ether, diethyl ether, chloroform, and ethyl acetate for completing the fermentation process to extract gums from the sample.
17. The HPLC quantification method recommends using high-purity solvents as it ensures consistent results and avoids contamination. Ensure the mobile and stationary phases are suitable to the method of extraction.

References

1. Dhingra D, Michael M, Rajput H et al (2012) Dietary fibre in foods: a review. J Food Sci Technol 49:255–266. https://doi.org/10.1007/s13197-011-0365-5
2. Barber TM, Kabisch S, Pfeiffer AF et al (2020) The health benefits of dietary fibre. Nutrients 12:3209. https://doi.org/10.3390/nu12103209
3. Tejada-Ortigoza V, Garcia-Amezquita LE, Serna-Saldívar SO (2016) Advances in the functional characterization and extraction processes of dietary fibre. Food Eng Rev 8:251–271. https://doi.org/10.1007/s12393-015-9134-y
4. Maphosa Y, Jideani VA (2016) Dietary fibre extraction for human nutrition – a review. Food Rev Int 32:98–115. https://doi.org/10.1080/87559129.2015.1057840
5. Nascimento RE, Carvalheira M, Crespo JG et al (2023) Extraction and characterization of cellulose obtained from banana plant pseudostem. Clean Technol 5:1028–1043. https://doi.org/10.3390/cleantechnol5030052
6. Hernández-Varela JD, Chanona-Pérez JJ, Benavides HAC et al (2021) Effect of ball milling on cellulose nanoparticles structure obtained from garlic and agave waste. Carbohydr Polym 255:117347. https://doi.org/10.1016/j.carbpol.2020.117347
7. Zhang QW, Lin LG, Ye WC (2018) Techniques for extraction and isolation of natural products: a comprehensive review. Chin Med 13:20. https://doi.org/10.1186/s13020-018-0177-x
8. Gan J, Xie L, Peng G et al (2021) Systematic review on modification methods of dietary fibre. Food Hydrocoll 119:106872. https://doi.org/10.1016/j.foodhyd.2021.106872
9. Serrano-Martínez VM, Pérez-Aguilar H, Carbonell-Blasco MP et al (2024) Steam explosion-based method for the extraction of cellulose and lignin from rice straw waste. Appl Sci 14:2059. https://doi.org/10.3390/app14052059
10. Wang J, Zhang J, Wang S et al (2023) Isolation and extraction of monomers from insoluble dietary fibre. Foods 12:2473. https://doi.org/10.3390/foods12132473
11. Pessoa A Jr, Mancilha IM, Sato S (1998) Acid hydrolysis of hemicellulose from sugarcane bagasse. Braz J Chem Eng 14. https://doi.org/10.1590/S0104-66321997000300014
12. Sun SF, Yang HY, Yang J et al (2022) The effect of alkaline extraction of hemicellulose on cocksfoot grass enzymatic hydrolysis recalcitrance. Ind Crops Prod 178:114654. https://doi.org/10.1016/j.indcrop.2022.114654
13. Martin-Sampedro R, Eugenio ME, Moreno JA (2014) Integration of a kraft pulping mill into a forest biorefinery: pre-extraction of hemicellulose by steam explosion versus steam treatment. Bioresour Technol 153:236–244. https://doi.org/10.1016/j.biortech.2013.11.088
14. Sipponen MH, Österberg M (2019) Aqueous ammonia pre-treatment of wheat straw: process optimization and broad spectrum dye adsorption on nitrogen-containing lignin. Front Chem 7:545. https://doi.org/10.3389/fchem.2019.00545
15. Galiwango E, Rahman NSA, Al-Marzouqi AH et al (2018) Klason method: an effective method for isolation of lignin fractions from date palm biomass waste. Chem Process Eng Res 57:46–58. https://doi.org/10.1016/j.heliyon.2019.e02937
16. Watkins D, Nuruddin M, Hosur M et al (2015) Extraction and characterization of lignin from different biomass resources. J Mater Res Technol 4:26–32. https://doi.org/10.1016/j.jmrt.2014.10.009
17. Sun J, Mao X (2016) An environmental friendly process for Antarctic krill (Euphausia superba) utilization using fermentation technology. J Clean Prod 127:618–623. https://doi.org/10.1016/j.jclepro.2016.04.020
18. Xie J, Xie W, Yu J et al (2021) Extraction of chitin from shrimp shell by successive two-step fermentation of Exiguobacterium profundum and Lactobacillus acidophilus. Front Microbiol 12:677126. https://doi.org/10.3389/fmicb.2021.677126
19. AACC International (2000) Approved methods of the AACC. Methods
20. Islam S, Miah MAS, Islam MF et al (2024) Enzymatic extraction of green banana resistant starch for future food preparation: structural, physicochemical and functional characterization. Future Foods 9:100308. https://doi.org/10.1016/j.fufo.2024.100308
21. Bashmil YM, Dunshea FR, Appels R et al (2024) Bioaccessibility of phenolic compounds, resistant starch, and dietary fibres from Australian green banana during in vitro digestion and colonic fermentation. Molecules 29:1535. https://doi.org/10.3390/molecules29071535
22. Zouambia Y, Ettoumi KY, Krea M et al (2017) A new approach for pectin extraction:

electromagnetic induction heating. Arab J Chem 10:480–487. https://doi.org/10.1016/j.arabjc.2014.11.011

23. Moczkowska M, Karp S, Niu Y, Kurek MA (2019) Enzymatic, enzymatic-ultrasonic and alkaline extraction of soluble dietary fibre from flaxseed – a physicochemical approach. Food Hydrocoll 90:105–112. https://doi.org/10.1016/j.foodhyd.2018.12.018
24. Cui SW, Chang YH (2014) Emulsifying and structural properties of pectin enzymatically extracted from pumpkin. LWT-Food Sci Technol 58:396–403. https://doi.org/10.1016/j.lwt.2014.04.012
25. Moczkowska M, Karp S, Niu Y et al (2019) Enzymatic, enzymatic-ultrasonic and alkaline extraction of soluble dietary fibre from flaxseed – a physicochemical approach. Food Hydrocoll 90:105–112. https://doi.org/10.1016/j.foodhyd.2018.12.018
26. Zabot GL, Schaefer Rodrigues F, Polano Ody L et al (2022) Encapsulation of bioactive compounds for food and agricultural applications. Polymers 14:4194. https://doi.org/10.3390/polym14194194
27. Pratiwi RD, El Muttaqien S, Gustini N (2023) Eco-friendly synthesis of chitosan and its medical application: from chitin extraction to nanoparticle preparation. ADMET and DMPK 11:435–455. https://doi.org/10.5599/admet.1999
28. Patel H (2023) Plant-derived gums: source, extraction, and food application. Pharma Innov J 2:3439–3447
29. Shiram S, Venugopal P, Tungare A et al (2021) Optimization of xanthan gum fermentation utilizing food waste. GRD J Eng 6:19–29
30. dos Santos BC, Bandiera JFV, Junges A et al (2024) Extraction and quantification of inulin prebiotic in leaves, tubers, and in vitro seedlings of Tropaeolum pentaphyllum. Sustain Chem Pharm 38:101486. https://doi.org/10.1016/j.scp.2024.101486
31. Eris FR, Pamela VY, Kusumasari S et al (2024) Extraction of inulin from Beneng tuber (*Xanthosoma undipes*) and its application to yogurt. Future Foods 9:100339. https://doi.org/10.1016/j.fufo.2024.100339
32. Pengkumsri N, Sivamaruthi BS, Sirilun S et al (2016) Extraction of β-glucan from Saccharomyces cerevisiae: comparison of different extraction methods and in vivo assessment of immunomodulatory effect in mice. Food Sci Technol (Campinas) 37:124–130. https://doi.org/10.1590/1678-457X.10716
33. Puligundla P, Lim S (2022) A review of extraction techniques and food applications of flaxseed mucilage. Foods 11:1677. https://doi.org/10.3390/foods11121677
34. Archana G, Sabina K, Babuskin S et al (2013) Preparation and characterization of mucilage polysaccharide for biomedical applications. Carbohydrate Polym 98:89–94. https://doi.org/10.1016/j.carbpol.2013.04.062
35. Cakmak H, Ilyasoglu-Buyukkestelli H, Sogut E et al (2023) A review on recent advances of plant mucilages and their applications in food industry: extraction, functional properties and health benefits. Food Hydrocoll Hlth 3:100131. https://doi.org/10.1016/j.fhfh.2023.100131
36. Lima BV, Oliveira MJ, Barbosa MA et al (2021) Immunomodulatory potential of chitosan-based materials for cancer therapy: a systematic review of in vitro, in vivo and clinical studies. Biomater Sci 9:3209–3227. https://doi.org/10.1039/D0BM01984D

Chapter 7

Extraction and Quantification of Health-Promoting Peptides

Manohar Lal, Meghana Maske, Bababode Kehinde, and Pragati Godara

Abstract

Health-promoting peptides are bioactive molecules derived from various natural protein sources, which may include dairy, meat, plant, and marine proteins. These peptides exhibit significant physiological health benefits such as antioxidant, antihypertensive, antimicrobial, and anti-inflammatory activities, which make them valuable for application in nutraceutical, pharmaceutical, and functional foods. The extraction and quantification of these peptides are critical steps for utilizing them for therapeutic use. This chapter provides an overview of the methods used for the extraction and quantification of health-promoting peptides. In enzymatic hydrolysis the utilization of proteolytic enzymes such as trypsin, pepsin, and alcalase is there, this is a most widely used method because of its specificity and efficiency to yield bioactive fragments. The quantification strategies emphasizing spectrophotometric assays, mass spectrometry (MS), etc. are the analytical tools which are essential for characterizing peptide sequences, molecular weights and biological acitivities. In this chapter, the recent advancements and challenges in peptide extraction and quantification have been highlighted.

Key words Health-promoting peptides, Peptide extraction, Enzymatic hydrolysis, Peptide quantification, Mass spectrometry

1 Introduction

Health-promoting peptides are short chains of amino acids, typically consisting of 2–20 residues which have beneficial physiological effects in addition to their basic nutritional value. The peptides are released from protein precursors mainly through enzymatic hydrolysis, microbial fermentation and other chemical processes. They are useful in functional foods and medications because they can alter biological processes including immunomodulatory, antimicrobial, anti-inflammatory, antihypertensive, and antioxidant effects [1]. One of the main causes of aging and various chronic illnesses such as cardiovascular diseases and neurodegenerative disorders, is oxidative stress. Bioactive peptides derived from food proteins, including casein, soy, and fish, can scavenge free radicals and prevent lipid peroxidation [2]. Peptides with histidine, tyrosine, and

Ashwani Kumar, Vidisha Tomer and Piyoosh Kumar Babele (eds.), *Phytochemicals: Extraction, Quantification, Isolation, and Characterization*, Methods and Protocols in Food Science, https://doi.org/10.1007/978-1-0716-4840-7_7,

proline residues have strong antioxidant effects because they have the ability to donate electrons and chelate metal ions [3]. It is well known that the hypertension is a major risk factor for cardiovascular diseases, and certain peptides function as angiotensin-converting enzyme (ACE) inhibitors. Peptides derived from dairy, egg, and marine sources, such as Val-Pro-Pro and Ile-Pro-Pro, have been found to lower the blood pressure by inhibiting ACE activity [4]. Peptides from milk, fish, and plant sources have been found to have antimicrobial activity by disrupting bacterial cell membranes or interfering with microbial metabolism in addition to this, the immunomodulatory peptides, derived from whey proteins, can enhance immune responses by stimulating macrophages and modulating cytokine production [5]. Peptides from soy, fish, and eggs can reduce inflammation by inhibiting pro-inflammatory cytokines such as TNF-α and IL-6. Bioactive peptides also support gut health by promoting beneficial gut microbiota and enhancing gut barrier function, which make them useful in the prevention of inflammatory bowel diseases (IBD) [6].

Advances in peptide extraction and quantification technologies are further expanding their applications in personalized nutrition. Effective extraction methods enable the large-scale production of peptides for medicines and functional foods. Standardized methods are required for food industry to extract bioactive peptides from various sources including whey, soy, and marine proteins to be utilized in health-promoting products [7]. Quantification is essential for determining the quantity and potency of peptides in health foods and pharmaceuticals. Consistency, efficiency, and regulatory compliances can be ensured only through accuracy in measurements, so an effective quantification technique allows the researchers and manufacturers to standardize peptide dosages for functional foods and therapeutic use [6]. The accuracy in measurement of peptide content is necessary to ensure its effectiveness in context of various health benefits, such as antihypertensive or cholesterol-lowering effects [2]. To ensure the accurate quantification of bioactive compounds, the spectrophotometric, chromatographic, and mass spectrometry techniques are commonly used [8].

The extraction and quantification of health-promoting peptides are essential to ensure their health benefits and consistency in functional food and pharmaceutical applications. Advancement in extraction technologies, combined with advanced quantification methods, are helpful for the large-scale production and commercialization of such health-promoting peptides. This chapter describes important techniques for obtaining and measuring health-promoting peptides, with an emphasis on accuracy and efficiency which are necessary for their proper utilization.

2 Materials

2.1 Extraction of Peptides

2.1.1 Enzymatic Hydrolysis [9]

(a) **Raw Materials**:

The proteins from dairy sources (casein and whey), meat and fish proteins (collagen and gelatin), muscle proteins, plant-based proteins (soy, pea, wheat gluten, etc.), microbial proteins (fermented milk, bacterial proteins, etc.) can be utilized as raw material.

(b) **Enzymes**:

- **Endopeptidases** (required to break the peptide bonds within the protein chain), the enzymes required includes: Trypsin (cleaves at Lys, Arg residues), pepsin (cleaves at hydrophobic residues), and alcalase (broad specificity).
- **Exopeptidases**: Cleave amino acids from the ends of peptides. The enzymes required includes: Carboxypeptidase (removes C-terminal residues) and Aminopeptidase (removes N-terminal residues).
- **Microbial Proteases**: It includes proteases from *Bacillus subtilis* (e.g., subtilisin) and Proteases from *Aspergillus oryzae*.

(c) **Apparatus**:

- pH meter.
- Magnetic stirrer.
- Water bath/shaker incubator.
- Centrifuge.
- Ultrafiltration or membrane filtration system.
- Freeze dryer.

(d) **Reagents**:

- Phosphate buffer (pH 7.0).
- Hydrochloric acid (HCl) or sodium hydroxide (NaOH) for pH adjustment.
- Distilled water.
- Ethanol (for enzyme deactivation, if needed).

2.1.2 Microbial Fermentation [10]

(a) **Microorganisms**:

- Any suitable microorganism known to produce the desired peptides through fermentation, examples include lactic acid bacteria (LAB) or specific strains of yeast.

(b) **Apparatus**:

- Bioreactors or fermenters equipped with agitation, temperature control, pH monitoring, and aeration capabilities.
- Centrifuges or filtration systems to separate the microbial biomass from the culture medium.

2.1.3 Acid Hydrolysis [11]

- Hydrochloric acid (6 N) (*see* **Note 1**).
- NaOH (5 M) (*see* **Note 2**).
- Whatman filter paper (0.45 μm).

2.1.4 Alkaline Hydrolysis [12]

- Hydrochloric acid (6 M) (*see* **Note 3**).
- NaOH solution (6 M) (*see* **Note 2**).
- Hydrolysis Vessel.
- pH Meter.
- Centrifuge and filtration unit.

2.2 Quantification of Peptides

2.2.1 UV-Vis Absorption [13]

(a) **Apparatus**:

- UV-Vis Spectrophotometer.
- Vortex Mixer.

(b) **Reagents**:

- Peptide Sample: Your extracted peptide solution.
- Standard peptide or protein: For calibration curve, e.g., Tyrosine, Tryptophan, or Bovine Serum Albumin (BSA) if using total protein-based quantification.
- Deionized or distilled water.
- Phosphate buffer (e.g., 50 mM, pH 7.0–7.4)—maintains pH for absorbance readings.

2.2.2 Fluorescence-Based Quantification [14, 20]

(a) **Apparatus**:

- Fluorescence Spectrophotometer: Capable of excitation at 390 nm and emission measurement at 475 nm.

(b) **Reagents**:

- Peptide samples: Dissolved in a suitable buffer or water.
- Fluorescamine reagent: A non-fluorescent compound that reacts with primary amines to yield a fluorescent product.
- Buffer solution: Typically, a borate buffer at pH 8.5 is used to facilitate the reaction.
- Standard peptides solution: A known concentration (0.1 to 10 μm) of a standard peptide or amino acid (e.g., glycine) to create a calibration curve.
- Fluorescamine reagent (*see* **Note 4**).

2.2.3 Matrix-Assisted Laser Desorption/Ionization (MALDI-TOF) Method [15]

(a) **Apparatus**:

- MALDI-TOF Mass Spectrometer: Calibrated and optimized for peptide analysis.
- MALDI Target Plate: Typically, a stainless steel or gold-coated plate.

(b) **Reagents**:

- Peptide samples: Dissolved in a suitable solvent, typically 0.1% trifluoroacetic acid (TFA) in water.
- Matrix solution: Commonly, α-cyano-4-hydroxycinnamic acid (HCCA) dissolved in 50% acetonitrile with 0.1% TFA.
- Internal standard: A peptide of known concentration and similar properties to the analyte peptides.
- Matrix solution (*see* **Note 5**).

2.2.4 Electrospray Ionization Mass Spectrometry (ESI-MS) [16]

(a) **Apparatus**:

- ESI-MS Instrument: Calibrated and optimized for peptide analysis.
- Liquid chromatography (LC) system: For separating peptide mixtures before mass spectrometric analysis.

(b) **Reagents**:

- Peptide samples: Dissolved in a suitable solvent, typically 0.1% formic acid in water or a water-acetonitrile mixture.
- Internal standard: A peptide of known concentration and similar physicochemical properties to the target peptides.

2.2.5 Ninhydrin Assay for Quantification of Peptides [17]

(a) **Apparatus**:

- Heating apparatus: Such as a water bath or heating block capable of maintaining 80–100 °C.
- Spectrophotometer: Capable of measuring absorbance at 570 nm.

(b) **Reagents**:

- Ninhydrin reagent: It is prepared by dissolving 550 mg of ninhydrin and 22 mg of hydrindant in 11 mL of Dimethyl Sulfoxide (DMSO), then mixing with 11 mL of acetate buffer (comprising 98.1 g potassium acetate and 111 mL glacial acetic acid, diluted to a final volume of 500 mL with water).
- Peptide sample: Dissolved in distilled water or an appropriate buffer.
- Test tubes or microplate wells: For reaction setup.
- Pipettes: For accurate liquid handling.

2.2.6 Automated Amino Acid Analyzer [18]

(a) **Apparatus**:
- Automated amino acid analyzer: Equipped with ion-exchange chromatography and post-column derivatization capabilities.

(b) **Reagents**:
- **Peptide sample**: Purified and accurately quantified.
- **Hydrolysis reagents**: Typically, 6 M hydrochloric acid (HCl) for acid hydrolysis.
- **Internal standards**: Known concentration of amino acids (100–500 μM) to calibrate the analyzer.

3 Methods

3.1 Extraction of Peptides

3.1.1 Enzymatic Hydrolysis [9]

1. **Preparation of protein solution**: Weigh the protein source and dissolve it in phosphate buffer (0.05 M, pH 7.0) at a concentration of 5–10% (w/v). Stir the solution at 40–50 °C to ensure complete dispersion of proteins.
2. **Enzyme addition and hydrolysis**: Select an enzyme based on the target peptide bioactivity. Adjust the solution to the optimal pH and temperature for enzyme activity (Trypsin: **pH 7.5–8.5, 37 °C,** Pepsin: **pH 2.0, 37 °C,** and Alcalase: **pH 8.0–9.0, 50 °C**). Add the enzyme at an **enzyme-to-substrate ratio (E/S) of 1–5% (w/w)**. Incubate the reaction mixture for **2–6 hrs**, depending on the degree of hydrolysis required. Stir continuously using a magnetic stirrer or orbital shaker to enhance enzyme-substrate interactions.
3. **Enzyme inactivation**: After hydrolysis, the enzymes need to be inactivated, which can be done by heating at 85–95 °C for 10–15 min.
4. **Peptide Separation and Purification**:
   - **Centrifugation**: Spin the hydrolysate at **10,000 rpm for 15 min** to remove insoluble particles.
   - **Ultrafiltration**: Separate peptides by molecular weight (e.g., <3 kDa, 3–10 kDa, >10 kDa).
   - **Precipitation (if required)**: Use ethanol (80%) or ammonium sulfate to precipitate unwanted proteins.
   - **Freeze Drying**: Lyophilize the purified peptide fraction for storage and further analysis.

3.1.2 Microbial Fermentation [10]

1. **Selection of microorganism**: Choose a suitable microorganism known to produce the desired peptides through fermentation. Examples include lactic acid bacteria (LAB) or specific strains of yeast.

2. **Inoculum preparation**: Start with a small culture of the selected microorganism and grow it in a nutrient-rich medium under controlled conditions to increase its biomass.
3. **Fermentation Process**:
   - Transfer the inoculum into a larger fermentation vessel containing a suitable growth medium.
   - Maintain optimal conditions such as temperature, pH, and oxygenation to promote peptide production.
   - Monitor the fermentation process closely to ensure optimal peptide synthesis.
4. **Harvesting**: Once fermentation is complete (typically monitored by growth kinetics or peptide production indicators), harvest the culture. This can involve centrifugation or filtration to separate the microbial biomass from the medium.
5. **Extraction of Peptides**:
   - Extract the peptides from the harvested biomass or the fermentation medium using methods like solvent extraction, enzymatic hydrolysis, or membrane filtration.
   - Purify the crude extract to isolate the desired peptides using techniques such as chromatography (e.g., HPLC) or ultrafiltration based on peptide size and properties.
6. **Characterization and analysis**: Analyze the purified peptides for their composition, structure, and bioactivity using methods like mass spectrometry, amino acid analysis, and bioassays.
7. **Storage and application**: Finally, store the purified peptides under appropriate conditions and test their application in various fields such as food, pharmaceuticals, or cosmetics.

3.1.3 Acid Hydrolysis [11]

1. **Sample Preparation**
   - Select a protein-rich raw material such as milk, soy, fish, or plant-based proteins.
   - Dry and grind the sample into a fine powder for uniform hydrolysis.
2. **Hydrolysis Process**
   - Weigh the protein sample and transfer it into a heat-resistant hydrolysis flask.
   - Add **6 M hydrochloric acid** (**HCl**) in a ratio of **1**:**20** (**w/v**) (**protein**: **acid solution**).
   - Seal the flask to prevent evaporation and degradation of volatile compounds.
3. **Incubation**
   - Heat the mixture at **110–120 °C for 12–24 h** in a hydrolysis oven.

- Ensure continuous stirring or shaking to facilitate uniform hydrolysis.

4. **Neutralization and Filtration**
   - After hydrolysis, cool the mixture to room temperature.
   - Neutralize with **NaOH** (**Sodium Hydroxide, 6 M**) to adjust the pH to 7.0.
   - Filter the solution using a **Whatman filter paper** (**0.45 μm**) to remove insoluble residues.
5. **Peptide Extraction and Purification**
   - Use **ultrafiltration** (**membrane filtration 3–10 kDa**) to separate peptides of different molecular weights.

3.1.4 Alkaline Hydrolysis [12]

1. **Sample Preparation**
   - Choose a protein-rich material such as plant proteins (e.g., soy, wheat gluten), animal proteins (e.g., casein, collagen), or marine proteins (e.g., fish muscle, seaweed).
   - Dry and grind the material into a fine powder to increase the surface area for hydrolysis.
2. **Hydrolysis Process**
   - Mix the protein powder with the NaOH solution at a ratio of 1:30 (w/v) (i.e., 1 g of protein to 30 mL of NaOH solution).
   - Conduct the hydrolysis at room temperature (25 °C) for 2 h.
3. **Incubation**
   - Alternatively, for more extensive hydrolysis, incubate the mixture at 105 °C for 20 h.
4. **Neutralization and Filtration**
   - After the desired hydrolysis time, cool the mixture to room temperature. Neutralize the solution to pH 7 by slowly adding 6 M hydrochloric acid (HCl) while stirring.
   - Filter the neutralized solution through filter paper (Whatman No. 1) to remove insoluble residues.
5. **Centrifugation**:
   - Centrifuge the filtrate at 10,000 rpm for 15 min to obtain a clear supernatant containing the peptides.
6. **Peptide Purification**:
   - Subject the supernatant to ultrafiltration using membranes with specific molecular weight cut-offs (e.g., 3 kDa or 10 kDa) to separate peptides based on size.

- Further purify the peptides using chromatographic techniques such as reversed-phase high-performance liquid chromatography (RP-HPLC).

3.2 Quantification of Peptides

3.2.1 UV-Vis Absorption [13]

Quantification of peptides using ultraviolet-visible (UV-Vis) absorption spectroscopy is a widely adopted technique due to its simplicity and rapid execution. This method leverages the intrinsic absorbance properties of peptide bonds and aromatic amino acids. Here is a detailed procedure for peptide quantification via UV-Vis absorption.

1. **Selection of Appropriate Wavelength**:
   - **Aromatic amino acids** (**Tryptophan and Tyrosine**): Peptides containing these residues exhibit strong absorbance at 280 nm. This characteristic allows for direct quantification without additional reagents.
   - **Peptides lacking aromatic residues**: For peptides devoid of tryptophan and tyrosine, absorbance at 205 nm is utilized, corresponding to the peptide bond's absorbance.
2. **Preparation of Peptide Samples**:
   - Dissolve the peptide in an appropriate buffer, i.e., phosphate-buffered saline (*see* **Note 6**) to a known concentration range, typically between 0.1 and 1.0 mg/mL.
   - Ensure the solution is free from particulates by filtering through a 0.22 μm filter.
3. **Calibration Curve Establishment**:
   - Prepare a series of standard peptide solutions with known concentrations within the expected sample range, i.e., 0.01–1.0 mg/mL.
   - Measure the absorbance of these standards at the selected wavelength (either 280 nm or 205 nm).
   - Plot the absorbance values against the corresponding concentrations to generate a calibration curve.
4. **Measurement of Sample Absorbance**:
   - Set the UV-Vis spectrophotometer to the predetermined wavelength (280 nm or 205 nm).
   - Zero the instrument using the buffer as a blank.
   - Measure the absorbance of the peptide samples.
5. **Calculation of peptide concentration**: Utilize the calibration curve to determine the concentration of the unknown peptide samples based on their absorbance readings.
   - **Determine the extinction coefficient** (ε): If the peptide sequence is known, the **molar extinction coefficient** can be calculated based on the amino acid composition. For example:

Tryptophan (W) = 5500 L $mol^{-1}cm^{-1}$

Tyrosine (Y) = 1490 L $mol^{-1}cm^{-1}$

Cysteine (C-C disulphide bond) = 125 L $mol^{-1}cm^{-1}$

For **unknown peptides**, an average extinction coefficient is used: for **280 nm**: $\varepsilon \approx 1{,}100$ L$mol^{-1}cm^{-1}$and for **205 nm**: $\varepsilon \approx 31{,}000$ L $mol^{-1}cm^{-1}$

- **Calculation of Concentration**

  Rearrange the Beer-Lambert equation to solve for C:

$$C = A/\varepsilon l$$

Where:

C = concentration of peptide.

A = absorbance.

ε = Molar absorptivity.

l = path length of cuvette (Typically 1 cm).

For Converting concentration from **mol/L** (**M**) to **mg/mL** using the peptide's **molecular weight** (**MW**):

$$\text{Concentration (mg/mL)} = c \times \text{MW}$$

Where

c = Molar concentration.

MW = Molecular weight of peptide.

6. **Considerations**:
   - **Buffer Compatibility**: Ensure the buffer used does not absorb significantly at the measurement wavelength to avoid interference.
   - **Sample Purity**: Contaminants absorbing at similar wavelengths can lead to inaccuracies; thus, sample purity is crucial.
   - **Path Length**: Use cuvettes with a known path length (commonly 1 cm) for accurate absorbance measurements.

3.2.2 Fluorescence-Based Quantification [14, 20]

Fluorescence-based quantification of peptides is a sensitive method that utilizes the interaction between peptides and specific fluorogenic reagents to produce measurable fluorescence signals. Here is a detailed procedure for quantifying peptides using fluorescamine, a reagent that reacts with primary amines to form fluorescent products.

A. **Preparation of Standard Curve**:

1. Prepare a series of standard peptide or amino acid solutions with concentrations ranging from 5 μg/mL to 1000 μg/mL in the buffer solution.

2. Pipette 10 μL of each standard solution into separate wells of a microplate or test tubes.
3. Add 90 μL of the buffer solution to each standard and sample, bringing the total volume to 100 μL.
4. Quickly add 100 μL of the fluorescamine reagent to each well or tube while mixing thoroughly.
5. Incubate the reaction mixtures at room temperature for 5 min.

B. **Fluorescence Measurement**:

1. Set the fluorescence spectrophotometer to an excitation wavelength of 390 nm and an emission wavelength of 475 nm.
2. Measure the fluorescence intensity of each standard and sample.
3. Plot the fluorescence intensity of the standards against their concentrations to generate a standard curve.
4. Use the standard curve to determine the concentrations of the peptide samples based on their fluorescence intensities.

C. **Considerations**:

1. **Sensitivity**: The fluorescamine assay is highly sensitive, detecting peptide concentrations as low as 5 μg/mL.
2. **Specificity**: Fluorescamine reacts with primary amines; thus, peptides with blocked N-termini or lacking lysine residues may yield lower signals.
3. **Interferences**: Ensure that the buffer and solvents used do not contain primary amines, which could interfere with the assay.
4. **Stability**: The fluorescence product is stable for a limited time; therefore, measure the fluorescence promptly after the reaction.

3.2.3 Matrix-Assisted Laser Desorption/Ionization (MALDI-TOF) Method [15]

Matrix-assisted laser desorption/ionization time-of-flight (MALDI-TOF) mass spectrometry is a powerful analytical technique primarily used for the identification and characterization of peptides and proteins. While MALDI-TOF is highly effective for qualitative analysis, its application in quantitative analysis presents challenges due to factors such as signal variability and matrix effects. However, with meticulous sample preparation and the use of internal standards, relative quantification of peptides can be achieved. Here is a detailed procedure for quantifying peptides using MALDI-TOF.

A. **Preparation of Internal Standard**:

1. Select an internal standard peptide that does not overlap with the mass range of the target peptides and has similar ionization efficiency.
2. Prepare a stock solution of the internal standard at a known concentration.

B. **Sample Preparation**:

1. Mix equal volumes of the peptide sample and internal standard solutions to achieve the desired concentration ratio.
2. Combine 1 μL of the sample mixture with 1 μL of the matrix solution.
3. Spot 1 μL of this mixture onto the MALDI target plate and allow it to air-dry completely.

C. **Calibration**: Apply calibration standards (e.g., Peptide Mix II) onto the target plate in designated spots. Perform external calibration using these standards to ensure mass accuracy.

D. **Data Acquisition**:

1. Insert the target plate into the MALDI-TOF mass spectrometer.
2. **Laser intensity**: Adjust to optimize signal without causing fragmentation.
3. **Ion mode**: Positive.
4. **Mass range**: Set according to the expected peptide masses.
5. Acquire spectra for both the samples and calibration standards.

E. **Data Analysis**:

1. Use the calibration data to correct the mass spectra of the samples.
2. Identify peaks corresponding to the target peptides and the internal standard.
3. Calculate the intensity ratio of the target peptide to the internal standard.
4. Determine the relative concentration of the target peptide based on this ratio and the known concentration of the internal standard.

F. **Considerations**:

1. **Matrix selection**: HCCA is commonly used for peptides due to its strong UV absorption and efficient energy transfer.

2. **Internal standard**: The internal standard should closely match the physicochemical properties of the target peptides to account for variability in ionization efficiency.
3. **Spotting technique**: Consistent spotting and drying techniques are crucial for reproducible results.
4. **Signal variability**: MALDI-TOF signals can vary due to inhomogeneous crystallization and matrix effects; averaging multiple spectra can improve accuracy.
5. **Quantification limitations**: Absolute quantification is challenging; however, relative quantification is feasible with proper controls and internal standards.

3.2.4 Electrospray Ionization Mass Spectrometry (ESI-MS) [16]

ESI-MS is a powerful analytical technique widely used for the quantification of peptides in proteomics research. The method involves ionizing peptides from a solution and analyzing their mass-to-charge ratios (m/z) to determine their concentrations. Here is a detailed procedure for quantifying peptides using ESI-MS.

A. **Analysis**:

1. Select an internal standard peptide that does not interfere with the target peptides' m/z values and has similar ionization efficiency.
2. Prepare a stock solution of the internal standard at a known concentration.
3. Mix equal volumes of the peptide sample and the internal standard solution to achieve the desired concentration ratio.
4. If necessary, perform desalting and concentration steps to remove impurities that may interfere with ionization.
5. Inject the prepared sample (*see* **Note 7**) into the LC system to separate the peptides based on their hydrophobicity.
6. Use a gradient elution with solvents such as water with 0.1% formic acid (solvent A) and acetonitrile with 0.1% formic acid (solvent B).
7. Couple the LC system directly to the ESI source of the mass spectrometer.
8. Set the ESI-MS parameters:
   - **Ionization mode**: Positive or negative, depending on the peptides' properties (*see* **Note 8**).
   - **Capillary voltage**: Optimize for maximum signal intensity (2.5–4.5 kV).
   - **Desolvation temperature**: Adjust to ensure efficient ionization (300–400 °C).

9. Acquire mass spectra over a suitable m/z range to detect the peptides and the internal standard.

B. **Data Analysis**:

1. Identify peaks corresponding to the target peptides and the internal standard in the mass spectra.
2. Calculate the intensity ratio of each target peptide to the internal standard.
3. Determine the concentration of the target peptides based on this ratio and the known concentration of the internal standard.

C. **Considerations**:

1. **Matrix effects**: Co-eluting substances can suppress or enhance ionization, affecting quantification accuracy. Employing chromatographic separation and using an internal standard can help mitigate these effects.
2. **Ionization efficiency**: Peptides with different sequences may ionize with varying efficiencies. Selecting an appropriate internal standard with similar properties to the target peptides is crucial for accurate quantification.
3. **Instrument calibration**: Regular calibration of the mass spectrometer is essential to ensure mass accuracy and reproducibility of the measurements.

3.2.5 Ninhydrin Assay for Quantification of Peptides [17]

The ninhydrin assay is a classical method utilized for the quantification of free amino acids, peptides, and proteins. This assay is based on the reaction between ninhydrin and amino groups, producing a colored complex measurable via spectrophotometry. Here is a detailed procedure for conducting the ninhydrin assay.

A. **Sample preparation**: Prepare peptide samples in distilled water or an appropriate buffer at known concentrations (*see* **Note 9**). Add 1 mL of the ninhydrin reagent to 5 mL of the peptide sample in a test tube. Seal the test tube with paraffin film or use a capped test tube to prevent solvent evaporation.

B. **Incubation**: Heat the reaction mixture at 80–100 °C for 4–7 min. Ensure consistent heating time for all samples to maintain assay accuracy. After incubation, cool the reaction mixture to room temperature using a cold water bath.

C. **Measurement**: Measure the absorbance of the resulting solution at 570 nm using a spectrophotometer. Use distilled water or an appropriate blank solution to zero the spectrophotometer.

D. **Quantification**: Generate a standard curve using known concentrations of a standard amino acid or peptide. Determine the concentration of unknown samples by comparing their absorbance values to the standard curve.

E. **Considerations**:
   1. **Reagent stability**: Prepare the ninhydrin reagent fresh or ensure it is stable at room temperature for up to 1 month in an ethylene glycol-sodium acetate solvent system without requiring a nitrogen atmosphere.
   2. **Sensitivity**: The assay is sensitive to primary amines, producing an intense blue color, while secondary amines, such as proline, yield a less intense red-brown color.
   3. **Interferences**: Ensure samples are free from contaminants that may react with ninhydrin, leading to erroneous results.

3.2.6 Automated Amino Acid Analyzer [18]

Automated amino acid analyzers are instrumental in quantifying peptides by determining their amino acid composition. The process involves hydrolyzing peptides into free amino acids, which are then separated and quantified. Here is a detailed procedure for this analysis:

A. **Sample hydrolysis**: Place the peptide sample in a hydrolysis vial. Add 6 M HCl to the vial, ensuring the sample is fully submerged. Seal the vial to prevent evaporation. Incubate at 110 °C for 18–24 h to hydrolyze peptide bonds into free amino acids.

B. **Post-hydrolysis processing**: Cool the hydrolyzed sample to room temperature. Remove HCl under reduced pressure (*see* **Note 10**) or by drying. Reconstitute the residue in a suitable buffer compatible with the amino acid analyzer.

C. **Calibration of the analyzer**: Prepare calibration standards using known concentrations of amino acids. Run these standards through the analyzer to generate calibration curves.

D. **Sample analysis**: Inject the reconstituted sample into the amino acid analyzer. Amino acids are separated via ion-exchange chromatography. Post-column derivatization, often with ninhydrin or o-phthalaldehyde (OPA), allows for detection. Detect and quantify amino acids using appropriate detectors (e.g., UV-Vis).

E. **Data analysis**: Compare sample chromatograms to calibration curves to determine amino acid concentrations. Calculate the peptide's quantity based on its known amino acid composition. The analyzer provides **peak area** (**A.A.**) for each amino acid, which is converted into concentration using a **standard calibration curve**. Prepare standard amino acid solutions of known concentrations (e.g., 10, 50, 100, 500 μm). Inject the standards into the AAA and record the peak areas. Generate a calibration curve:

$$C_{AA} = m(A_{AA}) + b$$

Where:

C_{AA} = Concentration of amino acid (μM).

A_{AA} = Peak area of amino acid.

m, b = slope and intercept from the calibration curve.

The total peptide concentration can be determined by the following formula:

$$C_{peptide} = \Sigma\ C_{AA}{}^{*}\ MW_{AA}/MW_{peptide}$$

Where:

$C_{peptide}$ = Estimated peptide concentration (mg/mL).

C_{AA} = Concentration of each amino acid (μM).

MW_{AA} = Molecular weight of amino acid.

$MW_{peptide}$ = Molecular weight of the intact peptide.

F. **Considerations**:

- **Hydrolysis conditions**: Optimize to prevent degradation of sensitive amino acids.
- **Derivatization reagents**: Choose based on the analyzer's specifications and sensitivity requirements.
- **Instrument calibration**: Regular calibration ensures accurate quantification.

4 Notes

1. **Hydrochloric Acid (6 N)**:

   Being HCl as strong acid we need calculate for 1N by following formula;

   $$\frac{\text{Gram Equivalent weight of acid}}{\text{Specific Gravity}} \times \frac{100}{\text{Percent Purity}}$$

   Gram equivalent weight of HCl = 36.45 g.

   Specific gravity of HCl = 1.18 g/mL.

   Percent purity of HCl = 35%.

   So, putting all these values in the above factor

   $$\frac{36.45}{1.18} \times \frac{100}{35} = 87.48 \text{ mL}$$

   For making 1000 mL of 1N HCL we require 87.48 mL of HCl.

And for making 1000 mL of 6N HCl we required $87.48 \times 6 = 524.88$ mL.

So, you have to take 524.88 mL of HCl in 1000 mL volumetric flask and make the volume to 1000 mL by adding distilled water and never forget the principal rule, i.e., always add acid to water, don't reverse it.

2. **NaOH (5 M)**: For making 1000 mL of 1 M NaOH solution, we must know the molecular weight of the NaOH, i.e., 40 g, so we need 40 g of NaOH pallets to be dissolved in 1000 mL of distilled water (the final volume makeup should be 1000 mL), we will get 1 M NaOH solution. Similarly, for making 5 M NaOH we need $40 \times 5 = 200$ g NaOH pallets to be dissolved in 1000 mL of distilled water.
3. **Hydrochloric Acid (6 M)**: For making 1000 mL of 1 M HCl solution we must know the molecular weight of the HCl, i.e., 36.46 g/mol, so we need 36.46 g of HCl to be dissolved in 1000 mL of distilled water (the final volume makeup should be 1000 mL), we will get 1 M HCl solution. Similarly, for making 6 M HCl we need $36.46 \times 6 = 218.76$ mL HCl to be dissolved in 1000 mL of distilled water.
4. Dissolve fluorescamine in an anhydrous solvent such as acetone to prepare a 3 mg/mL solution. Store the solution protected from light and prepare fresh before use, as fluorescamine is sensitive to moisture and light.
5. Dissolve HCCA in 50% acetonitrile with 0.1% TFA to achieve a saturated solution. Sonicate the solution for 5–10 min to ensure complete dissolution. Centrifuge at 12,000× *g* for 2 min to remove any undissolved particles.
6. **Preparation of Phosphate-Buffered Saline** [19]: In a beaker, add 800 mL of distilled water and then add the following salts while stirring, i.e., NaCl (8.0 g), KCl (0.2 g), Na_2HPO_4 (1.44 g), and KH_2PO_4 (0.24 g). Using a pH meter adjust the pH to 7.4 using 1NHCl (to lower pH) or 1N NaOH (to raise pH) if needed. Once everything is dissolved and pH is adjusted, bring the final volume to 1 liter with distilled water. Filter through a 0.22 μm filter paper. Store at room temperature.
7. Load sample into a syringe or static nanospray tip. Flow rate should be 3–10 μL/min and inject: 5–10 μL
8. The properties are as follows: Positive Mode: Most common for peptides (basic residues: Arg, Lys, Hist) produces $[M + H]^+$, $[M + 2H]^{2+}$, etc. The Negative Mode: Rarely used for peptides with strong acidic residues (Asp, Glu, phosphorylated peptides) or sulfated peptides
9. For Free amino acids: 1–200 μg/mL, for short peptides: 10–200 μg/mL and for longer peptides: 50–500 μg/mL.

10. **Rotary Evaporation (Rotavap)**: Cool hydrolysate to room temperature. Transfer hydrolyzed sample into a pre-weighed glass tube or flask. Add a few drops of water (or ethanol) helps to co-evaporate residual HCl. Evaporate under reduced pressure (Rotary evaporator: 40–50 °C, Pressure ~100 mbar or lower). After first round of drying, add 200 μL water, then dry again. Repeat 2–3 times to fully remove volatile HCl. Avoid high heat (>60 °C) to prevent degradation of sensitive amino acids like tryptophan, methionine, or cysteine. Use acid-resistant glassware (borosilicate). Work in a fume hood, as evaporated HCl gas is corrosive.

References

1. Korhonen H, Pihlanto A (2006) Bioactive peptides: production and functionality. Int Dairy J 16:945–960
2. Erdmann K, Cheung BW, Schröder H (2008) The possible roles of food-derived bioactive peptides in reducing the risk of cardiovascular disease. J Nutr Biochem 19:643–654
3. Sarmadi BH, Ismail A (2010) Antioxidative peptides from food proteins: a review. Peptides 31:1949–1956
4. Li Y, Li B, He C, He J (2021) Bioactive peptides as therapeutic agents for cardiovascular diseases: a review. J Funct Foods 86:104732
5. Sharma S, Singh R, Rana S, Sharma M (2019) Whey protein-derived bioactive peptides and their therapeutic potential. Nutr Metab 16:78
6. Udenigwe CC, Aluko RE (2012) Food protein-derived bioactive peptides: production, processing, and potential health benefits. J Food Sci 77:R11–R24
7. Agyei D, Danquah MK (2011) Industrial-scale manufacturing of pharmaceutical-grade bioactive peptides. Biotechnol Adv 29:272–277
8. Sánchez-Rivera L, Martínez-Maqueda D, Cruz-Huerta E, Miralles B, Recio I (2014) Peptidomics for discovery, bioavailability and monitoring of dairy bioactive peptides. Food Res Int 63:170–181
9. Aluko RE (2015) Enzymatic production of bioactive peptides from food proteins. Adv Food Nutr Res 76:1–20
10. Bartolomei M, Capriotti AL, Li Y, Bollati C, Li J, Cerrato A, Lammi C (2022) Exploitation of olive (Olea europaea L.) seed proteins as upgraded source of bioactive peptides with multifunctional properties: focus on antioxidant and dipeptidyl-dipeptidase—IV inhibitory activities, and glucagon-like peptide 1 improved modulation. Antioxidants 11:1730
11. Dullius A (2018) Peptide production by acid hydrolysis of protein-rich sources and their applications in food and health. Food Chem 245:30–42
12. Li H, Prairie N, Udenigwe CC, Adebiyi AP, Tappia PS, Aukema HM, Aluko RE (2011) Blood pressure lowering effect of a pea protein hydrolysate in hypertensive rats and humans. J Agric Food Chem 59:9854–9860
13. Anthis NJ, Clore GM (2013) Sequence-specific determination of protein and peptide concentrations by absorbance at 205 nm. Protein Sci 22:851–858
14. Thermo Fisher Scientific (n.d.) Fluorescamine Protein Assay
15. Tamura K, Matsumoto K (2005) Quantitative determination of peptides using matrix-assisted laser desorption/ionization time-of-flight mass spectrometry. BioTechniques 38:481–486
16. Chowdhury SK, Katta V, Chait BT (1990) A nelectrosprayionization mass spectrometer with new features. Rapid Commun Mass Spectrom 4:81–87
17. Starcher B (2001) A ninhydrin-based assay to quantitate the total protein content of tissue samples. Anal Biochem 292:125–129
18. Sykora C, Schmid W (2000) High-performance amino acid analysis using modern automated analyzers. J Chromatogr B 749:61–68
19. Sambrook J, Russell DW (2001) Molecular cloning: a laboratory manual, 3rd edn. Cold Spring Harbor Laboratory Press
20. Sedgwick AC (2018) Peptide-based fluorescent probes for biological imaging. Chem Soc Rev 47:884–906

Chapter 8

Extraction of Lectin FRIL from Leguminous Plant *Lablab purpureus*

Subhash Chandra Pani and Immanuel Dhanasingh

Abstract

FRIL, a mannose/glucose binding lectin, from *lablab purpureus* or hyacinth beans exhibits therapeutic and biomedical applications including antibacterial, antifungal, and anti-SARS-Cov-2 activities, due to its higher affinity toward carbohydrates. In addition to its antimicrobial activity, FRIL protein was able to preserve primitive hematopoietic and neural progenitors from differentiation in suspension culture for a month without changing the medium. In this chapter, we illustrate the procedure to extract this valuable protein directly from the seeds of *Lablab purpureus*. The obtained pure protein techniques include homogenization, ammonium sulfate precipitation and anion exchange chromatography. The obtained pure protein was biochemically characterized using hemagglutination and hemagglutination-inhibition assays.

Key words *Lablab purpureus*, Antimicrobial activity, SDS-PAGE, Ammonium sulfate precipitation, Hemagglutination assay, Anion exchange chromatography

1 Introduction

Lectins are widely distributed proteins that can bind to carbohydrates and glycoconjugates both selectively and reversibly. Lectins from leguminous plants are the subject of much research because of their widespread existence in nature and the fact that they have been found in a variety of biological sources. Legume lectins are an important tool in glycomic research because of their capacity to identify glycans and glycoconjugates on cells and other intracellular structures. Lectins exhibit various therapeutic and biomedical applications due to their variable affinity toward carbohydrates attached to the envelope of viruses, bacteria, and fungi [1]. In addition, lectins not only play an important role in disease diagnosis, but also in activating the host's immune response and preventing pathogen adherence and invasion [2]. Therefore, lectins are frequently employed in biomedical uses, such as the regulator of

Ashwani Kumar, Vidisha Tomer and Piyoosh Kumar Babele (eds.), *Phytochemicals: Extraction, Quantification, Isolation, and Characterization*, Methods and Protocols in Food Science, https://doi.org/10.1007/978-1-0716-4840-7_8,

immune responses, healing of wounds by identifying glycans on cell surfaces, glycoconjugates, or free sugars [3].

Among different kinds of lectin proteins available, the mannose-binding plant lectin FRIL [Flt3 receptor-interacting lectin], also known as DLL-I [*Dolichos lablab* lectin], derived from hyacinth beans (*Dolichos lablab*), is of therapeutic interest in research community. This is because FRIL possess unique ability to maintain primitive hematopoietic progenitors in a dormant state, showcasing its potential in clinical applications related to stem cell preservation and therapy [4, 5]. One of the major issues with hematopoietic stem cells is that they quickly proliferate and differentiate hindering major research activities. As a differentiation inhibitor, FRIL has shown to selectively interact with NIH 3T3 fibroblasts that have been transfected with the Flt3 tyrosine kinase receptor, keeping human cord blood stem cells and progenitors quiescent for extended periods of time in vitro. Therefore, treatment of primitive hematopoietic and neural progenitors in suspension culture with purified FRIL alone can preserve the cells for 1 month without medium change, thereby implicating the therapeutic significance of this protein molecule [4–8].

In addition to being a stem cell preservation factor, FRIL has been shown to possess antiviral activity against diverse influenza virus strains and SARS-CoV-2 [9]. The viral envelopes contain complex-type N-glycans that are neutralized by FRIL, thereby inhibiting viral entry into the cells. As demonstrated recently, at low nanomolar concentrations, FRIL neutralized 11 typical strains of the human and avian influenza, and when administered intranasally, it protected mice against a fatal H1N1 infection [9]. These works imply that the FRIL can also be applied for COVID-19 and influenza prevention and/or therapy. These applications indicate the need for a standardized protocol for the extraction of FRIL protein from the seeds of hyacinth beans.

The FRIL monomer (269 amino acids) consists of a typical lectin fold containing a sandwich of large and small β-sheets at either end (Fig. 1a). It has been reported frequently that the FRIL monomer undergoes proteolytic digestion in vivo resulting in two chains named α-subunit (N-terminus) and β-subunit (C-terminus), mainly responsible for oligomerization and mannose-binding, respectively (Fig. 1a, b). FRIL exists as a homo-tetramer in solution (Fig. 1a) as verified using gel-filtration chromatography, dynamic light scattering, and negative-stain electron microscopy [9]. The purified mannose/glucose-specific FRIL lectin from hyacinth bean has always resolved into five bands within the molecular mass range of 10–20 kDa when viewed under SDS-PAGE gel (Fig. 1b) [6]. The top four bands correspond to the differentially digested and N-glycosylated α-subunit while the lowest band is designated as β-subunit (Fig. 1b) [4, 6].

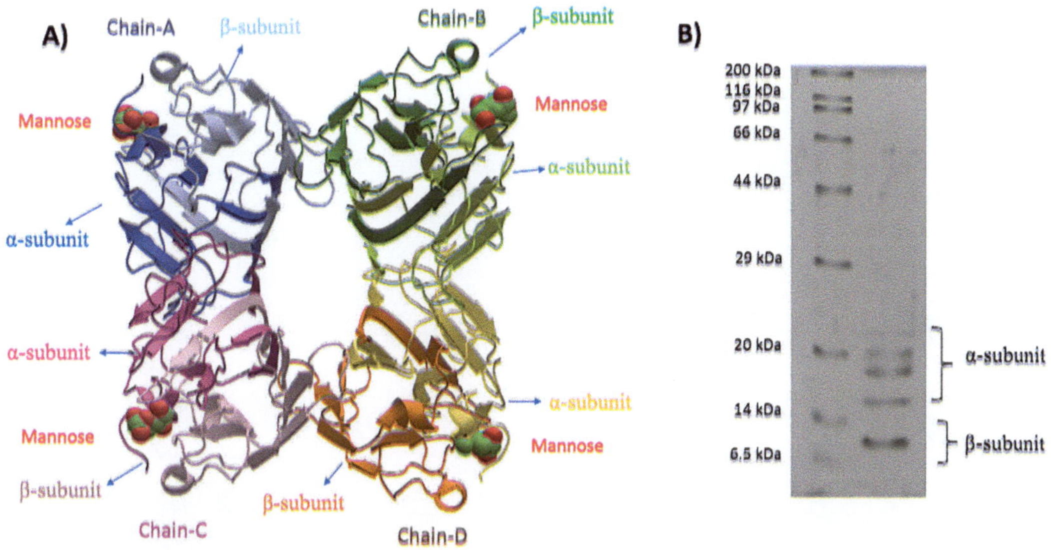

Fig. 1 FRIL Protein. (**a**) FRIL protein exists as a homotetramer (PDB: 1QMO). Each monomer exists as an N-terminal α-subunit and C-terminal β-subunit. The bound ligand molecule is Mannose represented in ball representation. (**b**) SDS-PAGE of FRIL protein shows a characteristic four-banded α-subunit (between 22 and 15 kDa) and single-banded β-subunit (~9 kDa)

Owing to the therapeutic significance of FRIL protein, here we have established a standardized protocol for the FRIL protein extraction directly from the hyacinth bean seeds. Following the multi-step protocol as described below, we were able to isolate and purify approximately 30 mg FRIL protein from 10 g of seeds (Fig. 2d). The purified FRIL protein had the characteristic five bands corresponding to α- and β-subunits when run on a SDS-PAGE gel (Fig. 2d). The obtained FRIL protein was further successfully characterized for its activity using hemagglutination and hemagglutination inhibition assays (Fig. 3). The purified FRIL protein was able to successfully agglutinate the human red blood cells as against the only buffer negative control (Fig. 3) while in the presence of glucose as an inhibitor, the agglutination of the blood was inhibited (Fig. 4). These results indicate that the extracted FRIL protein from hyacinth bean is pure and is able to retain its activity.

2 Materials

Prepare all solutions using ultrapure water and analytical-grade reagents. Store all prepared reagents at room temperature (unless otherwise specified). While disposing of waste products, follow waste disposal protocols as per rules and regulations. We do not add sodium azide to our reagents and buffers.

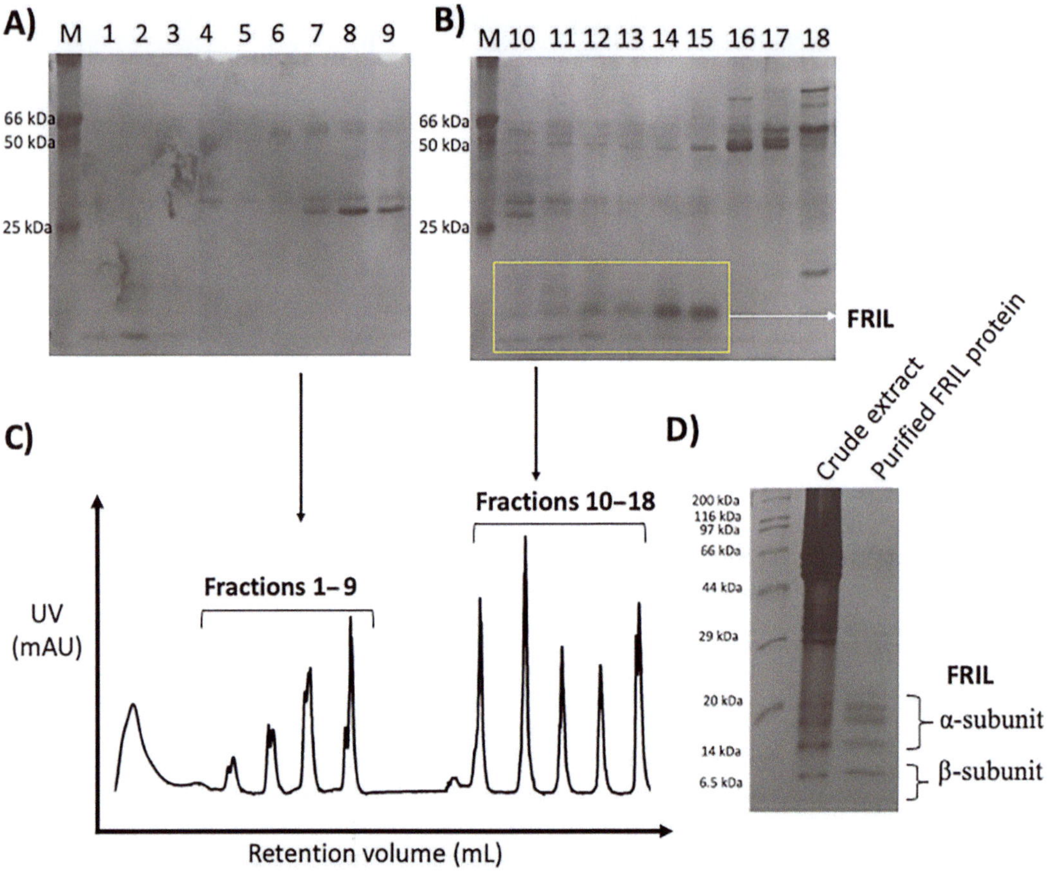

Fig. 2 FRIL protein purification using anion exchange chromatography. (**a**), (**b**) SDS-PAGE of the eluents from anion exchange chromatography (**c**) The crude seed extract subjected to ammonium sulfate precipitation was then subjected to anion exchange chromatography. The protein fractions bound to the bead were eluted using gradients of sodium chloride. The elution profile peaks obtained in **C** (fractions 1–18) were subjected to SDS-PAGE electrophoresis **A** and **B**. The yellow box in **B** indicates the characteristic FRIL protein bands. (**d**) SDS PAGE gel of the crude hyacinth seed extract and purified lectin FRIL protein, with the characteristic five bands corresponding to α- and β-subunits

2.1 Common Buffers

1. Sodium-Phosphate buffer pH 8.0: Prepare 500 mL of 0.5 M sodium phosphate monobasic solution and 500 mL of 0.5 M sodium phosphate dibasic solution, mix the sodium monobasic and dibasic solutions to obtain the pH according to the phosphate buffer table (*see* **Note 1**).
2. Phosphate buffered saline (PBS) pH 7.4: Prepare 500 mL of 1X PBS with pH 7.4 (*see* **Note 2**).

2.2 Sodium-Dodecyl Sulfate Polyacrylamide Gel Electrophoresis (SDS-PAGE)

1. 1.5 M Tris–HCl, pH 8.8 for resolving gel: Add about 200 mL water to a beaker. Weigh 54.51 g Tris–HCl and transfer to the beaker. Adjust the pH to 8.8 with HCl. Make up to 300 mL with water and store at 4 °C (*see* **Note 3**).

Hemagglutination assay

	Negative control	Positive control (Seed extract)	**FRIL Protein**
Buffer (PBS 1X)	50 μL	-	-
Crude extract (50 μg)	-	50 μL	-
FRIL Protein (50 μg)	-	-	50 μL
Blood (RBC 20%)	30 μL	30 μL	30 μL
Agglutination	No	Yes	Yes
Slide image		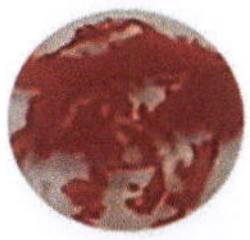	

Fig. 3 FRIL protein characterization using hemagglutination assay. The droplets of negative control (buffer only), positive control (seed extract), and the isolated FRIL protein were subjected to agglutination in presence of RBCs. The fractions containing seed extract and FRIL protein showed agglutination

2. 1 M Tris–HCl, pH 6.8 for stacking gel: Weigh 12.11 g of Tris–HCl and dissolve in 50 mL of water by mixing well. Finally, adjust the pH to 6.8. Make up to 100 mL with water and store at 4 °C.
3. Thirty percent acrylamide/Bis solution (29.2:0.8): Weigh 29.2 g of acrylamide and 0.8 g Bis acrylamide and transfer to a beaker containing about 40 mL of water. Mix while stirring it for about 30 min. Make up to 100 mL with water and filter through a 0.45 μm filter (*see* **Note 4**). Store at 4 °C, in a bottle wrapped with aluminium foil (*see* **Note 5**).
4. Ammonium persulfate (APS): Prepare 10% APS solution in water (*see* **Note 6**).
5. N, N, N, N ′-Tetramethyl-ethylenediamine (TEMED): Store at 4 °C (*see* **Note 7**).
6. SDS-PAGE running buffer: Prepare 10× buffer (0.25 M Tris–HCl, 1.92 M glycine). Weigh 30 g Tris–HCl and 144 g glycine, mix, and make up to 1 L with water. Add 10 g of SDS (10%)

Hemagglutination inhibition assay

	Negative control	Fructose	Sucrose	Glucose (inhibitor)
Buffer (PBS 1X)	100 µL	-	-	-
Fructose (50 mM)	-	50 µL	-	-
Sucrose (50 mM)	-	-	50 µL	-
Glucose (50 mM)	-	-	-	50 µL
FRIL Protein (50 µg)	-	50 µL	50 µL	50 µL
Blood (Whole blood)	30 µL	30 µL	30 µL	30 µL
Agglutination	No	Yes	Yes	No
Slide image			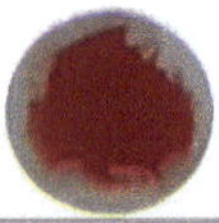	

Fig. 4 FRIL protein characterization using hemagglutination-inhibition assay. The droplets of negative control (buffer only), isolated FRIL protein and different sugars such as sucrose, fructose, and glucose were subjected to agglutination in presence of RBCs. The fractions containing negative control and FRIL protein's substrate molecule: Glucose failed to show agglutination while agglutination happened in presence of sucrose and fructose

carefully as it lathers. Stir slowly using a magnetic stirrer for 30 min to avoid foam formation. Dilute 100 mL of 10× native buffer with 900 mL water to give 1× buffer.

7. SDS sample buffer (5×): Mix 2.5 mL of 1 M Tris–HCl pH 6.8, 1 mL of 10% SDS, 0.01 g of bromophenol blue, 5.5 mL of 70% glycerol, 2 mL of β-mercaptoethanol. Leave one aliquot at room temperature and store the remaining aliquots at −20 °C.
8. In order to prepare 18% resolving gel, the above solutions were mixed as per the protocols mentioned in subheading 3.1.
9. Coomassie blue staining buffer: To prepare 100 mL of Coomassie brilliant blue staining buffer, add 0.1 g of Coomassie brilliant blue powder in 100 mL of destaining buffer.
10. Destaining buffer: To prepare 500 mL of destaining solution, measure 200 mL of 100% methanol and mix 50 mL of glacial acetic acid into it. Make up the final volume to 500 mL by adding 250 mL of distilled water.

3 Methods

All the procedures should be carried out at room temperature unless otherwise specified.

3.1 18% Sodium Dodecyl Sulfate-Polyacrylamide Gel Preparation (SDS-PAGE)

1. Resolving gel (5 mL): Pour 0.6 mL of water into a mixing tube with 3 mL of 30.8% previously prepared acrylamide stock. Add 1.3 mL of Tris–HCl pH 8.8 stock, followed by 50 μL of 10% SDS stock and 50 μL of 10% APS stock. Mix the solutions thoroughly and add 2 μL of TEMED. Pour the mixture into a 1.0 mm gel cassette.
2. Stacking gel (3 mL): Pour 2.1 mL distilled water into a mixing tube with 0.5 mL of 30.8% acrylamide stock solution. Add 0.38 mL of Tris–HCl pH 6.8 from stock, followed by 30 μL of 10% SDS and 30 μL of 10% APS. Mix the solution gently and add 3 μL of TEMED. Cast the stacking gel solution after complete settling of the resolving gel buffer. Insert a 10-well gel comb immediately and make sure no bubbles enter the well.

3.2 Extraction of FRIL from Lablab purpureus Seeds Using PBS Buffer

1. Measure 20 g of *Lablab purpureus* (Hyacinth bean) seeds without seed coat and rinse 2–3 times with deionized water. Spread it on a blotting paper and dry them at room temperature overnight (*see* **Note 8**).
2. Next day, homogenize the dried seeds into fine powder form using a blender. Again, homogenize the seed powder more finely in powder form using a pre chilled mortar and pestle (*see* **Note 9**).
3. Prepare 200 mL of 0.01 M PBS, pH 7.4 containing 1 mM of phenylmethylsulfonyl fluoride (PMSF) and store it at 4 °C. Resuspend all the finely ground seed powder obtained above in 100 mL of pre chilled PBS buffer prepared earlier. Continue to stir the solution with a magnetic stirrer for 6 h at 4 °C (*see* **Note 10**).
4. Centrifuge the above suspension at 10,000× *g* for 45 min at 4 ° C. Collect the supernatant (seed extract) and store at −80 °C in 10 mL aliquots. Resuspend the pellet with the remaining 100 mL PBS buffer again while stirring overnight at 4 °C (*see* **Note 11**).
5. Repeat step-4 to isolate the remaining seed extract.
6. Run SDS-PAGE 18% acrylamide gel and stain with Coomassie blue staining buffer for 1 h followed by destaining buffer overnight to visualize the protein fractions present in supernatant and pellet (*see* **Note 12**).

3.3 Ammonium Sulphate Precipitation of the Crude Extract

1. Take one aliquot containing 10 mL of frozen seed extract from −80 °C and thaw it in a beaker filled with tap water at room temperature. Centrifuge the thawed seed extract at 10,000× *g* for 30 min at 4 °C. Discard the pellet while keeping the supernatant in a 100 mL beaker with a magnetic bar placed into it (*see* **Note 13**).
2. The supernatant is subjected to ammonium sulfate saturation between 0% and 80%. Collect the precipitate formed between 40% and 60% ammonium sulfate saturation (Refer Ammonium Sulfate Nomogram table). Note that ammonium sulfate should be added slowly at every saturation percentage with continuous stirring on a magnetic stirrer to avoid froth formation which may cause protein denaturation. Add ammonium sulfate till required saturation is achieved (*see* **Notes 14** and **15**).
3. Centrifuge the seed extract in ammonium sulfate solution (from the previous step) at 10,000× *g* for 40 min at 4 °C. Separate the supernatant and pellet. Resuspend the pellet with 8–10 mL of 50 mM phosphate buffer pH 7.4 and mix properly. Collect 100 μL of both precipitant and supernatant to load on the 18% SDS-PAGE gel to verify if FRIL protein can be visualized under pellet fraction (Fig. 2).
4. Now in order to remove ammonium sulfate completely, dialyze the 10 mL of pellet fraction against the dialysis buffer: 50 mM phosphate buffer, pH 7.4, for 48 h at 4 °C with at least six buffer changes of 500 mL dialysis buffer (*see* **Note 16**).
5. 0%–60% saturation of ammonium sulfate was found to be more appropriate for the precipitation of FRIL protein.

3.4 Anion Exchange Chromatography

1. Prepare 20 mM phosphate buffer pH 8.0 for equilibration of the anion exchange column (*see* **Notes 17** and **18**).
2. Wash the anion exchange bead with 10 column volumes of deionized water.
3. Load the equilibration buffer into the column through a pump to stabilize pH and ionic strength of the resins. Equilibrate the beads in the column using an equilibration buffer with a flow rate of 1 mL/min (*see* **Note 19**).
4. Next day, collect the dialyzed crude extract sample from the previous step from 4 °C and centrifuge it at 10,000× *g* for 30 min. Filter the samples with a 0.45 μm syringe filter. Load the supernatant into the pre-equilibrated column and set the UV to zero (*see* **Note 20**). Keep the flow rate 0.4 - 0.6 mL/ min.
5. Since FRIL protein has a characteristic PI value of 6.3, we attempted to purify the protein using anion exchange chromatography. The filtered crude extract is allowed to pass through

the beads. The flow through contains an unbound fraction of the crude extract. Upon gradient increase in Sodium chloride, the bound protein starts to elute out depending on their affinity to the beads. Prepare the gradient elution using sodium chloride from 50 to 500 mM concentration.

6. The multiple fractions were eluted separately based on the peak intensity and elution flow volume (*see* **Notes 21** and **22**).
7. Collect 80 μL from each sample (Flow through, Wash, and Elution) and dilute it with 20 μL of 5× SDS loading dye. Observe the bands in SDS-PAGE. The fraction that contains the signature multiple α and β subunit bands corresponding to the molecular weight of FRIL is isolated for characterization.

3.5 Characterization of the Purified Protein

3.5.1 Hemagglutination Assay

1. After purification of the protein with anion exchange chromatography, immediately go for dialysis with 20 mM phosphate buffer pH 8.0 for the removal of NaCl concentration.
2. Collect 2 mL of human whole blood in a tube and centrifuge it immediately at 2000 rpm for 10 min to separate plasma and serum from RBC (*see* **Note 23**).
3. Collect the RBC in a separate tube and make it to 20% with PBS pH 7.4.
4. Take one glass slide and add 50 μL PBS as control at one end. Add 50 μL crude seed extract in the middle (50 μg) and add 50 μL isolated fraction of FRIL protein (50 μg) in the other end. Then add 30 μL of 20% RBC in each droplet. Keep the slide at room temperature for 15 min and observe agglutination (Fig. 3) (*see* **Note 24**).

3.5.2 Hemagglutination Inhibition Assay

1. Prepare 50 mM sugar solutions of Glucose, Sucrose, and Fructose (*see* **Note 24**).
2. Take one 12-well microtiter plate and add 100 μL PBS pH 7.4 as control in the 1st well. Add 50 μL of 50 mM sugar solutions such as fructose, sucrose, and glucose in consecutive wells. Incubate the plate for 30 min at 37 °C inside the incubator (*see* **Note 25**).
3. Add 30 μL of human whole blood in each well and observe agglutination visually (Fig. 4).

4 Notes

1. Weigh 29.99 g of sodium phosphate monobasic and mix in 400 mL water. Mix it properly and make the final volume 500 mL with water (0.5 M Stock).

   Weigh 35.49 g of sodium phosphate dibasic powder and dissolve it in 400 mL water. Mix it and make the final volume to 500 mL with water (0.5 M stock).

Mix the monobasic and dibasic solution as per the following composition to prepare phosphate buffer pH 8.0.

46.6 mL of dibasic sodium phosphate (0.5 M stock) with 3.4 mL of monobasic sodium phosphate (0.5 M stock). Adjust the pH to 8.0 while finally making the volume to 500 mL.

2. Mix the following ingredients: NaCl—4 g (137 mM), KCl—0.1 g (0.0027 mM), Dibasic sodium phosphate—0.72 g (0.01 M), and Monobasic sodium phosphate—0.12 g (0.0018 M), make up the final volume to 500 mL with water while adjusting the pH to 7.4.

   Autoclave both the buffers separately. Do not add sodium azide to the buffers. Adjust the pH of the Phosphate buffer according to the phosphate buffer table.

3. Sodium Dodecyl Sulphate-Polyacrylamide Gel electrophoresis is a technique used for separating the proteins based on their molecular weight. It contains two layers in the gel, the top (stacking gel) and the bottom (resolving gel). The resolving gel has more acrylamide percentage and higher pH (8.8). The stacking gel is where you load the samples and the separation of the protein samples based on molecular weight happens at the resolving gel.
4. Wear a mask and gloves while handling acrylamide since it is a neurotoxin. Use a hood while filtering the solution.
5. Acrylamide can be stored inside 4 °C for up to 1 month. Don't make acrylamide in larger amounts. We prefer to prepare fresh solutions once a month.
6. We prefer to use freshly prepared APS solutions each time.
7. Storing TEMED inside 4 °C reduces its pungent smell.
8. Keeping the seeds at room temperature for drying will remove the excess moisture along the seed coat.
9. Keep the blender as well as the mortar and pestle at 4 °C before use. Temperature maintenance is critical for keeping the protease activity to the minimum.
10. Adding PMSF will prevent the activity of proteases in the crude extract.
11. We found more concentration of protein while extracting. So, we collected it in 2 batches.
12. As our protein was observed as 5-subunit bands and it comes between 12 and 20 kDa, we used 15% gel. Increasing the percentage of gel means increasing the concentration of acrylamide in resolving gel which will lower the pore size of the gel and will help to observe lower molecular weight proteins.
13. Make sure to keep the supernatant inside the beaker full of ice to avoid protein degradation.

14. Salt precipitation can be a useful tool to purify proteins by precipitation. When high concentrations of small, highly charged ions such as ammonium sulfate are added, these groups compete with the proteins to bind to the water molecules. This removes the water molecules from the protein and decreases its solubility. The removal of water molecules causes the protein to fold tighter. The reduction in protein-water interactions allows for more hydrophobic interactions between protein molecules, which causes aggregation and precipitation of the protein.
15. Cooling the ammonium sulfate solution to 4 °C for a longer period is an important step that facilitates protein precipitation and fractionation, alters the kinetics of ammonium sulfate crystallization, and assists in accomplishing desirable crystal characteristics and protein fractions [10].
16. Dialysis is a process for exchanging the buffer of a protein. In general, the protein solution is placed inside a semi-permeable membrane (dialysis bag) which is suspended in a larger volume of the solution to be exchanged into. The dialysis continues until the concentration of solute remains equal inside and outside the dialysis membrane.
17. The molecules separated based on their charge are eluted using a solution of varying ionic strength [11]. The pI of the protein FRIL is 6.3. In such a case, anion exchange chromatography should be performed. DEAE Sepharose will be used for the anion exchanger.
    (a) If the pI is greater than pH of the buffer, then the protein will be positively charged whereas the bead inside the column will be negatively charged. In such a case, Cation exchange chromatography will be performed.
    (b) If the pI is less than the pH of the buffer, then the protein will be negatively charged whereas the bead inside the column will be positively charged. In such a case, an anion exchange chromatography will be performed.
    (c) As the pI of the FRIL protein is 6.3, we should perform anion exchange chromatography where the stationary phase (DEAE Sepharose) will be positively charged and the mobile phase (Protein sample) will be negatively charged [4].
18. Prepare 20 mM phosphate buffer from 0.1 M stock, always make a fresh phosphate buffer of working concentration.
19. In our lab, we are using a Bio-Rad Model EP-1 Econo pump (Biologic LP) for controlling the flow rate across the column.
20. Protein crude lysates should always be filtered with a 0.22 μm syringe filter before loading into the bead, to avoid any kind of contaminants from entering inside the anion exchanger.

21. When protein passes through a UV detector, the peak will be observed in the system. Flowthrough is the unbound protein that doesn't bind to anion exchanger. Washing the bead using the same equilibration buffer will remove other unbound non-specific proteins.
22. In anion exchange chromatography, as the mobile phase is negatively charged and the stationary phase is positively charged while adding NaCl concentration as elution, the Cl^- will compete with negatively charged proteins to bind with the positively charged resins, thereby eluting the protein.
23. Make sure to maintain a temperature of 4 °C while isolating RBC. RBCs will be settled as a pellet after centrifugation. Carefully discard the supernatant containing plasma and one buffy layer coat (serum) and dilute the RBC pellet into 1x PBS buffer and make it to 20%.
24. Lectins can bind with erythrocytes by recognizing the cell surface carbohydrates which result in the formation of a cross-linking activity between them. This is known as hemagglutination assay [12].
25. Weigh 0.045 g of D-Glucose and dissolve in 5 mL PBS Buffer pH 7.4.

    Weigh 0.045 g of D-Fructose and dissolve in 5 mL PBS buffer pH 7.4.

    Weigh 0.085 g of Sucrose and dissolve in 5 mL PBS buffer pH 7.4.
26. The addition of Lectins into the sugar solution will allow the sugar to bind with the lectins. Therefore, a 30-min incubation period is necessary. If the binding affinity between lectins and protein is high, then the blood coagulation reaction will be inhibited [13].

References

1. Naik S, Kumar S (2022) Applications of plant lectins in biotechnology and therapeutics: plant lectins application. J Microbiol Biotechnol Food Sci 11:e4224. https://doi.org/10.55251/jmbfs.4224
2. Costa ACM, Malveira EA, Mendonça LP et al (2022) Plant lectins: a review on their biotechnological potential toward human pathogens. Curr Protein Pept Sci 23:851–861. https://doi.org/10.2174/1389203724666221014142740
3. Coelho LCBB, Silva PMDS, de Menezes Lima VL et al (2017) Lectins, interconnecting proteins with biotechnological/pharmacological and therapeutic applications. Evid Based Complement Alternat Med:1594074. https://doi.org/10.1155/2017/1594074
4. Colucci G, Moore JG, Feldman M et al (1999) cDNA cloning of FRIL, a lectin from Dolichos lablab, that preserves hematopoietic progenitors in suspension culture. Proc Natl Acad Sci 96:646–650. https://doi.org/10.1073/pnas.96.2.646
5. Yao H, Xie X, Li Y et al (2008) Legume lectin FRIL preserves neural progenitor cells in suspension culture in vitro. Clin Dev Immunol:531317. https://doi.org/10.1155/2008/531317
6. Gnanesh Kumar BS, Pohlentz G, Schulte M et al (2014) N-glycan analysis of mannose/glucose specific lectin from Dolichos lablab

seeds. Int J Biol Macromol 69:400–407. https://doi.org/10.1016/j.ijbiomac.2014.05.073

7. Kollet O, Moore JG, Aviram R et al (2000) The plant lectin FRIL supports prolonged in vitro maintenance of quiescent human cord blood CD34(+)CD38(−/low)/SCID repopulating stem cells. Exp Hematol 28:726–736. https://doi.org/10.1016/S0301-472X(00)00163-6
8. Li J, Xie XY, Wang DM et al (2007) FRIL maintains quiescence of hematopoietic stem cells through regulation of cell cycle related factors. Zhonghua Xue Ye Xue Za Zhi 28:37–40
9. Liu YM, Shahed-Al-Mahmud M, Chen X et al (2020) A carbohydrate-binding protein from the edible lablab beans effectively blocks the infections of influenza viruses and SARS-CoV-2. Cell Rep 32:108016. https://doi.org/10.1016/j.celrep.2020.108016
10. Tadayon A, Rohani S, Bennett MK (2002) Estimation of nucleation and growth kinetics of ammonium sulfate from transients of a cooling batch seeded crystallizer. Ind Eng Chem Res 41:6181–6193. https://doi.org/10.1021/ie020163r
11. Cheriyedath S (2024) How does ion exchange chromatography work? Available from: https://www.news-medical.net/life-sciences/How-Does-Ion-Exchange-Chromatography-Work.aspx. Accessed 20 Mar 2024
12. St-Pierre G, Sato S (2022) Hemagglutination assay for the estimation of a carbohydrate-binding activity and determination of its carbohydrate-binding specificity. In: Nishihara S, Angata K, Aoki-Kinoshita KF et al (eds) Glycoscience protocols (GlycoPODv2), 2nd edn. JCGG, Japan
13. Naik S, Kumar S (2020) Biochemical characterization of lactose binding Entadin lectin from Entada rheedii seeds with cytotoxic activity against cancer cell lines. ACS Omega 5:16430–16439. https://doi.org/10.1021/acsomega.0c00577

Chapter 9

Isolation and Characterization of Polyphenols

Priyanka Suthar, Anika Panwar, Ajay Kumar, Vishant Sharma, and Satish Kumar

Abstract

Plants contain phenolic compounds, which are of great importance and have drawn increasing attention recently because of their bioactive properties. Polyphenols' antioxidant properties make them one of the most sought-after phytochemicals. These substances might be free or matrix-bound, monomers or highly polymerized structures, aglycones or glycosides, or both. Additionally, their stability varies significantly, and there is uneven distribution within the plant. Thus, a single standard technique for isolating and characterizing polyphenols cannot be recommended. Therefore, methods need to be tuned according to the study's objective, sample type, and target analytes. This chapter has discussed the general protocols for preparing samples, isolation, and characterization approaches that range from conventional chromatographic procedures to more modern modified approaches.

Key words Polyphenols, Chromatography, Isolation, Identification, LC-MS, Protocols

1 Introduction

Phytochemicals, often known as "secondary metabolites," comprise a broad category of organic complexes produced by plants. All higher plants contain secondary metabolites called polyphenols, which are crucial factors in determining the nutritional value and sensory appeal of vegetables, fruits, and other plants [1]. The polyphenols have an aromatic ring with several hydroxyl groups, and the structures of polyphenols range from a basic phenolic monomer unit to a complex polymer [2]. The classification of polyphenols into categories such as phenolic acids, flavonoids, lignans, stilbenes, and tannins is based on the total number of phenol rings and structural elements that hold them together [3]. They come from a variety of sources, including fruits, vegetables, nuts, seeds, bark, roots, leaves, and herbs; they also come from whole grain items, processed foods, coffee, tea, and red wine. A wide range of biological activities is characteristic of these substances. For

Ashwani Kumar, Vidisha Tomer and Piyoosh Kumar Babele (eds.), *Phytochemicals: Extraction, Quantification, Isolation, and Characterization*, Methods and Protocols in Food Science, https://doi.org/10.1007/978-1-0716-4840-7_9,

millennia, people have known that fruits, vegetables, and herbs are good for their health. It has been established that polyphenols lower morbidity and/or decrease the progression of cancer, neurological, and cardiovascular illnesses. Polyphenols' antioxidant activity is closely linked to their mode of action. In addition, plant polyphenols have anti-allergic/inflammatory, anti-mutagenic, and anti-thrombotic qualities that contribute to health [4]. Since the chemical structure of polyphenols is closely linked to all of the aforementioned properties, researchers have concentrated on appropriate extraction, isolation, and characterization of these chemicals from natural sources as well as food by-products. The discovery and characterization of recently discovered bioactive phenolic compounds are of great interest, which is not surprising. But because of their intricate structure, there is now a need for novel analytical methods. These methods are necessary for their isolation, separation, and structural clarification [5]. Phenolic compounds are primarily separated and isolated via chromatographic methods as summarized in Table 1. However, the selection of an isolation technique depends upon the kind of target component

Table 1
Characterization techniques used for the identification of polyphenols in food samples

Characterization technique	Polyphenols identified	Sample	References
NMR	Catechin, gallocatechin, catechin-3-0-gallate, gallocatechin-3-0-gallate, epicatechin, epigallocatechin, epicatechin-3-0-gallate, epigallocatechin-3-0-gallate, epiafzelechin and epigallocatechin-3-O-[3′-O-methyl] gallate	Tea	[6]
	Tricaffeoylhexaric acid, tetracaffeoylhexaric acid, 3,5-dicaffeoylquinic acid, caffeoyl dihexoside, and chlorogenic acid	*Pallenis spinosa* [L.] Cass	[7]
	Polyphenols, epicatechin, phloridzin-phloretin, and *p*-coumaric, chlorogenic	Apple cider juice	[8]
	Flavanone-*C*-glycosides [[*R*]/[*S*]-eriodictyol-6-*C*-β-d-glucopyranoside and [*R*]/[*S*]-eriodictyol-8-*C*-β-d-glucopyranoside]	*Aspalathus linearis*	[9]
LC-MS/ chromatography	3 hydroxycinnamic acid conjugates, procyanidin	Apricot puree	[10]
	Free polyphenols: procyanidin B1, procyanidin B2, catechin, epicatechin, and protocatechuic acid, procyanidin A2 Bound polyphenols: epicatechin and procyanidin B2 Others: dihydroxybenzoic acid isomers	Seedless dates	[11]
	Ellagitannins, punicalagin and punicalagin	Pomegranate peel	[12]
	p-hydroxybenzoic acid, *p*-coumaric acid, pyrogallol, vanillin, ferulic acid, and caffeic acid	Cinnamon	[13]
	p-coumaric acid, trans-ferulic acid, and syringic acid	Barley husk	[14]

found in plant extracts or fractions. The important parameters that must be considered for the target molecule[s] are solubility (hydrophobic nature or hydrophilic nature), charge, acid-base characteristics, stability, and molecular size. Collecting details regarding a known phenolic compound's chromatographic behaviour in the scientific literature is easy, enabling the choice of the best isolation technique. However, designing a separation technique for crude extract is more complex because the kind of phenolic compounds present is unidentified or has not been specified before [15]. Qualitative chemical testing for substances such as alkaloids, flavonoids, phenolics, and steroids, along with preliminary thin-layer chromatography [TLC] or high-performance liquid chromatography [HPLC] profiling, can aid in selecting a suitable isolation protocol. Two broad categories can be used to group the chromatographic approaches used to isolate different phenolic compounds: Conventional chromatographic techniques and recent chromatographic techniques. The conventional chromatographic methods comprise column chromatography [CC], size-exclusion chromatography [SEC], preparative thin-layer chromatography [PTLC], and ion-exchange chromatography, whereas dry column chromatography [DCC], chromatotron, high-performance thin layer chromatography [HPTLC], vacuum liquid chromatography [VLC], and high-performance liquid chromatography [HPLC] may be considered to be modern/recent chromatographic techniques [16]. Further, for characterization of the isolated phenolic compounds, approaches, i.e., nuclear magnetic resonance [NMR], mass spectrometry [MS], and more often both, are used for the structure identification of isolated secondary phytochemicals following phytochemical separation utilizing chromatographic techniques as summarized in Table 2. NMR offers a very potent supplementary approach for the characterization and quantification of secondary metabolites in plant extracts, while being less sensitive than MS. MS, however, when combined with liquid chromatography, is a popular analytical method for identifying phytochemicals such as polyphenols due to its exceptional sensitivity and broad detection range [5]. Table 1 provides overview of various characterization techniques used for polyphenols identification. The chapter contains further information regarding the majority of the aforementioned techniques commonly utilized for the isolation or purification, and characterization of phenolic compounds.

2 Materials

2.1 Isolation Techniques

2.1.1 Preparation of Sample

This preparatory step may involve multiple equipment [i.e., Soxhlet apparatus, accelerated solvent extractor, microwave-assisted or ultrasound-assisted extractor, or supercritical fluid extraction devices], solvents [i.e., ethanol, ethyl acetate, methanol, acetone, and water], in addition to chemicals [i.e., constituting stationary

Table 2
Isolation techniques used for the purification of polyphenols in food samples

Flavonoid class	Isolation techniques	Compounds isolated	Plant source	Quantity	References
Flavonols	Column chromatography on polyamide and Sephadex LH-20	Kaempferol Quercetin 5-deoxy kaempferol Fisitin	*Astragalus beckari*	1. Kaempferol: 5 mg 2. Quercetin: 7 mg 3. 5-Deoxy kaempferol: 2 mg 4. Fisitin: 1.5 mg	[17]
	Silica gel column chromatography and HPLC	Tricin Luteolin Thalictoside Icariin I Baohuoside I Quercitrin Icariin Epimedin C Epimedin B	*Epimedium hunanense*	–	[18]
	Column chromatography	Kaempferol-3-O-rhamnoside Quercetin-3-O-rhamnoside	*Pometia pinnata*	1. Kaempferol-3-O-Rhamnoside: 39.0 mg 2. Quercetin-3-O-Rhamnoside: 17.4 mg	[19]
Flavones	Silica gel chromatography		*Artemisia vestita*		[20]

		Pectolinarigenin		1: 43.3 mg	
		Jaceosidin		2: 39.5 mg	
		Cirsilineol		3: 37.7 mg	
		Cirsimaritin		4: 60.9 mg	
		Hispidulin		5: 10.2 mg	
		6-methoxytricin		6: 4.2 mg	
		Acacetin		7: 3.3 mg	
		Apigenin		8: 2.8 mg	
	Silica gel chromatography	Chrysoeriol-7-O[200-O-E-feruloyl]-b-D-glucoside	*Allium vineale*	1. Chrysoeriol-7-O[200-O-E-feruloyl] b-D-glucoside: 50 mg	[21]
		Chrysoeriol		2. Chrysoeriol: 35 mg	
		Isorhamnetin-3-b-D-glucoside		3. Isorhamnetin-3-b-D-glucoside: 40 mg	
	Column chromatography and thin layer chromatography	5-Hydroxy-3′,4′,6,7-tetramethoxyflavone	*Artemisia monosperma*	175 mg	[22]
				70 mg	
		Eupatorine		55 mg	
		Ladanein		640 mg	
		Arcapillin		280 mg	
		Cirsilineol		60 mg	
		Eupatilin		415 mg	
		Jaceosidin		20 mg	
		Circimaritin		180 mg	
		Hispidulin			

(continued)

Table 2
(continued)

Flavonoid class	Isolation techniques	Compounds isolated	Plant source	Quantity	References
Phenolic acids class					
Benzoic acid [Hydroxybenzoic acid, Gallic acid, Protochatechuic acid]	High-speed counter-current chromatography [HSCCC] + Preparative liquid chromatography	Gallic acid	*Punica granatum husk*	Gallic acid = 48 mg [by HSCCC] Gallic acid = 15 mg [by liquid chromatography]	[23]
	High-speed countercurrent chromatography [HSCCC]	Protocatechuic acid 4-hydroxybenzoic acid Myricetin Caffeic acid	*Ribes nigrum*	Protocatechuic acid = 0.8 mg 4-hydroxybenzoic acid = 0.5 mg Myricetin = 2.5 mg Caffeic acid = 1.0 mg	[24]
	Silica gel column chromatography + Sephadex LH-20 chromatography + Thin-layer chromatography + High performance liquid chromatography	Gallic acid or 3,4,5-trihydroxy benzoic acid	*Terminalia nigrovenulosa bark*	Gallic acid or 3,4,5-trihydroxy benzoic acid = 107 mg	[25]
Cinnamic acid [Hydroxycinnamic acid, Chlorogenic acid, Coumaric acids, Ferulic acids, Caffeic acids, Sinapic acid]	Preparative high-speed countercurrent chromatography [HSCCC]	Caffeic acid	*Eupatorium adenophorum*	Caffeic acid = 63.2 mg	[26]
	Preparative Thin-layer chromatography	*p*-coumaric acid	*Cynodondactylon*	*p*-coumaric acid = 0.479 mg	[27]
	Counter-current chromatography	Caffeic acid Ferulic acid *p*-coumaric acid	*Saccharum officinarum*	Caffeic acid = 4.8 mg Ferulic acid = 12.9 mg *P*-coumaric acid = 65.2 mg	[28]

Lignans	Column chromatography on	6-allyl-5-methoxy-1,3-	*Peperomia*	2.4 mg	[29]
	silica gel, Lobar column,	benzodioxol-4-ol	*pellucida*	0.8 mg	
	LiChroprep RP-18, and	Pachypostaudin B		3.4 mg	
	Sephadex LH-20 and thin	Pellucidin A		61.2 mg	
	layer chromatography	Dillapiole		4.4 mg	
		Apiol			
	Silica gel column	Micrantherin A	*Schisandra*	13 mg	[30]
	chromatography and HPLC	Gomisin M_2	*chinensis*	100 mg	
		Schizandrin		15 mg	
Stilbenes	HPLC	[E]-resveratrol	*Vitis vinifera*	81.7 mg	[31]
		[E]-ε-viniferin		197.2 mg	
		[E]-piceatannol		13.4 mg	
		Vitisin B		14.8 mg	
		Ampelopsin A		17.5 mg	
		[E]-δ-viniferin		3.3 mg	
		[E]-ω-viniferin		1.1 mg	
		Pallidol		2.13 mg	
		Scirpusin A		2.8 mg	
		[E]-trans-cis-miyabenol C		6.2 mg	
Tannins	HPLC	Proanthocyanidins	*Rubus*	5–46 mg/100 g	[32]
		Ellagitannins	*fructicosus*	150–270 mg/100 g	
	HPLC	Hydrolysable tannins and condensed tannins	*Terminalia* sp.	126–822 mg/100 g	[33]
	Solvent extraction	Total tannins	*Swietenia mahagony*	1. 29.9% DM	[34]
	HPLC	Condensed tannins	Carya	4.66 mg/g DW	[35]
		Hydrolysable tannins	illinoinensis	0.22 mg/g DW	

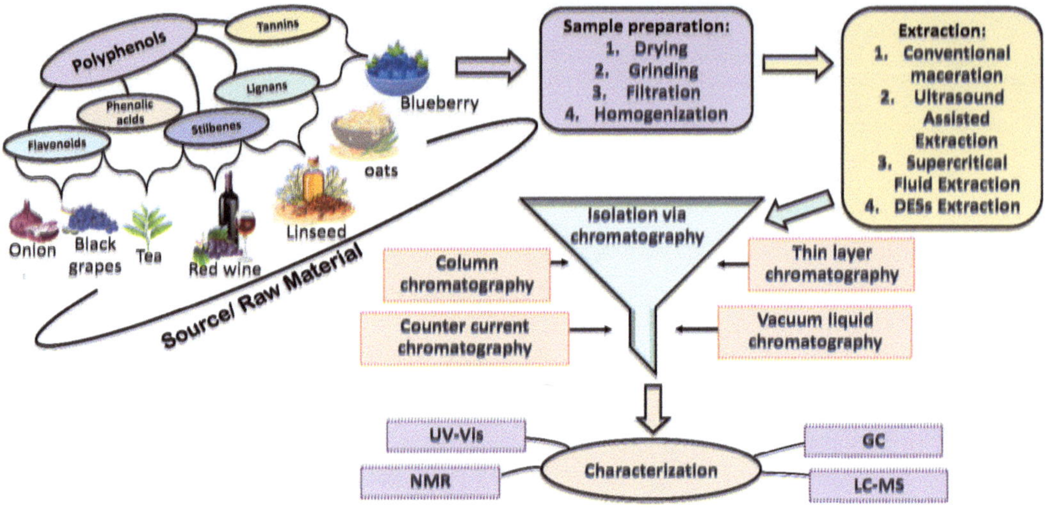

Fig. 1 Schematic illustration of pre-requisite steps for identification and characterization of polyphenols

and mobile phases] for drying, grinding, sieving, homogenization, extraction, derivatization, and hydrolysis. Enzymatic hydrolysis requires the presence of enzymes such as β-glycosidases and/or sulfatases [16]. The schematic illustration of the pre-requisite steps for isolation and identification of polyphenols is depicted in Fig. 1.

2.1.2 For Adsorption and Size Exclusion Column Chromatography

Stationary Phases

(a) *Silica gel [$SiO_2.H_2O$]*: Silica gel is a three-dimensional polymer comprising silicon oxide tetrahedral elements. This porous substance has a significant surface area of 5–800 m^2 per gram. In CC, silica gel particles typically range from 40 to 200 μm, with pore diameters ranging from 40 to 300 Å (*see* **Note 1**).

(b) *Bonded-phase silica gel*: Bonded-phase silica gel is the chemically modified silica gel with altered surface/physical characteristics and chromatographic performance. Surface modification aims to safeguard active silanol groups and link them to organic ligands on the adsorbent surface, resulting in specific surface interactions. Silyl chlorides can block silanol groups, resulting in non-polar [reverse phase] or intermediately polar [normal bonded phase] chromatographic support. The stationary phase for reversed phase and intermediate phase (*see* **Note 2**).

(c) *Alumina*: Alumina is a polymer of high porosity made from aluminium oxide, having either acidic, neutral, or even basic surface. Acidic alumina, i.e., pH ~ 4.0, is effective in separating carboxylic acid groups, basic alumina, i.e., pH ~ 10.0, in isolating alkaloids, and neutral alumina, i.e., pH ~ 7.0, in separating non-polar molecules like steroids.

(d) *Polyacrylamide beads*: Polyacrylamide beads, also known as Biogel P, are formulated by co-polymerization of acrylamide and N, N-methylene-bis-acrylamide. These beads are utilized as a stationary phase to purify natural compounds such as tannins and polysaccharides.

(e) *Carbohydrate [i.e., Dextran]*: Carbohydrates are crosslinked to a three-dimensional network and converted to beads that are generally utilized for chromatographic separation of labile bioactive complexes. For example: Sephadex®, i.e., created by cross-linking a water-soluble dextran with epichlorohydrin. Sephadex® media, particularly the G-10 and G-15, are extensively used for isolating phenolic compounds. Sephadex® LH-20, which is a hydroxypropylated form of Sephadex® G-25, is a popular gel utilized in polyphenols separation, particularly through non-polar or intermediate polarity chemicals.

Mobile Phases

The mobile phase includes solvents of distinct polarities, i.e., non-polar solvents, which include *n*-hexane, pentane, chloroform, dichloromethane, and polar solvents, i.e., methanol and ethyl acetate, whereas for the reversed-phase adsorption chromatography: methanol, water, and acetonitrile are broadly utilized.

2.1.3 For Counter-Current Chromatography

(a) *Apparatus*:

- Centrifugal partition instrument.
- Coil planet centrifuge.

(b) *Reagents*:

Bi-phase solvent systems: Consisting of phases, i.e., aqueous and organic phase or two non-aqueous or two aqueous phases. Generally utilized solvents in CCC are *n*-hexane, methanol, chloroform, and water.

2.1.4 For Thin Layer Chromatography

(a) *Apparatus*:

- Separation chamber with lid.
- Glass micro pipettes.

(b) *Reagents*:

- Silica plates [20 * 20 cm].
- Solvent for the polar phase.
- Solvent for non-polar phase.

2.1.5 For Vacuum Liquid Chromatography

(a) *Apparatus*:

- Sintered glass funnel with fritted disk grade i.e. G3 [15–40 μ], G2 [40–90 μ] or G1 [90–150 μ].
- Laboratory-scale water pump.

- Water aspirator for providing a 20–70 mmHg vacuum essential for suctioning.

(b) *Sample and column size*: The sample size and column size for different scale separations are listed below:
   (a) For <100 mg sample: 1.0 × 4 cm column size.
   (b) For 0.5–1.0 g sample: 2.5 × 4 cm.
   (c) For 1–10 g sample: 5 × 5 cm.

2.2 Characterization Techniques

2.2.1 Nuclear Magnetic Resonance (NMR)

1. Mixture of various standards: arginine, alanine, coumaric acid, catechin, epicatechin, citric acid, chlorogenic acid, fructose, gallic acid, glutamic acid, glucose, proline, serine, sucrose, naringin, malic acid, tartaric acid, malvidin-3, 5-di glucoside, gentisic acid, and resveratrol.
2. 0.1 M Phosphate buffer; pH 3.4.
3. Methanol, acetone, and ethanol used for polyphenol extraction with a 7:3 solvent ratio [solvent: water v/v].
4. Centrifugal evaporator.
5. Sonicator.
6. C18 stationary phase.
7. NMR spectroscopy [36].

2.2.2 For Gas Chromatography

1. *Extraction solvents*: The solvent used to extract varies depending on the interest of the analyte. There is no universal single solvent for all the analyte groups.
2. *Stock standard solutions [1000 mg/L]*: Prepare stock standard solutions by accurately weighing about 0.0100 g of pure compound. Dissolve the compound in a 10 mL volumetric flask with isooctane or hexane (*see* **Note 3**). The stock standard solution is then transferred to screw-caped bottles. Different types of standards used are:
3. *Composite stock standard*: For the preparation of composite stock solution, 1 mL stock solution with stock solution concentration 1000 mg/L is transferred, followed by the addition of solvent, mixing, and diluted properly in a volumetric flask.
4. *Calibration standards*: A calibration standard is prepared using at least five different concentrations using a standard stock solution and diluted with appropriate solvent (*see* **Note 4**). *Internal standard*: An internal standard used for spiking. For this, a standard stock solution of 1000 mg/L was diluted to 50 ng/μL. The spiking volume of extract used is10 μL/mL. The spiking concentration of internal standards is kept constant for all the samples.
5. *Surrogate standard*: Surrogate compounds are used to monitor the performance of this method (*see* **Note 5**).

2.2.3 For Liquid Chromatography Mass Spectrometry (LC-MS)

(a) Distilled water.
(b) Ethanol, methanol, formic acid [HPLC grade].
(c) Volumetric flasks.
(d) Standards with purity above 98%: ascorbic acid, catechol, p-coumaric acid, caffeic acid, ellagic acid, ferulic acid, α-tocopherol, gallic acid, quercetin, p-hydroxybenzoic acid, pyrogallol, vanillin, and syringic acid.
(e) Stock solution: 5 mg/L in ethanol (except ascorbic acid and catechol, which are prepared as 25 mg/L and 50 mg/L, respectively).
(f) Calibration solution: Ethanol-water (50:50, v/v). Dilution prepared by using automatic pipettes and glass volumetric flasks stored at −20 °C.
(g) Internal standard: 1 mg/L curcumin is prepared, and 0.1 mL of curcumin solution is further used.
(h) Gravity column C18.
(i) Lyophilizer.
(j) Rotatory evaporator.
(k) Magnetic stirrer.
(l) LC-MS with tandem gold triple quadrupole [37].

3 Methods

3.1 Isolation Technique for Polyphenols

3.1.1 Column Chromatography [CC]

CC is an isolation method that is employed for separating impurities while isolating active compounds and extracting intermediates from various materials. Both liquid and solid materials can be separated and purified using this method. The stationary phase is set within a narrow tube [called a column], and passing compounds are separated and adsorbed using a liquid mobile phase as depicted in Fig. 2a. The chemical makeup of compounds causes them to be adsorbed, and the stationary phase's differential adsorption is the basis for elution. Silica gel adsorption chromatography and size exclusion chromatography are the most commonly used column chromatography techniques [38].

3.1.2 Silica Gel Adsorption Chromatography

Adsorption is a fundamental liquid chromatographic mode that relies on the phenomenon of adsorption to achieve separation. The analytic molecules are maintained due to their contact with the surface of the stationary phase. The primary basis for separation, therefore, is variation in adsorption affinities of the analyte molecule for surface of stationary phase. Several factors influence analyte adsorption on the stationary phase, including hydrogen bonding, complexation, van der Waal forces, acid-base characteristics, dipole-dipole interactions, as well charge transfer [15].

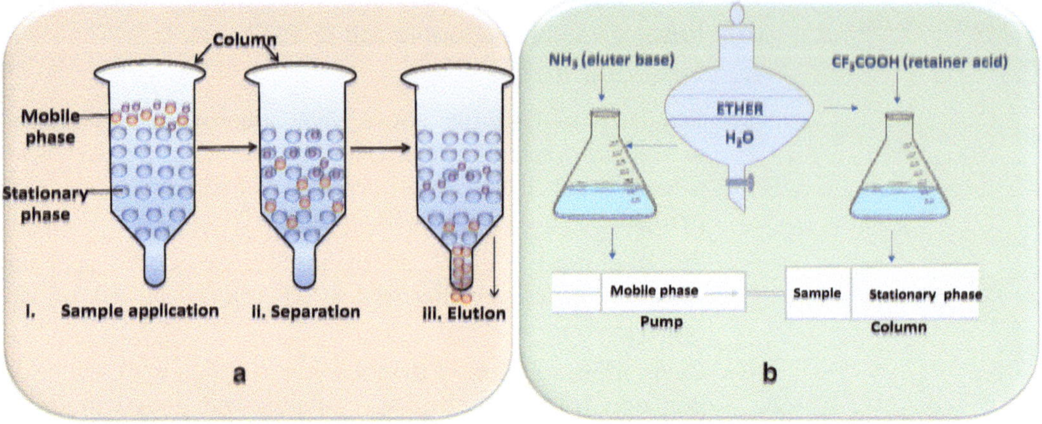

Fig. 2 Schematic illustration of: (**a**) Column chromatography, and (**b**) pH zone refining counter current chromatography. (Adapted from Ito (2012))

Silica gel's surface contains hydroxyl groups that act as active sites. Hydroxyl groups generate stronger hydrogen bonds with different substances. Stronger hydrogen bonds lead to prolonged retention of compounds on silica gel. Silica gel exhibits significant absorption of polar molecules with carboxylic acid, amine, or amide functionalities. Non-polar chemicals have minimal to no retention and are only weakly absorbed because they lack the hydrogen bonding sites or functional groups, i.e., polar [38]. The general silica gel chromatography protocol, as described by Reid and Sarker, can be summarized as follows [39]:

Preparation of Column

1. Pick a glass column with thick walls that is roughly three times as large as the amount of silica gel required for separation.
2. For easy separation, utilize a proportion of 100 g of silica gel/1 g of crude sample. A proportion of as high as 500 g silica gel/1 g of raw material is utilized for more challenging separations. Bed height, i.e., 20–30 cm with a headspace of 40–60 cm, is considered ideal to accommodate the mobility phase.
3. To assist the stationary phase, place a piece of glass wool or a sintered glass frit at the column's bottom. Transfer the stationary phase into the column while vibrating or tapping, followed by opening the outlet.
4. Allow the liquid flow while allowing the stationary portion to settle.
5. Add the stationary phase continuously till the bed is finished.

Equilibration of the Column

1. Loading the column with the original mobile phase, without disturbing the bed.
2. Allow the solvent to flow spontaneously through the column exit using gravity.
3. Continue the equilibration until the stationary phase of the bed is homogenous, without any dry areas or air pockets.
4. Reduce the solvent's quantity to slightly beyond the stationary phase.
5. Stop the flow by closing the exit valve.

Sample Application

1. Allow at least a portion of the mobile phase to dissolve the sample. If D is the column's diameter [cm], then the volume [measured in mL] shouldn't be more than $0.4 \times D^2$.
2. Apply the sample solution with caution using a Pasteur pipette, being cautious without disrupting the bed.
3. To allow the sample to enter the bed, partially release the outflow valves. Following full adsorption of the sample, the sides of the glass columns are gently cleaned by applying a tiny quantity of mobile phase and letting it flow again.
4. To make sure the sample has completely adhered to the bed, this step can be repeated.
5. Fill the column with the original mobile phase required for separation.

Elution and Fraction Collection of Samples

1. Let the original mobility phase run.
2. Gather the amounts of fractions. Fractions can be gathered either automatically or manually. For complex separations, modest fractions [5–10 mL/100 mL column] are recommended. Fractions of greater quantities can be gathered for particular solvent polarities in crude separations.
3. Load the new polar mobile phase and further collect the sample as described before.
4. Repeat with different polarity solvents.

3.1.3 Size-Exclusion Chromatography (SEC)

The size-exclusion chromatography (SEC) or gel permeation chromatography (GPC) is another chromatographic method that separates molecules in solution based on their size when organic solvents are utilized. The non-adsorbing porous particles utilized as stationary phases have pores that are around the same size as the molecules. Beads with different pore diameters are made from a variety of materials. Large molecules cannot easily enter the matrix, whereas intermediate-sized molecules can only enter a fraction of it. Consequently, molecules elute from the column in decreasing order of size, with the smallest molecules eluting last and the larger molecules eluting first. This kind of column has a high sample recovery rate since the packing ingredients are inert [39]. The general protocol for size exclusion chromatography, as described by Reid and Sarker, can be summarized as follows [39]:

Preparation of Column

1. Select a glass column that is between 40 and 100 cm long and has a diameter of 15–25 mm, ideally composed of sintered glass frit. The necessary bed volume determines the column's size. In practical terms, 1 mg of extract must be separated using 1 mL of swelling gel.
2. Suspend the stationary phase in methanol to cause the gel [Sephadex® LH-20] to swell overnight. To obtain a pourable slurry, use enough methanol. Make sure the gel has ample time to fully swell.
3. One full pouring session should be used to finish the packing process. Slowly pour gel into the column after partially opening the exit valve.
4. If required, additional nonpolar solvents like DCM [Dichloromethane] or chloroform [$CHCl_3$] can be employed, particularly when working with extracts that are nonpolar or medium polarity.

Equilibration of Column

1. To give the packing bed time to settle, change the flow rate from 1 to 5 mL/min. It could take several hours.
2. After finishing, let the methanol level fall until it is slightly higher than the packing material.

Sample Application

1. Verify that the sample volume does not exceed 5% of column volume before dissolving it in methanol. The viscosity of the sample solution and the mobile phase should be comparable.
2. Apply the sample solution gently with a Pasteur pipette.
3. To enable the sample to enter the bed, open the outflow valve. It's important to take care not to disrupt the bed or let it get too dry.

4. Fill the bed with an additional 10 mL of methanol, then flow through to confirm that the bed has been completely cleaned of the sample solution.

Elution and the Collection of Samples

1. Add methanol to the column and connect a pumping system or solvent reservoir.
2. Open a column outlet and adjust the flow to 2–4 mL/min. A contact duration of roughly 50–100 min should be provided by the flow rate.
3. Gather the fractions, usually around 5 mL per fraction.

3.1.4 Counter-Current Chromatography [CCC]

CCC, i.e., type of liquid-liquid partition chromatography, involves two non-miscible liquids. In the absence of an adsorption matrix, one phase is retained as a stationary phase. A hydrodynamic or turbulent mixture is used to efficiently equilibrate a second phase upon completion of the stationary phase. Unlike conventional chromatographic systems, this approach uses gravity or a centrifugal force for holding a stationary phase within the column, eliminating the need for solid support [3, 38]. The pH-zone refining centrifugal partition chromatography is one of the CCC techniques that have been developed for isolation. Additionally, a new technique using a deep eutectic solvent with high-speed CCC has been developed to use a more sustainable solvent for isolation [5].

pH-Zone Refining Counter-Current Chromatography

A pH-zone refinement is a large-scale process used to separate ionizable analytes as depicted in Fig. 2b. The approach elutes extremely rigorous rectangular peaks joined together with minimal overlap. In contrast, contaminants are more concentrated and eluted between the major peaks according to their hydrophobicity and pKa content. The method has a ten-fold higher sample loading capacity than traditional HSCCC in a similar separation column [40].

The protocol for pH-zone refining counter-current chromatography, as reported by it, can be summarized as below [41]:

Retainer and Eluter Selection

1. Organic acids, mainly trifluoroacetic acid [TFA], are the highly employed retainers for acidic analytes, while carbon-based bases like triethylamine [TEA] are utilized for basic analytes. To maintain extremely polar molecules like sulfonic acid dyes and sulfuric acid is combined with a basic ligand [a hydrophobic counterion] like dodecyl amine in the upper organic phase. Depending on the acidity of the analyte,

inorganic bases such as ammonia, sodium carbonate, and sodium hydroxide are the eluters for acidic analytes.

2. Among them, ammonia [i.e., NH_3] is highly employed in carboxylic acid isolation, while sodium carbonate [i.e., Na_2CO_3] and sodium hydroxide [i.e., NaOH] are used for polyphenols. The alkaline analytes are eluted with inorganic acids i.e., hydrochloric acid [HCl].
3. Typically, equimolar doses of retainer and eluter [10–20 mM each] are employed for separation.

Two-Phase Solvent Selection

Choosing an appropriate two-phase based solvent system for the pH zone-refining CCC is more constrained compared to the traditional HSCCC approach. Solvent selection for acidic or basic samples is often confined to two-phase systems like [Methyl tert.-butyl ether, i.e., MTBE/water, and 1-butanol/water for hygroscopic materials. An efficient approach for selecting a two-phase solvent solution is described below:

1. Choose a two-phase solvent system with a K value near one for the target compound. This will result in a sharp retainer border that moves similarly to the $K = 1$ compound.
2. Use a small amount of material in a test tube to equalize the two separate phases (2 mL each).
3. Create a hydrodynamic condition on the left side of each sharp retainer border by adding the proper eluter to the testing tube. Add NH_3 to the test tube to bring the pH up to greater than 10 in case the sample contains acid. In case the material being tested is a base, add HCl to the test tube until the pH is less than 2.
4. Next, find the K values. Proceed to the next step if K is significantly less than one.
5. Place the retainer within the test tube. As a result, the right end of the steep peak enters a hydrodynamic state.
6. Add TFA to an acidic sample to bring the pH down to less than 2. To raise the pH of a basic sample to more than 10, add TEA.

Determination of K Value

Then determine the compound's K value. This two-phase solvent system is capable of being efficiently utilized for pH-zone refining CCC if K is significantly more than one.

Sample Solution

1. For as long as solubility allows, the 320 mL semi-preparative column may separate up to 5 g of sample, and the final volume

of pure fractions rises with sample size (*see* **Note 6**). The retained amount of the stationary phase within a column can be used to calculate the highest amount of the sample solution.

2. As long as the stationary phase has retention greater than 50%, about 100 mL of the sample solution made up of equivalent quantities of individual phases can be charged without any issues in the previously described separation column.
3. It is recommended to use roughly equal amounts of each phase: A higher organic phase with the retainer and the lower phase without the eluter. If the sample is not completely dissolved, one can either add more solvents or remove the undissolved particles.
4. Measure the sample's pH and K values before injection. To raise the value of K to the required level [i.e., $K > 3$], add a retainer to the sample solution in case the analyte is not sufficiently dispersed into the upper phase. This modification is required when there is a significant amount of salt in the sample, which alters the analyte's pH and K value (*see* **Note** 7).

Elution Curve Monitoring

1. With UV-absorbing analytes, the effluent can be monitored similarly to the standard HSCCC method; the only difference is that the attenuation needs to be changed to account for the effluent's high absorbance.
2. The effluent can be examined for analytes lacking a chromophore using an ELSD [evaporative light scattering detector] or a flow-through pH meter.

3.1.5 Thin Layer Chromatography

The TLC remains widely utilized today, offering several advantages over mass spectrometry-based techniques due to its low-cost equipment, simplicity, and accessibility in most laboratories. First described in 1938, TLC has now replaced paper chromatography as it is highly efficient and does not demand complex sample preparation or derivatization, making it a convenient option. TLC, unlike other chromatography techniques [HPLC], does not require complicated instrumentation; hence it is an easy chromatographic technique offering low cost, minimum sample preparation, and high sample throughput [42]. There has been a wide application of TLC in various fields, including the food industry, for quality control. TLC has a wide range of applications for analyzing numerous substances such as lipids, carbohydrates, amino acids, and natural pigments.

The chromatography technique is based on the principle of mixture separation by capillary action, in which the mobile phase is used to drag up the sample placed in the stationary phase. The solute and mobile phase compete with one another for binding sites on the stationary phase during compound separation [43]. Furthermore, each compound is characterized by Rf [Retention factor], which is determined using the following formula:

$$\mathrm{Rf} = \frac{\text{Compound displacement from start}}{\text{Solvent discplacement fom start}}$$

The Protocol for TLC Is As Follows

1. Initial step includes the application of a stationary phase on plates, allowing them to dry and stabilize, or precoated plates of different thicknesses available commercially.
2. Sorbent activated by drying, followed by sample application at the marked spot as a single spot or streak [1–2 cm] at one end of the plate.
3. TLC plate placed in a separation chamber filled with mobile phase filled with the chamber touching plate below the spot of sample.
4. Solvent ascends up the plate via capillary action, separating the sample (compounds).
5. Further plates were taken out from the chamber, followed by drying of the plates.
6. Separated sample detected/made visible in different methods (UV light, acid treatment, ninhydrine test, iodine vapor treatment).

3.1.6 Vacuum Liquid Chromatography

The term vacuum liquid chromatography (VLC) was first coined by NM Targett, J.P. Kilcoyne, and B. Green. VLC was first employed to separate diterpenes in Australia. Since the separation is done on aluminium oxide or TLC-grade silica gel, and after every fraction, the column is dried, VLC is regarded as a preparative thin layer chromatography (PTLC). In PTLC, plates are first dried and then run again to enhance the process of separation. After properly cleaning the column with methanol and scraping off the band or disintegrated polar material from the top of the adsorbent column, VLC columns thus packed can be used again for the same or a comparable separation. Both small and huge amounts of mixtures can be separated using gradient elution, which is incredibly effective [44].

Adsorbents and Operations

Aluminium oxide or silica gel (silica gel H), which are TLC grade binder-free, are typically utilized in VLC; however, additional adsorbents may also be studied based on their TLC profiles. Packing of VLC, the column is straightforward. As in Column Chromatography (CC), in the sintered funnel, dry adsorbent is placed and then gently tapped with a rubber tube to allow it to settle under gravity. First, the layer can be pressed using a rubber stopper or a glass. To properly pack a tight and consistent column, the adsorbent layer should be further compressed with a stopper after the vacuum has been applied. To check the column's homogeneity, release the vacuum and immediately pour a non-polar solvent (like hexane) on top of the adsorbent before applying the vacuum again. If the solvent doesn't flow through the column evenly, it has to be repacked. It is necessary to vacuum-dry the column before loading samples [45].

Loading of Sample

Slurry-based solid loading can also be accomplished in VLC, although loading of a liquid sample is preferable because it is simpler or quicker. The loading sample can be loaded as a liquid if it is liquid; if it is solid, it has to be dissolved in a slightly polar solvent. Using a dropper, the appropriately dissolved sample ought to be added to the surface of the adsorbent. It is essential to take into account that the column should not have any vacuum when adding samples. The vacuum should be applied once the entire sample has been introduced. The sample is placed onto the adsorbent in this manner as a homogeneous, attractive band. The better the separation, the smaller the band. To remove the polar solvent that was utilized to dissolve the material, the column is placed under vacuum for 5–10 min. After that, the column is operated under vacuum using the appropriate solvent mixes. To avoid solvent loss from evaporation, a moderate vacuum of 20–70 mmHg should be employed. Standard round-bottom flasks with joints can be used to collect fractions. Each fraction is followed by the addition of solvent to the column's top and vacuuming. After every fraction, the column should be sucked dry to prevent channelling [46]. Table 2 contains various isolation techniques used in food samples for purification of polyphenols.

3.2 Characterization Techniques for Polyphenols

3.2.1 NMR

NMR, i.e., Nuclear Magnetic Resonance, is a powerful tool for biologists to study the structure, dynamics, and interactions of biological macromolecules. NMR was initially established in scientific fields like physics, biology, and chemistry; however, at present, it has gained significant importance in various interdisciplinary fields. Before conducting the NMR study, an understanding of key parameters is essential. Standard NMR spectrometers have classes: One has having high field that ranges from 200 to 600 MHz [^{1}H], and another has a low field that typically ranges from 40 to 80 MHz [^{1}H]. Both classes operated on the same principles. High fields employ cryogenically cooled superconducting electromagnets, require huge space for the laboratory, and therefore, are usually costly for both maintenance and purchase, with high-quality spectra. In contrast, low-field spectrometers include permanent magnets, require less space, are less expensive, and portable, and therefore are also referred to as benchtop spectrometers. Further, the detection sensitivity also varies among the spectrometers. For the same amount of volume, better resolution with greater signal intensity is possible with high-field magnets. Also, by reducing the thermal noise by using cryogenically cooled RF coils, the signal-to-noise ratio can be enhanced. The higher sensitivity per unit mass of sample is achieved by using a small probe diameter (mm probe, standard) [47]. In NMR spectroscopy, the peak intensity is directly proportional to the number of atomic nuclei, and hence, it is used to classify plant samples according to species, origin, age, processing, and various other parameters based on the NMR profiles.

Protocol for Characterization by NMR

1. The extraction solvent used for the characterization of polyphenols using NMR involves the preparation of methanol, ethanol, and acetone in a 7:3 ratio (v/v; solvent: water). The prepared extract was then kept in freeze-drier overnight and re-extracted with varied volumes of extraction solvents (5 mL and 2.5 mL), followed by sonication for 15 min. The obtained supernatant was recovered using a centrifuge and dried in a centrifugal evaporator using nitrogen. Similarly, for ethyl acetate extract, 0.5 mL standard mixture in a 1:1 and 1:5 ratio was mixed with a solvent (v/v), standard mix: solvent, and sonicated for 15 min. Further, the mixture is shaken up to 2 h to obtain the organic layer and aqueous layer, and dried by using a centrifugal evaporator with nitrogen [36]. 2 mL of methanol was run through C18 solid phase extraction under vacuum for the activation. Under similar conditions, 2 mL of distilled water was then run for the equilibration of the column. 1 mL of a standard mixture having a pH of 7 was loaded on the cartridge. Nitrogen gas was used to elute the non-retained components. The column was washed with 4 mL methanol: water [10:90, v/v], subsequently followed by 4 mL ethyl acetate. The column was then acidified with 2 mL of 0.01 M HCl. For polyphenols like anthocyanins, catechins, and flavonols, ethyl acetate was used for elution. To remove the polymeric fraction, 4 mL of solvent containing acetone and water in a 66: 34, v/v ratio was used in the column. The final extracts obtained were subjected to a centrifugal evaporator for drying by removing the extraction solvent. The dried extract was then reconstituted in potassium phosphate buffer having 0.1 M concentration and pH 3.4 for NMR spectroscopy analysis [36].
2. ^{1}H NMR spectra were obtained using NMR spectroscopy with a 400.13 MHz 1H resonance frequency. The internal probe temperature for all the spectra was 298 K. Each sample of 0.6 mL was mixed with 0.2 mL distilled water containing the

primary reference standard (5.8 mM sodium trimethylsilyl propionate) and placed in NMR tubes. By using pre-saturation solvent, the larger water signal in the spectra is suppressed with an 11.5 s relaxation delay. Typically, 128 transients with a spectral width of 6000 Hz were collected into 32 K data points. The baseline and manual spectra were collected, and chemical shifts were referenced to at d 0.00. Before the Fourier transform, the broadening of the exponential of 0.3 Hz was applied to FIDs and zero-filled by a factor 2. The quantification of the recovered compound is evaluated by ^{1}H NMR peak integration. The resonances obtained were integrated with resonances obtained from an internal standard [36]. The following equation is used to calculate the amount of each compound [mmol/L]:

$$Cc = Ic/Is \times Ns/Nc \times Cs$$

c = Compound; s = Standard; I = Peak area integral; N = No. of protons; C = Concentration

3. This calculation was carried out before and after extraction using 1H NMR spectra. The performance of the NMR-based method is validated by using repeatability, recovery, limit of quantification, and limit of detection are based on the standard mixture quantification. Recovery is obtained by calculating the mean standard deviation from the peak integration of peaks. Under the experimental conditions, the LODs were calculated using a 3× level of noise, and the LOQs were calculated by using a 10× level of noise criteria [36].

3.2.2 Gas Chromatography

Gas chromatography (GC) is widely used in the characterization of polyphenols owing to their aromatic nature. This technique is the most efficient and reliable separation tool for multiple-component mixtures like hydrocarbons, essential oils, polyphenols, etc. The instrumentation of GC includes the flame ionization detector and an electron capture detector, providing high sensitivity when the sample volume is low [48]. The application of GC has expanded widely; therefore, various evolutions have been made to invent more powerful instruments along with the data analysis approaches needed for the study of complex samples. Therefore, GC has now evolved into one-dimensional (1D) GC and multidimensional GC. The identification of the analyte is made by adding multivariant detectors like time-of-flight mass spectrometry (TOFMS). The modulator is another part of GC instrumentation, which is known as the heart of GC × GC, to facilitate comprehensive separations. These modulators are classified in three categories, i.e., thermal, valve-based, and flow. Further, many other developments have been made in detectors like the introduction of high-resolution time of flight mass spectroscopy (HRTOFMS), which provides high chemical selectivity relative to MS with unit mass resolution. However, there is still a need to tackle the obstacles like varied retention time shifting and variation in the peak intensities. Therefore, advances in such techniques have been made for their effective utility in both industries and institutional research laboratories to obtain chemical information [49].

Protocol for Characterization by GC

1. The extraction solvent must be exchanged before gas chromatographic analysis with 2-propanol, with a compatible detector. For exchange, the extract is concentrated to 1 mL by using a macro-Snyder column and allowed to cool and drain the apparatus for a minimum of 10 min. Afterward, the micro-Snyder column is removed and its lower joint is rinsed with a small volume of 2-propanol into a concentrator tube. The extract volume is then adjusted to 1 mL. Prepare the calibration standards: Operating conditions for the GC were established for the single-column and dual-column approaches. Afterwards, the instrumental conditions were optimized for better resolution and sensitivity of target analytes (*see* **Note 8**). Further, each calibration standard of 2 μL volume is used to inject (*see* **Note 9**).
2. Gas chromatographic analysis of sample extracts: Inject a 2-μL aliquot of the concentrated sample extract. Record the volume injected to the nearest 0.05 μL and the resulting peak size in area units. The same GC operating conditions used for the initial calibration must be employed for sample analyses. A calibration verification standard is injected before analyzing the sample to verify the calibration. Sample injected until both calibration verification as well as standard interspersed with sample meet quality control requirements of instruments (*see* **Note 10**). For each analyte, the quantifiable calibration factor must not be above a ±15% difference compared to the initial calibration curve (*see* **Note 11**). The retention time for each analyte in the calibration standard is compared with the absolute retention time windows. The retention time in the mid-concentration standard analyzed during initial calibration is the centre of the absolute retention time for each analyte. Both retention times are compared (*see* **Note 12**). Tentative identification of an analyte occurs when a peak from a sample extract falls within the absolute retention time window. Each tentative identification must be confirmed using either a second GC column of dissimilar stationary phase or using another technique such as GC/MS (*see* **Note 13**).
3. The results are calculated from either internal or external calibration. The derivatization or underivatization of phenols is

applied to the analysis for both the standard. For the selection of an appropriate base line, proper quantitation is important to determine the peak height and area. Dilution of the extract is applied when the responses exceed the calibration range and samples are re-analyzed. Peak height measurements are preferred over peak area integration as overlapping peaks cause error in area integration (*see* **Note 14**).

3.2.3 Liquid Chromatography Coupled with Mass Spectroscopy [LC-MS]

LC-MS, i.e., liquid chromatography coupled with mass spectroscopy or tandem spectrometry [LC-MS/MS], is also an effective tool for high and low molecular weight polyphenol characterization in food samples. Recently, ultra-high-performance LC has been used, having sub-2 μm particle-packed columns or porous shell columns having sub-3 μm porous particles, which provide up to 5–10 times faster separation of polyphenols from complex food matrix compared to conventional LC [50].

Protocol for Characterization by LC-MS

1. Bursal et al. used a 25 g dried plant sample, milled, and 100 mL of distilled water was added to obtain the water extract [37]. The extract was then boiled using a magnetic stirrer for 20 min, followed by filtration and freezing of the filtrate. The frozen filtrate was lyophilized under 5 μm Hg pressure at −50 ° C. For the preparation of ethanolic extract, 25 g of dried samples were ground in a mill and mixed with 100 mL of ethanol using a magnetic stirrer for 1 h. The extract was filtered, and ethanol was removed by rotary evaporator at 50 °C. All the extracts were stored at −20 °C. 100 mg of both water and ethanolic extracts were then dissolved in 5 mL of 50:50 (v/v) water: ethanol solvent, and 1 mL of which was then transferred to a 5 mL volumetric flask. In another flask, 100 μL of curcumin was added in a solvent prepared from ethanol and water (5:50, v/v). 1.5 mL aliquot was then transferred to an autosampler vial and a 10 μL sample injected into the LC. During the entire experiment, the samples in the autosampler were kept at 15 °C. C18 gravity column with 5 μm particle size, equipped with Tandem Gold Triple quadrupole mass spectroscopy, is used for the experiment. For mobile phase, methanol [A, 0.5% formic acid] in water [B, 0.5% formic acid] with 0 to 1.00 min program gradient and 50% A and 50% B with 0.3 mL/min flow rate at 30 °C column temperature and 10 μL kept as injection volume. For the instrument analysis, triple quadruple mass spectroscopy was selected, and optimum ESI parameters were 600 V shield voltage, 5000 V needle voltage, and 2.40 mTorr CID gas pressure, respectively. Further, 300 °C and 50 °C were kept as drying gas and API housing temperatures, respectively. The gas pressure for the

nebulizer is 55 psi, and 40 psi for the drying gas pressure. The internal standard used for the experiment is curcumin. For validation, recovery, repeatability, LOD, and LOQ were done for the experiments. Experiments were repeated to measure the precision of each compound [37].

4 Notes

1. While using water or methanol [mobile phase] for silica gel [stationary phase], the mobile phase's pH needs to be less than 7.0 when silica gel is used with water. Silica gel dissolves easily at any pH higher than 7.0.
2. Chlorodimethylalkylsilanes or chloroalkoxysilanes of varying chain lengths are applied to silica gel to create the nonpolar [reversed phase] stationary phase. Although most chromatographers prefer C_8 and C_{18}, the integrated alkyl chain length may contain C_2, C_4, C_6, C_8, or C_{18}. Because monochlorosilanes guarantee monomolecular surface coverage, the packing material performance is significantly more reproducible. Nitro [NO_2], Cyanopropyl [CN], diol, and aminopropyl [NH_2] are among the stationary phases that are currently accessible.
3. In case where the purity of the compound is greater than 96%, the weight is used without applying a correction to the calculation for the standard stock solution.
4. For the dilution of standards, the solvents used should be the same as the final solvent mixture in the sample extract used for analysis. The analyte must be divided into two or more calibration standards for determining the analyte. Further, the analyst must consider GC column conditions to be used in laboratory, and form of phenols i.e., underivatized, derivatized, methylated etc. and then prepare the mixtures as applicable to those circumstances.
5. Surrogates are mixed with all the blanks used in methods, samples, calibration standards, and matrix spikes.
6. It is advised to dilute the sample even further if the sample solution takes a long time to settle.
7. Measure the sample solution's pH regularly before putting it in the column.
8. The same operating conditions are used for both sample and calibration analyses.
9. Different injection volumes may be used when the analyst can demonstrate adequate sensitivity for the target compound.
10. After every 10 samples, analyze the calibration verification standards. In case all sample standards fail the QC limit,

samples are processed only after re-injection to reduce the unnecessary repeat analysis.

11. In case where the criterion is exceeded, check the GC system and perform maintenance before verifying calibration and sample analysis. Further, if the maintenance does not meet the QC requirements based on last calibration, it is recommended to initiate new calibration.
12. It is important that each analyte in the standard must fall within its retention time window. If not, GC must be adjusted so that result of all the analyte falling in their respective retention time window or establish the new retention time window.
13. Additional confirmation is not required while using the dual-column technique; however, the analyte must meet the identification criteria for both columns.
14. It is recommended to change the column if partially overlapping peaks are found.

References

1. Kontogianni VG (2014) Novel techniques towards the identification of different classes of polyphenols. In: Polyphenols in plants. Academic Press, London, pp 159–185
2. Ignat I, Volf I, Popa VI (2011) A critical review of methods for characterisation of polyphenolic compounds in fruits and vegetables. Food Chem 126(4):1821–1835
3. Plaza M, Domínguez-Rodríguez G, Castro-Puyana M, Marina ML (2018) Polyphenols analysis and related challenges. In: Polyphenols: properties, recovery, and applications. Woodhead Publishing, pp 177–232. https://doi.org/10.1016/B978-0-12-813572-3.00006-3
4. Gorzynik-Debicka M, Przychodzen P, Cappello F, Kuban-Jankowska A, Marino Gammazza A, Knap N et al (2018) Potential health benefits of olive oil and plant polyphenols. Int J Mol Sci 19(3):686
5. Khalaf RA, Alhusban AA, Al-Shalabi E, Al-Sheikh I, Sabbah DA (2019) Isolation and structure elucidation of bioactive polyphenols. Stud Nat Prod Chem 63:267–337
6. Davis AL, Cai Y, Davies AP, Lewis JR (1996) 1H and 13C NMR assignments of some green tea polyphenols. Magn Reson Chem 34(11): 887–890
7. Adoui N, Souilah N, Bendif H, Sut S, Dall'Acqua S, Flamini G et al (2023) Characterization of polyphenols and volatile compounds from understudied Algerian *Pallenis spinosa* by HS-SPME-GC-MS, NMR and HPLC-MSn approaches. Appl Sci 13(18): 10113
8. Del Campo G, Santos JI, Iturriza N, Berregi I, Munduate A (2006) Use of the 1H nuclear magnetic resonance spectra signals from polyphenols and acids for chemometric characterization of cider apple juices. J Agric Food Chem 54(8):3095–3100
9. Krafczyk N, Glomb MA (2008) Characterization of phenolic compounds in rooibos tea. J Agric Food Chem 56(9):3368–3376
10. Cocconi E, Stingone C, Zanotti A, Trifiro A (2016) Characterization of polyphenols in apricot and peach purees by UHPLC coupled to HRMS Q-Exactive™ mass spectrometer: an approach in the identification of adulterations. J Mass Spectrom 51(9):742–749
11. Sirisena S, Zabaras D, Ng K, Ajlouni S (2017) Characterization of date [*Deglet Nour*] seed free and bound polyphenols by high-performance liquid chromatography-mass spectrometry. J Food Sci 82(2):333–340
12. Abid M, Yaich H, Cheikhrouhou S, Khemakhem I, Bouaziz M, Attia H, Ayadi MA (2017) Antioxidant properties and phenolic profile characterization by LC–MS/MS of selected Tunisian pomegranate peels. J Food Sci Technol 54:2890–2901
13. Gülçin I, Kaya R, Gören AC, Akincioglu H, Topal M, Bingol Z et al (2019) Anticholinergic, antidiabetic and antioxidant activities of cinnamon (*Cinnamomum verum*) bark extracts: polyphenol contents analysis by

LC-MS/MS. Int J Food Prop 22(1): 1511–1526

14. Hajji T, Mansouri S, Vecino-Bello X, Cruz-Freire JM, Rezgui S, Ferchichi A (2018) Identification and characterization of phenolic compounds extracted from barley husks by LC-MS and antioxidant activity in vitro. J Cereal Sci 81: 83–90
15. Sarker SD, Nahar L (2012) An introduction to natural products isolation. Humana Press, New Jersey, pp 1–25
16. Santos-Buelga C, Gonzalez-Manzano S, Dueñas M, Gonzalez-Paramas AM (2012) Extraction and isolation of phenolic compounds. Nat Prod Isol 864:427–464
17. Hasan A, Sadiq A, Abbas A, Mughal E, Khan KM, Ali M (2010) Isolation and synthesis of flavonols and comparison of their antioxidant activity. Nat Prod Res 24(11):995–1003
18. Liang HR, Vuorela P, Vuorela H, Hiltunen R (1997) Isolation and immunomodulating effect of flavonol glycosides from Epimedium hunanense. Planta Med 63(4):316–319
19. Utari F, Itam A, Syafrizayanti S, Putri WH, Ninomiya M, Koketsu M, Tanaka K, Efdi M (2019) Isolation of flavonolrhamnosides from Pometia pinnata leaves and investigation of α-glucosidase inhibitory activity of flavonol derivatives. J Appl Pharm Sci 9(8):53–65
20. Yin Y, Gong FY, Wu XX, Sun Y, Li YH, Chen T, Xu Q (2008) Anti-inflammatory and immunosuppressive effect of flavones isolated from Artemisia vestita. J Ethnopharmacol 120(1):1–6
21. Demirtas I, Erenler R, Elmastas M, Goktasoglu A (2013) Studies on the antioxidant potential of flavones of Allium vineale isolated from its water-soluble fraction. Food Chem 136(1): 34–40
22. Abu-Niaaj L, Katampe I (2018) Isolation and characterization of flavones from Artemisia monosperma. Pharm J 10(5):1018–1023
23. Lu JJ, Wei Y, Yuan QP (2007) Preparative separation of gallic acid from Chinese traditional medicine by high-speed counter-current chromatography and followed by preparative liquid chromatography. Sep Purif Technol 55(1):40–43
24. He D, Gu D, Huang Y, Ayupbek A, Yang Y, Aisa HA, Ito Y (2009) Separation and purification of phenolic acids and myricetin from black currant by high-speed countercurrent chromatography. J Liq Chromatogr Relat Technol 32(20):3077–3088
25. Seo DJ, Lee HB, Kim IS, Kim KY, Park RD, Jung WJ (2013) Antifungal activity of gallic acid purified from *Terminalia nigrovenulosa* bark against *Fusarium solani*. Microb Pathog 56:8–15
26. Wei Y, Gao Y, Zhang K, Ito Y (2010) Isolation of caffeic acid from *Eupatorium adenophorum* Spreng by high-speed countercurrent chromatography and synthesis of caffeic acid-intercalated layered double hydroxide. J Liq Chromatogr Relat Technol 33(6):837–845
27. Karthikeyan R, Devadasu C, Srinivasa Babu P (2015) Isolation, characterization, and RP-HPLC estimation of p-coumaric acid from methanolic extract of durva grass (*Cynodondactylon* Linn.) (Pers.). Int J Anal Chem 2015: 201386
28. Fang YT, Li Q, Cao AC, Li Y, Wei Y (2017) Isolation and purification of phenolic acids from sugarcane (*Saccharum officinarum* L.) rinds by pH-zone-refining counter-current chromatography and their antioxidant activity evaluation. Food Anal Methods 10(7): 2576–2584
29. Kartika IGA, Bang IJ, Riani C, Insanu M, Kwak JH, Chung KH, Adnyana IK (2020) Isolation and characterization of phenylpropanoid and lignan compounds from *Peperomia pellucida* [L.] Kunth with estrogenic activities. Molecules 25(21):4914
30. Li F, Zhang T, Sun H, Gu H, Wang H, Su X et al (2017) A new nortriterpenoid, a sesquiterpene and hepatoprotective lignans isolated from the fruit of *Schisandra chinensis*. Molecules 22(11):1931
31. Sáez V, Pastene E, Vergara C, Mardones C, Hermosín-Gutiérrez I, Gómez-Alonso S et al (2018) Oligostilbenoids in *Vitis vinifera* L. Pinot Noir grape cane extract: isolation, characterization, in vitro antioxidant capacity and anti-proliferative effect on cancer cells. Food Chem 265:101–110
32. Smeriglio A, Barreca D, Bellocco E, Trombetta D (2017) Proanthocyanidins and hydrolysable tannins: occurrence, dietary intake and pharmacological effects. Br J Pharmacol 174(11): 1244–1262
33. Pfundstein B, El Desouky SK, Hull WE, Haubner R, Erben G, Owen RW (2010) Polyphenolic compounds in the fruits of Egyptian medicinal plants (*Terminalia bellerica*, *Terminalia chebula* and *Terminalia horrida*): characterization, quantitation and determination of antioxidant capacities. Phytochemistry 71(10):1132–1148
34. Yuliana P, Laconi EB, Wina E, Jayanegara A (2014) Extraction of tannins and saponins from plant sources and their effects on in vitro methanogenesis and rumen fermentation. J Indones Trop Anim Agric 39(2):91–97

35. Fraga-Corral M, García-Oliveira P, Pereira AG, Lourenço-Lopes C, Jimenez-Lopez C, Prieto MA, Simal-Gandara J (2020) Technological application of tannin-based extracts. Molecules 25(3):614
36. Savage AK, van Duynhoven JP, Tucker G, Daykin CA (2011) Enhanced NMR-based profiling of polyphenols in commercially available grape juices using solid-phase extraction. Magn Reson Chem 49(S1):S27–S36
37. Bursal E, Köksal E, Gülçin İ, Bilsel G, Gören AC (2013) Antioxidant activity and polyphenol content of cherry stem (Cerasus avium L.) determined by LC–MS/MS. Food Res Int 51(1):66–74
38. Susanti I, Pratiwi R, Rosandi Y, Hasanah AN (2024) Separation methods of phenolic compounds from plant extract as antioxidant agents candidate. Plants 13(7):965
39. Reid RG, Sarker SD (2012) Isolation of natural products by low-pressure column chromatography. Nat Prod Isol 864:155–187
40. Ito Y (2005) Golden rules and pitfalls in selecting optimum conditions for high-speed counter-current chromatography. J Chromatogr A 1065(2):145–168
41. Ito Y (2013) pH-zone-refining counter-current chromatography: origin, mechanism, procedure and applications. J Chromatogr A 1271(1):71–85
42. Nielsen SS (ed) (2010) Food analysis. Springer, New York
43. Bele AA, Khale A (2011) An overview on thin layer chromatography. Int J Pharm Sci Res 2(2):256
44. Maurya A, Kalani K, Verma SC, Singh R, Srivastava A (2018) Vacuum liquid chromatography: simple, efficient and versatile separation technique for natural products. Org Med Chem Int J 7:1–3
45. Coll JC, Bowden BF (1986) The application of vacuum liquid chromatography to the separation of terpene mixtures. J Nat Prod 49(5):934–936
46. Bucar F, Wube A, Schmid M (2013) Natural product isolation – how to get from biological material to pure compounds. Nat Prod Rep 30(4):525–545
47. Giberson J, Scicluna J, Legge N, Longstaffe J (2021) Developments in benchtop NMR spectroscopy 2015–2020. In: Annual reports on NMR spectroscopy, vol 102. Academic Press, pp 153–246. https://doi.org/10.1016/bs.arnmr.2020.10.006
48. Al-Rubaye AF, Hameed IH, Kadhim MJ (2017) A review: uses of gas chromatography-mass spectrometry (GC-MS) technique for analysis of bioactive natural compounds of some plants. Int J Toxicol Pharmacol Res 9(1):81–85
49. Prebihalo SE, Berrier KL, Freye CE, Bahaghighat HD, Moore NR, Pinkerton DK, Synovec RE (2018) Multidimensional gas chromatography: advances in instrumentation, chemometrics, and applications. Anal Chem 90(1):505–532
50. Lucci P, Saurina J, Núñez O (2017) Trends in LC-MS and LC-HRMS analysis and characterization of polyphenols in food. TrAC Trends Anal Chem 88:1–24

Chapter 10

Isolation and Characterization of Carotenoids

Pragya Pandey, Nidhi Dangi, Prixit Guleria, Pratibha Rathee, and Amit Lather

Abstract

This chapter presents protocols for the purification and analytical characterization of major dietary carotenoids such as lycopene, β-carotene, lutein, zeaxanthin, and β-cryptoxanthin from plant and microbial sources. Methods include flash chromatography, column chromatography, and TLC for purification, followed by TLC, UV-Visible spectrophotometry, and HPLC for compound identification and quantification. Purification methods include column chromatography, flash chromatography, and TLC. Characterization is performed using TLC, UV-Visible spectrophotometry, and HPLC. Emphasis is placed on minimizing pigment degradation and ensuring analytical suitability for research and quality control.

Key words Carotenoids, Lycopene, β-Carotene, Lutein, Zeaxanthin, β-Cryptoxanthin, Flash chromatography, Thin-layer chromatography, HPLC, UV-visible spectrophotometry, Purification, Analytical methods

1 Introduction

Carotenoids are naturally occurring pigments that are widely distributed in plants, algae, and some microorganisms. These compounds are responsible for the vibrant yellow, orange, and red coloration seen in many fruits and vegetables. Chemically, carotenoids are tetraterpenoids, composed of eight isoprene units with a long conjugated double-bond system. This conjugated structure gives them their distinct color and contributes to their antioxidant activity. Carotenoids are broadly classified into carotenes, which include α-carotene, β-carotene, and lycopene, and xanthophylls, such as lutein, zeaxanthin, and β-cryptoxanthin [1, 2].

In photosynthetic organisms, carotenoids perform essential roles including light harvesting, photoprotection, and stabilization of pigment–protein complexes. In human health, their significance is widely recognized. Carotenoids such as β-carotene and β-cryptoxanthin act as provitamin A, contributing to vision,

Ashwani Kumar, Vidisha Tomer and Piyoosh Kumar Babele (eds.), *Phytochemicals: Extraction, Quantification, Isolation, and Characterization*, Methods and Protocols in Food Science, https://doi.org/10.1007/978-1-0716-4840-7_10,

immune function, and cell growth [1, 2]. Lutein and zeaxanthin are primarily accumulated in the retina and are believed to protect against age-related macular degeneration [2, 3]. These health-promoting effects have led to increasing use of carotenoids as bioactive ingredients in the food, nutraceutical, pharmaceutical, and cosmetic industries.

Carotenoids are often embedded within complex biological matrices, localized in organelles such as chromoplasts or associated with lipids and proteins. Their efficient recovery from these matrices is challenging due to their sensitivity to oxidation, light, and elevated temperatures. Natural sources such as tomatoes, carrots, algal biomass, and microbial cultures are widely explored for carotenoid extraction. Each matrix presents specific challenges and requires tailored handling protocols to ensure good yield and compound stability [4].

While extraction methods are discussed in detail in Chapter 2, it is important to note that proper extraction forms the foundation for subsequent purification steps. Carotenoids must be isolated in a form that retains their structural integrity and functionality. This requires optimized protocols that minimize degradation during handling and separation. Factors such as the polarity, molecular structure, and physical state of the target carotenoid must be considered when selecting appropriate procedures [5].

Conventional extraction techniques using organic solvents remain common, but greener alternatives are increasingly favored. Methods such as ultrasound-assisted extraction (UAE), supercritical fluid extraction (SCFE), and deep eutectic solvent (DES)-based extraction have demonstrated improvements in extraction efficiency and compound recovery [5, 6]. Once carotenoids are extracted, they must be purified to separate them from lipids, chlorophylls, and other co-extracted compounds. Purity is critical when carotenoids are used as analytical standards or functional ingredients.

Purification techniques include silica gel column chromatography, flash chromatography, preparative high-performance liquid chromatography (prep-HPLC), and centrifugal partition chromatography (CPC). Each technique offers specific advantages depending on the nature of the carotenoid and the level of purity required. For example, flash chromatography allows efficient recovery of lycopene from tomato extract based on polarity differences. Preparative HPLC is well suited for precise separation and quantification of structurally similar xanthophylls such as lutein and zeaxanthin [6, 7].

Once purified, carotenoids must be characterized to verify their identity and determine purity levels. Analytical methods such as thin-layer chromatography (TLC), UV-Visible spectrophotometry, and high-performance liquid chromatography (HPLC) are commonly used for this purpose. These techniques provide important qualitative and quantitative information and are essential in quality control and research settings [7, 8].

This chapter presents standardized laboratory-scale protocols for the purification and characterization of selected carotenoids.

2 Materials

This section lists the essential apparatus and reagents required for the purification and characterization of carotenoids from various biological matrices. Primary extraction steps are described in Chapter 2.

2.1 Isolation/ Purification Techniques

2.1.1 Lycopene from Tomato [9]

(a) *Apparatus*:
- Rotary evaporator.
- Glass chromatography column (2 × 30 cm).
- Silica gel (60–120 mesh).
- TLC plates and chamber.
- UV cabinet.
- Amber vials.
- Nitrogen gas source.

(b) *Reagents*:
- Petroleum ether.
- Acetone.
- Hexane.
- Lycopene (analytical-grade standard).

2.1.2 β-Carotene from Carrot [10]

(a) *Apparatus*:
- Rotary evaporator.
- Glass chromatography column (2 × 25 cm).
- Silica gel (60–120 mesh).
- TLC plates and chamber.
- UV cabinet.
- Amber vials.
- Nitrogen gas source.

(b) *Reagents*:
- Petroleum ether.
- Acetone.
- Hexane.
- β-Carotene standard (analytical grade).

2.1.3 Lutein from Algal Biomass [11]

(a) *Apparatus*:
- Rotary evaporator.
- Glass chromatography column (1.5 × 20 cm).
- Silica gel (60–120 mesh).
- TLC plates and chamber.
- UV cabinet.
- Amber vials.
- Nitrogen gas source.

(b) *Reagents*:
- Methanol.
- Acetone.
- Hexane.
- Lutein (analytical-grade standard).

2.1.4 β-Cryptoxanthin and Zeaxanthin from Microbial Biomass [12]

(a) *Apparatus*:
- Flash chromatography unit.
- Rotary evaporator.
- Glass chromatography column (1.5 × 20 cm).
- Silica gel (60–120 mesh).
- TLC plates and chamber.
- UV cabinet.
- Amber vials.
- Nitrogen gas source.

(b) *Reagents*:
- Dichloromethane.
- Ethyl acetate.
- Hexane.
- Petroleum ether.
- β-Cryptoxanthin and Zeaxanthin standards.

2.2 Characterization Techniques

2.2.1 Thin-Layer Chromatography (TLC)

(a) *Apparatus*:
- TLC plates (silica gel 60 F254).
- TLC chamber.
- UV cabinet (254 and 366 nm).
- Capillary tubes.
- Micropipettes.

(b) *Reagents*:
- Hexane.
- Acetone.
- Petroleum ether.
- Authentic carotenoid standards (β-carotene, lycopene, lutein, zeaxanthin).

2.2.2 UV–Visible Spectrophotometry

(a) *Apparatus*:
- UV–Vis spectrophotometer.
- Quartz cuvettes (1 cm).
- Amber vials.
- Syringe filters (0.45 μm).

(b) *Reagents*:
- Hexane (non-polar carotenoids).
- Ethanol or acetone (polar carotenoids).
- Standard carotenoid solutions.

2.2.3 High-Performance Liquid Chromatography (HPLC)

(a) *Apparatus*:
- HPLC system with diode-array detector (DAD).
- C18 column (250 × 4.6 mm, 5 μm).
- Syringe filters (0.45 μm).
- Vortex mixer.

(b) *Reagents*:
- Methanol (HPLC grade).
- Acetonitrile.
- Dichloromethane.
- Milli-Q or HPLC-grade water.
- Analytical-grade carotenoid standards.

3 Methods

This section describes laboratory protocols used to purify and analyze carotenoids from crude plant and microbial extracts.

3.1 Isolation/ Purification Techniques

After carotenoids are extracted from natural matrices, the resulting crude extracts often contain a mixture of pigments, lipids, chlorophylls, and other non-target compounds. To isolate individual carotenoids such as lycopene, β-carotene, lutein, or β-cryptoxanthin in analytically or industrially useful purity, further purification is necessary. The choice of purification technique depends on factors such as the chemical structure, polarity, thermal stability, and intended application of the target compound.

Lycopene, a non-polar carotenoid found abundantly in tomatoes, can be selectively purified from crude extracts using silica gel-based flash chromatography. This method allows efficient separation from co-extracted compounds such as chlorophylls and lipids by exploiting lycopene's low polarity and red chromophore. β-Carotene present in carrot tissues is often mixed with other carotenoids, lipids, and polar pigments. The techniques like, column chromatography using silica gel allows its separation based on polarity differences. TLC is typically used for monitoring and confirming the purity of the collected fractions. Lutein, a polar xanthophyll carotenoid is found in many yellow flowers and microalgae. It is commercially isolated from the flowers of *Tagetes erecta* and some microalgae species, where, it typically coexists with β-carotene and chlorophyll derivatives. Because of its polarity and heat sensitivity, its purification requires careful handling with polar solvents and a suitable chromatographic system that allows distinction from non-polar carotenoids. It is generally purified by column chromatography. β-cryptoxanthin and zeaxanthin are polar xanthophyll carotenoids found in citrus fruits, papaya, corn, green leafy vegetables and microbial sources such as *Rhodotorula* spp. These pigments also exist with other carotenoids and typically require both extraction and post-treatment purification steps due to co-extraction with lipids and other intracellular components. Flash chromatography and TLC are commonly used techniques for selective separation and identification. The simple methods like crystallization, and recrystallization can also be used for their purification. More sophisticated options, such as preparative high-performance liquid chromatography (prep-HPLC) and centrifugal partition chromatography (CPC), are used when higher resolution or compound recovery is required [13, 14].

In this section, selected methods are presented for purification of carotenoids from crude extracts, aligned with the upstream extraction techniques already described.

3.1.1 Purification of Lycopene from Tomato Extract [9]

1. Dissolve the crude tomato extract in 5 mL of petroleum ether.
2. Pack a glass chromatography column (2 × 30 cm) with silica gel (60–120 mesh) using hexane as the slurry solvent.
3. Load the sample slowly on top of the column.
4. Elute with petroleum ether: acetone (95:5, v/v) at a controlled flow rate, collecting 10 mL fractions.
5. Monitor fractions using TLC (hexane: acetone, 9:1 v/v) under UV or visible light.
6. Pool fractions containing lycopene, as confirmed by Rf comparison with standard and its intense red color.

7. Concentrate the pooled fractions using a rotary evaporator at 35 °C under reduced pressure (*see* **Note 1**).
8. Store the purified pigment in amber vials flushed with nitrogen at −20 °C.

3.1.2 Purification of β-Carotene from Carrot Extract [10]

1. Dissolve the crude extract in 5 mL of petroleum ether (*see* **Note 2**).
2. Pack a chromatography column (2 × 25 cm) with silica gel (60–120 mesh), pre-equilibrated with hexane.
3. Carefully apply the sample to the column without disturbing the bed.
4. Elute using hexane: acetone (90:10, v/v) and collect 10 ml fractions.
5. Monitor collected fractions using TLC (hexane: acetone, 85: 15, v/v). Visualize under visible light and compare with a β-carotene standard.
6. Pool β-carotene-positive fractions and concentrate using a rotary evaporator at ≤40 °C (*see* **Note 3**).
7. Store the purified pigment in amber vials under nitrogen at −20 °C.

3.1.3 Purification of Lutein from Algal Biomass [11]

1. Dissolve the crude algal extract in 5 mL of chilled methanol (*see* **Note 4**).
2. Prepare a glass column (1.5 × 20 cm) packed with silica gel (60–120 mesh), pre-conditioned with hexane: acetone (7:3, v/v).
3. Load the sample on top of the column and elute with hexane: acetone (7:3, v/v) at a steady flow rate.
4. Collect 5 mL fractions and analyze by TLC using silica plates and the same solvent system. Compare with lutein standard under visible or UV light.
5. Pool lutein-positive fractions and evaporate at ≤35 °C using a rotary evaporator (*see* **Note 5**).
6. Transfer to amber glass vials and store at −20 °C under nitrogen.

3.1.4 Purification of β-Cryptoxanthin and Zeaxanthin from Microbial Biomass [12]

1. Dissolve the dried microbial extract in 5 mL of dichloromethane (*see* **Note 6**).
2. Prepare a flash chromatography column (1.5 × 20 cm) packed with silica gel (60–120 mesh), pre-washed with hexane: ethyl acetate (8:2, v/v).
3. Load the extract onto the column and elute with a gradient of hexane: ethyl acetate from 90:10 to 70:30, v/v (*see* **Note 7**).

4. Collect 5 mL fractions and monitor them using TLC (hexane: ethyl acetate, 85:15 v/v). Compare retention factor (Rf) with standards for β-cryptoxanthin and zeaxanthin (*see* **Note 8**).
5. Pool the identified fractions and evaporate under reduced pressure at ≤35 °C (*see* **Note 7**).
6. Store in amber glass vials at −20 °C under nitrogen.

3.2 Characterization Methods

After purification, it is essential to confirm the identity and purity of carotenoids through reliable analytical techniques. These methods provide structural, qualitative, and quantitative insights that help validate the effectiveness of the isolation and purification process. Common techniques include thin-layer chromatography (TLC) for quick pigment profiling, UV-Vis spectrophotometry for concentration estimation, and high-performance liquid chromatography (HPLC) for detailed separation and quantification.

TLC is a simple and rapid technique for the preliminary identification and purity assessment of carotenoids. Different carotenoids show distinct Rf values and color intensities under visible or UV light. It is particularly useful for comparing extracted pigments with reference standards (*see* **Note 9**). UV-Vis spectrophotometry is widely used for estimating carotenoid concentrations due to its simplicity, sensitivity, and non-destructive nature. Each carotenoid has a characteristic absorption maximum (λmax), allowing for rapid quantification in crude or purified extracts. HPLC is a powerful and precise technique for separating and quantifying individual carotenoids, even in complex mixtures. It is especially useful for confirming compound identity, evaluating purity, and detecting minor carotenoid isomers.

Selection of the appropriate method depends on factors such as the carotenoid's polarity, sensitivity to light and temperature, matrix complexity, and the intended application (e.g., food, pharmaceutical, analytical standards). This section outlines the materials and step-by-step protocols used for characterization of lycopene, β-carotene, lutein, and zeaxanthin from various natural sources.

3.2.1 Thin-Layer Chromatography (TLC) for Carotenoid Identification [15]

1. Draw a pencil line 1.5 cm from the bottom of the TLC plate.
2. Spot 5 μL of the sample and standard carotenoid solutions onto the line using capillary tubes.
3. Develop the plate in a TLC chamber pre-saturated with hexane: acetone (9:1, v/v).
4. Allow the solvent front to migrate up to 8 cm from the baseline.
5. Remove the plate, air dry, and observe under visible light and UV at 254 nm.
6. Measure the Rf values and compare color and position with standard spots.

3.2.2 UV-Visible Spectrophotometry for Carotenoid Quantification [1, 13, 16]

1. Prepare a standard curve by diluting a known concentration of carotenoid standard (e.g., β-carotene or lycopene) in an appropriate solvent (hexane or ethanol).
2. Record absorbance at the compound-specific λmax (e.g., 450 nm for β-carotene, 472 nm for lycopene).
3. Prepare sample solutions similarly and filter through 0.45 μm syringe filters.
4. Measure absorbance using a UV-Vis spectrophotometer and compare with the standard curve to determine concentration.
5. Perform all measurements using quartz cuvettes, ensuring solvent blanks are used to zero the instrument.

3.2.3 High-Performance Liquid Chromatography (HPLC) for Carotenoid Profiling [16]

1. Filter both sample and standard carotenoid solutions through 0.45 μm syringe filters.
2. Inject 20 μL of each solution into an HPLC system equipped with a reverse-phase C18 column (250 × 4.6 mm, 5 μm particle size).
3. Use a mobile phase of acetonitrile: methanol: dichloromethane (70:20:10, v/v/v) at a flow rate of 1.0 mL/min.
4. Set the column temperature at 25 °C and the detection wavelength at 450 nm.
5. Identify carotenoids by comparing retention times and UV-Vis spectra with those of authentic standards.
6. Quantify each compound using a standard curve generated from known concentrations (*see* **Note 10**).

4 Notes

1. Lycopene is light-, heat-, and oxygen-sensitive. Use foil-wrapped glassware, chilled solvents, and keep evaporation ≤40 °C under dim light [9, 16].
2. Carrots have high moisture and fiber. Pre-drying improves extraction. Petroleum ether must be handled in a fume hood. Monitor Soxhlet to avoid overheating or loss [15].
3. β-Carotene is light- and oxygen-sensitive. Use chilled solvents and limit exposure to air. TLC can confirm identity by Rf and color [1].
4. Algal cells have rigid walls. Freeze-drying and ultrasonic treatment improve yield. Work under low light and use chilled methanol to prevent carotenoid loss [8].
5. Lutein is more polar and heat-sensitive than other carotenoids. Use polar solvents and controlled evaporation. Always compare bands with standards [11].

6. Microbial carotenoids are mostly intracellular. Lyophilization preserves them. Flash chromatography with TLC helps separate β-cryptoxanthin and zeaxanthin [12].
7. Microbial carotenoids may co-elute with lipids. Gradient flash chromatography and TLC with authentic standards help achieve separation [15].
8. Rf values vary with solvent and TLC plate conditions. Always dry plates properly and compare with standards for consistency [13].
9. Use quartz cuvettes above 300 nm. Solvent mismatch or turbidity affects results. Filter samples and use the same solvent for blank and sample [1, 16].
10. In HPLC, resolution depends on column, mobile phase, and wavelength. C18 columns with diode-array detection improve carotenoid profiling [13].

References

1. Saini RK, Keum YS (2018) Carotenoid extraction methods: a review of recent developments. Food Chem 240:90–103. https://doi.org/10.1016/j.foodchem.2017.07.099
2. Li BX, Rognon GT, Mattinson T et al (2018) Supplementation with macular carotenoids improves visual performance of transgenic mice. Arch Biochem Biophys 649:22–28. https://doi.org/10.1016/j.abb.2018.07.003
3. Johnson EJ (2019) Role of lutein and zeaxanthin in visual and cognitive function throughout the lifespan. Nutrients 11(1):43. https://doi.org/10.3390/nu11010043
4. Asaduzzaman M (2022) Lycopene—a review: chemistry, source, health role, extraction, applications. Annu Res Rev Biol 37(2): 87–101. https://doi.org/10.9734/ARRB/2022/v37i230490
5. Viñas-Ospino A, López-Malo D, Esteve M et al (2023) Green solvents: emerging alternatives for carotenoid extraction. Foods 12(1): 23–34. https://doi.org/10.3390/foods12010023
6. Badea G, Gatea F, Litescu-Filipescu SC et al (2025) Optimization of green ultrasound-assisted extraction of carotenoids from tomato waste using NADESs. Molecules 30(4): 911–926. https://doi.org/10.3390/molecules30040911
7. Gkioni M, Andriopoulos V, Koutra E et al (2022) Ultrasound-assisted extraction of *Nannochloropsisoculata* with ethanol and betaine: 1,2-propanediol eutectic solvent for antioxidant pigment-rich extracts. Antioxidants 11(3):563. https://doi.org/10.3390/antiox11030563
8. Khoo HE, Prasad KN, Kong KW, Jiang Y, Ismail A (2011) Carotenoids and their isomers: color pigments in fruits and vegetables. Molecules 16(2):1710–1738. https://doi.org/10.3390/molecules16021710
9. Strati IF, Gogou E, Oreopoulou V (2015) Enzyme and high pressure assisted extraction of carotenoids from tomato waste. Food Bioprod Process 94:668–674. https://doi.org/10.1016/j.fbp.2014.09.002
10. Qamar S, Shaikh S, Choudhary R et al (2020) Lycopene extraction and stabilization using green solvents: a comparative study. Food Biosci 35:100582. https://doi.org/10.1016/j.fbio.2020.100582
11. Matos ÂP, Nunes RG, da Silva TL et al (2020) Microalgae as a natural source of carotenoids: applications and extraction techniques. Mar Drugs 18(12):658. https://doi.org/10.3390/md18120658
12. Sharma R, Ghoshal G (2020) Optimization of carotenoid production by *Rhodotorula mucilaginosa* using agro-industrial waste in bioreactor: a statistical approach. Biotechnol Rep 25: e00407. https://doi.org/10.1016/j.btre.2020.e00407
13. Rodriguez-Amaya DB et al (2019) Advances in purification and quantification of carotenoids. Foods 8(10):524. https://doi.org/10.3390/foods8100524
14. Manzoor MF, Ahmad N, Ahmed Z et al (2019) A novel technique for the extraction

and purification of carotenoids from fruit and vegetable waste using green solvents. J Clean Prod 225:630–641. https://doi.org/10.1016/j.jclepro.2019.03.344

15. Parasar DP, Ramakrishnan E, Kabilan S et al (2020) Characterization of β-cryptoxanthin and other carotenoid derivatives from *Rhodotorula taiwanensis*. Chem Biodivers 17(4): e2000198. https://doi.org/10.1002/cbdv.202000198

16. Cucu T, Huvaere K, Van Den Bergh MA et al (2012) A simple and fast HPLC method to determine lycopene in foods. Food Anal Methods:1221–1228. https://doi.org/10.1007/s12161-011-9354-6

Chapter 11

Isolation and Characterization of Fatty Acids

Ashwani Kumar, Devendra Kumar Bhatt, Mukul Kumar, and Prashant Sahni

Abstract

Fatty acids are a concentrated source of energy, are necessary components of cell walls, and possess several physiological functions like gene regulation and biological signalling. The role of fatty acids in healthy fetal development, prevention of cardiovascular diseases, and Alzheimer is also observed. The ester of some fatty acids is used as food flavorings. They are also used as inactive ingredients in drug preparation. These applications of fatty acids have increased their demand in the purified form. To isolate fatty acids, the fats and oils are first saponified to extract glycerol and produce free fatty acids. These are further isolated by urea fractionation, low temperature crystallization, molecular distillation, chromatographic separation, enzymatic purification, and supercritical extraction. A combination of purification methods produces fatty acids with high purity. The isolated fatty acids can be further characterized using chromatographic techniques like thin layer chromatography, gas chromatography, high pressure liquid chromatography, and mass spectrometry.

Key words Fatty acids, Saponification, Chromatographic techniques, Purification

1 Introduction

Fatty acids are primary components of lipids that are made up of carbon chains with a methyl group at one end of the molecule and the carboxyl group at the other end [1]. They are classified as saturated and unsaturated based on the double bonds present in their structures. Fatty acids are a source of energy, a necessary component of cell walls, regulates genes and signalling molecules. The major fatty acids of dietary importance are palmitic, stearic, oleic, linoleic, α-linolenic, eicosapentaenoic (EPA), and docosahexaenoic acid (DHA), and they are further classified as essential and non-essential fatty acids. Among these, two fatty acids, i.e., linoleic and α-linolenic are considered essential as they cannot be produced by the body and their dietary intake is a must. In addition to these essential fatty acids, the other members of the ω-3 and ω-6 family, especially, EPA and DHA are reported to have various health

Ashwani Kumar, Vidisha Tomer and Piyoosh Kumar Babele (eds.), *Phytochemicals: Extraction, Quantification, Isolation, and Characterization*, Methods and Protocols in Food Science, https://doi.org/10.1007/978-1-0716-4840-7_11,

benefits. These fatty acids help in healthy fetal development, prevent cardiovascular diseases, and Alzheimer. Their role in weight management is also reported [2]. As per the health organizations, the intake of ω-3 and ω-6 fatty acids should be 1%–3%, and 3%–5% of total energy intake, respectively. Butyric acid, a short chain fatty acid, induces morphological and biochemical differentiation in a range of cells and thus exhibits anticarcinogenic properties. Methyl butyrate, an ester of butyric acid has pleasant aroma and is used for food flavorings. In addition to their dietary applications, fatty acids also have industrial applications especially as inactive ingredients in drug preparations. The co-formulations of fatty acids with drugs improve intestinal permeability. In case of orally administered drugs, the presence of fatty acids or triglycerides at low levels in intestinal lumen promote the intestinal uptake of small hydrophilic and lipophilic molecules. The conjugation of fatty acids with proteins and peptides helps in sustained release and site-specific delivery of drugs and thus improves the therapeutic outcomes [3]. Butyric acid is supplemented in animal feeds to prevent pathogenic bacterial colonization. It is also used for the production of butanol [4].

The application of fatty acids in nutricare/health care products and drugs have increased their demand in the purified forms. To obtain the fatty acids in purified form we first need to extract the oil. The major methods of oil extraction are either mechanical methods or solvent extraction methods. The major mechanical extraction methods are screw press expeller and hydraulic pressure extraction. However, the yield is more in solvent extraction methods. Both polar (ethanol, methanol, acetone, ethyl acetate, etc.), and non-polar solvents (hexane, chloroform, diethyl ether, toluene, etc.) can be used for the extraction of oil [5]. Better extraction results have been observed with the mixture of polar and non-polar solvents [6]. A ternary solution composed of chloroform, methanol, and water in the ratios 2:2:1.8 v/v/v is proposed by Bligh and Dyer to extract lipids from fresh tissues [7]. Further advancements in the extraction methods such as microwave-assisted extraction, ultrasonic extraction, and enzyme-assisted extraction also improve the yield of oil and fats. Organic solvents can be replaced with green solvents like bio derived solvents (terpenes, fatty acids methyl esters of plant origin), ionic liquids (1-butyl-3-methylimidazolium tetrafluoroborate, 1-butyl-3-methylimidazolium methyl sulfate, 1-butyl-3-methylimidazolium chloride, etc.), deep eutectic solvents (quaternary ammonium salt + metal chloride or metal chloride hydrate, etc.), natural deep eutectic solvents [(sugars: glucose, sucrose, fructose), (organic acids: lactic acid, malic acid, and citric acid), choline chloride, urea, etc.], fluorous solvents (polyethers and perfluorocarbons, etc.), and supercritical solvents (carbon dioxide, water, etc.) [6, 8]. The methods of the extraction of oils and fats have been discussed in detail in Chapter 3.

On the successful extraction of the oils and fats the next step is isolation and purification of desired fatty acids. The technologies for the purification of the fatty acids depend on chain length, molecular weight, melting point, degree of crystallization, degree of unsaturation, and number, position, and geometric configuration of double bonds in the molecule [6, 9]. For purification, the oil is first converted to free fatty acids and then the fatty acids are separated by methods like, winterization (low temperature crystallization), urea complexation, molecular distillation, chromatographic separation, enzymatic purification, and supercritical extraction. In general, a combination of purification methods is applied to obtain fatty acids of high purity. To further ensure the purity and type of fatty acids they are characterized by various techniques like thin layer chromatography (TLC), gas chromatography (GC), and high pressure liquid chromatography (HPLC). The chromatographic techniques work on the principle of interaction of the molecules to stationary phase or mobile phase. The other factors that affect separation are the shape of the stationary phase and the state of the stationary phase and mobile phase [10].

TLC is an inexpensive, quick, and sensitive liquid-solid adsorption chromatographic separation technique that works on the principle of capillary action. In this separation technique, the stationary phase is solid and the mobile phase is a liquid. To prepare a TLC plate, a thin layer of inert substance like silica, alumina, bentonite, cellulose, charcoal, dextran gel, and polyamide is supported on a flat, nonreactive surface like glass, plastic, or aluminium. The mobile phase is generally a mixture of two or more solvents and only in a rare case a single solvent is used. The commonly used solvents are petroleum ether, benzene, and carbon tetrachloride. The components having lower affinity towards the stationary phase moves faster and vice versa. More polar analytes interact more strongly with the stationary phase [11].

GC is employed for the separation and analysis of thermally stable volatile compounds. In this chromatographic technique, the mobile phase is a gas, and the stationary phase is either a solid like silica, alumina, charcoal, porous polymers, and molecular sieves or an immobilized liquid like polydimethylsiloxane, phenyl methyl polysiloxane, trifluoropropyl methyl polysiloxane, polyethylene glycol, and squalane supported on inert support made up of silica, alumina, etc. In GC, where the stationary phase is a solid material, it is known as gas-solid chromatography (GSC) and when the stationary phase is a liquid. It is known as gas-liquid chromatography (GLC). The separation of solutes in GSC is based on adsorptive properties of stationary phase, while, in GLC, the separation is based on portioning of the solutes in and out of the liquid film. Major parts of a GC system include: (1) Flow control section, (2) Sample injection port, (3) Column, (4) Column oven, (5) Detector, and (6) Data processor. Helium, nitrogen, and hydrogen are used as carrier gas. The columns used are of two types, i.e., packed column and capillary column. Packed column is about 2–6 m in length, has an internal diameter of 2–4 mm and is made up of glass, stainless steel, copper, or aluminium. It is filled with a particulate solid support like diatomaceous earth having particle diameter ranging from 37–44 μm to 250–354 μm. The particles are very porous having surface area ranging from 0.5 to 7.5 m^2/g. The capillary columns are a very thin type of column having length of 15–100 m and internal diameters of approximately 150–300 μm. Typically, it is a silica tubing whose outer surface is coated with polyimide resin to give it robustness. It has three major classes, i.e., (1) Wall coated open tubular (WCOT), (2) Support coated open tubular (SCOT), and (3) Porous layer open tubular (PLOT). The most common type of columns is chemically bonded WCOT type. The detectors used are Flame ionization detector (FID), Thermal conductivity detector (TCD), Electron capture detector (ECD), and Mass spectrometry (MS). Among these, FID is the most commonly used type of detector. The other commonly used detector is MS. MS analyzes and identifies the chemical nature and quantity of a substance by measuring mass-to-charge ratio. For MS analyses, the sample must be first ionized into gaseous ions. The ionization is achieved by gas chromatography and the ions are then characterized by their mass to charge ratios by mass spectrometry. Gas chromatography-mass spectrometry (GC-MS) has better selectivity and efficiency for the identification of fatty acids compared to GC-FID. GC-MS has well established databases for the identification of fatty acids and unknown compounds can be identified via molecular weight determination without any prior comprehensive information of the sample [12, 13].

HPLC is used for separating components in a liquid mixture based on the differential migration. The major parts of a basic HPLC system are: (1) Mobile phase reservoir, (2) Solvent degasser, (3) Solvent pump, (4) Sample injection port, (5) Column, and (6) Detector. The length of the column varies from 30 to 250 mm, diameter is about 2.1–4.6 mm, and is filled from inside with uniform fully porous spherical particles of 3–5 μm diameter. The various types of column used are normal phase columns, reverse phase columns, ion exchange columns, and size exclusion columns. Normal phase columns contain a polar stationary phase like silica gel, and the mobile phase consists of less polar organic solvents like n-hexane and ethyl acetate. Normal phase HPLC is suitable for separating lipophilic compounds, long chain alkanes, and the compounds having limited solubility in aqueous solutions. In reversed phase, HPLC a nonpolar stationary phase and polar mobile phase is used for the separation of organic compounds. The stationary phase is surface modified silica or a polymer that is covalently bound to alkyl or aromatic ligands. The mobile phase is a polar solvent, often a mixture of water and an organic solvent such as methanol, or acetonitrile. In the ion exchange column, the separation occurs due to the charge difference between the molecules of the sample and the charged stationary phase. The stationary phase used in this chromatography is usually an acid with either positive or negative charge and the mobile phase is a polar aqueous buffer like salt water. This method is used for the separation of carbohydrates, proteins, and amino acids. The size exclusion chromatography separates the sample based on particle size. In this, the stationary phase is a porous material like zeolite, polysaccharides, and polymers that allows only the small particles into pores. The retention time and elution profile of sample components depends on the pore size in the stationary phase. The rate of diffusion into pores varies as per each molecule and large molecules pass through the column faster. The detectors used in HPLC are UV/vis, photodiode array, fluorescence, refractive index, evaporative light scattering, and mass spectrometric detectors. Among these, the most commonly used detector is UV/vis due to its good stability, ease of operation, and high sensitivity. The column used for the separation of fatty acids is reverse phase column and the detector used is UV/vis detector [12].

This chapter discusses the fatty acid extraction methods optimized by various researchers and the methods and the optimized conditions for the characterization of fatty acids.

2 Materials

2.1 Isolation and Purification of Saturated Fatty Acids from Palm Oil [14]

(a) *Apparatus*:
- Two-neck round bottom flask.
- Separating funnel.
- Rotary evaporator.
- Deep freezer.

(b) *Reagents*:
- *Hydrolysis solution*: 1.75 M ethanolic potassium hydroxide and ethanol (150 mL:90% v/v) (*see* **Notes 1** and **2**).
- 6 N Hydrochloric acid (HCl) (*see* **Note 3**).
- Hexane.
- Distilled water.
- Anhydrous sodium sulphate.
- Methanol.

2.2 Isolation and Purification of α-Linolenic Acid from Perilla frutescens var. Japonica Oil [15]

(a) *Apparatus*:
- Medium pressure liquid chromatography (MPLC).
- Gas Chromatography:
  (a) *Detector*: Flame ionization detector.
  (b) *Column*: 30 mm × 0.25 mm × 0.25 μm.
- Refrigerator.
- Deep freezer.

(b) *Reagents*:
- n-hexane (C_6H_{14}) (isocratic: 100 *v/v*).
- Chloroform ($CHCl_3$) (isocratic: 100 *v/v*).
- Dichloromethane (CH_2Cl_2) (isocratic: 100 *v/v*).
- Potassium hydroxide (KOH).
- Urea.
- Sodium hydroxide (NaOH).
- Hydrochloric acid.
- Iodomethane.
- Dimethylsulfoxide.

2.3 Isolation of α-Linolenic Acid from Flaxseed [16]

(a) *Apparatus*:
- Coffee grinder.
- Refrigerator.
- Deep freezer.

(b) *Reagents*:
- Chloroform.
- Methanol.
- Potassium hydroxide.
- Nitrogen.
- Sulphuric acid.
- Distilled water.

2.4 Separation and Purification of ω-6 Linoleic Acid from Crude Tall Oil [12]

(a) *Apparatus*:
- Separating funnel.
- Centrifuge.
- Distillation unit.
- Air compressor.
- Buchner funnel.
- Deep freezer.

(b) *Reagents*:
- Urea.
- Ethanol.
- Methanol.
- Hexane (ACS grade).
- Sodium chloride.
- Sodium hydrogen carbonate.
- Acetone (HPLC grade).
- Sulphuric acid (99.9%).
- Ethanoic urea solution (ethanol: urea = 96:4).

(c) *Others*:
- Glass wool.
- Whatman filter paper.

2.5 Optimization of Novel Method for Isolation of High Purity Food Grade α-Linolenic Acid from Linum usitatissimum Seeds [17]

(a) *Apparatus*:
1. Grinder.
2. Hot plate.
3. Rotary evaporator.
4. Temperature controlled reactor with shaker.
5. Separating funnel.
6. Orbital agitator.
7. Pressure vessel.
8. Cold water bath.
9. Glass chromatography column.
10. Desiccator.

(b) *Reagents*:
- Extraction solvent: 2:1 solution of methanol and chloroform.
- Saponification solution (1 L) (*see* **Note 4**).
- Hexane.
- Petroleum ether.
- Ethanol.
- Hydrochloric acid.
- Acetyl chloride.
- Silver nitrate.
- Silica.

(c) *Others*:
- Glass wool.

2.6 Acid Ether and Supercritical Extraction of Docosahexaenoic Acid from Schizochytrium limacinum Using Vegetable Oils as Entrainer [18]

(a) *Apparatus*:
- Grinder.
- 1 mm sieve.
- Water bath.
- Vacuum evaporator.
- Supercritical fluid extraction unit.
- Glass tubes.

(b) *Reagents*:
- Anhydrous ethanol.
- Hydrochloric acid.
- Anhydrous ether.
- Soybean or canola oil.

2.7 Isolation of High Purity Alpha-Linolenic Acid from Perilla Seed Oil by Counter Current Chromatography (CCC) [19]

A. *High Performance Counter Current Chromatography (HPCCC)*:
- *Spectrum HPCCC*: This contains two columns, i.e., an analytical column having 0.8 mm internal diameter and tubing of 22.5 mL and a semi preparative column having 1.6 mm internal diameter and 133.5 mL of tubing with adjustable rotation speed of 0–1600 rpm. This is also equipped with a constant temperature regulator with SH 150–1500, a high performance liquid chromatography (HPLC) system containing two P-1000 pumps, a UV detector (UV-2500), EuroChrom workstation, fraction collector.
- *Midi HPCCC: Midi*: HPCCC is used for the scale up of the process. This consists of preparative columns with 4 mm internal diameter and 100 mL tubing with the rotation speed of 0–1400 rpm and equipped with column temperature regulator SH 150–2100, a preparative HPLC system with two P 2.1 pumps and 2.1 L UV detector.
- *Solvent System*: Analytical grade n-heptane/methanol/water (10:9:1, v/v) (*see* **Note 5**).

B. *pH-Zone Refining Counter Current Chromatography*:

In this, the outlet of spectrum HPCCC is connected with a pH detector combined with a flow cell for the online pH detection.
- *Solvent system: Hexane/methanol/water (10:5:5 v/v).*
- *Retainer acid*: Trifluoroacetic acid (TFA) for acidic samples and Triethylamine for basic samples.
- *Eluter*: Ammonia for acidic and HCl for basic samples (*see* **Note 6**).

2.8 Extraction and Separation of Butyric Acid from *Clostridium tyrobutyricum* Fermentation Broth in PEG/Na_2SO_4 Aqueous Two-Phase System [20]

(a) *Apparatus*:
- Centrifuge.
- Microporous barrel type filter.
- Separating chamber.

(b) *Reagents*:
- Sodium sulphate.
- PEG 6000.

2.9 Isolation of Fatty Acids and Fatty Alcohols from Wax Esters in a Wax-Ester Rich Marine Oil Without Formation of Fatty Acid Methyl Esters (FAMEs) [21]

(a) *Apparatus*:
- Mega Bond Elute (5 g) aminopropyl Solid Phase Extraction (SPE) Column.
- SPEVisiprep™ Vacuum Manifold.

(b) *Reagents*:
- Heptane (99.8%).
- Isopropanol (100%).
- Diethyl ether (≥99.8%).
- Ethanol (96%).
- Acetic acid (99.8%).
- Hydrochloric acid (37%).
- Chloroform (99.0%–99.4%).
- Phosphoric acid (99.99%).
- Potassium bicarbonate (≥99.8%).
- Copper (II) sulphate pentahydrate ≥98.0%.
- Sulphuric acid.
- Toluene.

2.10 Characterization of Fatty Acids by TLC

(a) *Apparatus*:
- TLC plate: Glass, plastic, aluminium (generally 20 cm × 20 cm × 20 cm).
- Adsorbent: Silica, alumina, etc.
- Binder: Calcium sulphate.
- Sample spreader.
- Hot air oven.
- Sample spotting device: Melting point capillary tube, micropipette, microliter syringe, etc.
- Development chamber.

(b) *Reagents*:
- Water.
- Dichloromethane.
- Developer: Bomothymol blue.

2.11 Characterization of Fatty Acids by GC

(a) *Apparatus*:
- Gas chromatograph.

(b) *Reagents*:
- Stationary phase: Solid or immobilized liquid (silica, alumina, charcoal, polymethylsiloxane, phenyl methyl siloxane, etc.).
- Carrier gas: Helium, nitrogen, hydrogen.

2.12 Characterization of Fatty Acids by HPLC [12]

(a) *Apparatus*:
- High pressure liquid chromatography.

(b) *Reagents*:
- Stationary phase: Reversed phase C18 columns.
- Mobile phase: Hexane, acetonitrile, etc.

3 Methods

3.1 Isolation and Purification of Saturated Fatty Acids from Palm Oil [14]

1. Take 25 g of oil sample and place in a 250 mL two-neck round bottom flask.
2. Add 150 mL of hydrolysis solution containing 1.75 M ethanolic potassium hydroxide and ethanol, and carry out hydrolysis at 70 °C for 2 h.
3. Separate the unsaponifiable matter by adding 100 mL of water with 50 mL of hexane to the mixture.
4. Acidify soap containing aqueous alcohol to pH 1 by adding approximately 60 mL of 6 N HCl.
5. Recover the free fatty acids in a nonpolar solvent like hexane.
6. Wash the extracted free fatty acids in a separating funnel with 25 mL distilled water for three to four times to a neutral pH.
7. Discard the lower (aqueous) layer and dry the upper layer containing FFAs with anhydrous sodium sulphate.
8. Evaporate the hexane under reduced pressure using a rotary evaporator at 35 °C.
9. Dissolve the dehydrated FFA's in 95% methanol (1:15) and crystallize at −15 °C for 24 h.

3.2 Isolation and Purification of α-Linolenic Acid from Perilla frutescens *var.* Japonica Oil [15]

(a) *Extraction of Oil*:
1. Roast the seeds at 170 °C for 10 min.
2. Press the roasted oil seeds in an oil press.
3. Separation of oil by MPLC using a suitable solvent (n-hexane, chloroform, dichloromethane, etc.) (*see* **Note 7**).
4. Concentrate the final residue layer to obtain Perilla oil.

(b) *Saponification*:

5. Dissolve 20 mL each of Perilla oil and the solvent layer (dichloromethane) in a solvent containing 10 mL of methanol mixed with 2 g of potassium hydroxide.
6. Heat the mixture at 65–70 °C for 1 h until a homogenous mixture is obtained.

(c) *Urea Complexation Treatment*:

7. Dissolve 2 g of this mixture in 10 mL methanol and treat this with 2 g of urea (*see* **Note 8**).
8. Cool the mixture to 10 °C for 24 h in a refrigerator.
9. Crystallize the annexations of treated mixture and remove by filtration.

(d) *Acidification*:

10. Acidify the filtrate with HCl to a pH of 2–3.

(e) *Recovery of Fatty Acids*:

11. Add 10 mL of n-hexane and 5 mL of ethyl ether to the separated fatty acids, stir, and remove the residues by filtration.
12. Dry the filtrate using anhydrous sodium sulphate and evaporate with nitrogen gas.

(f) *Synthesis of Fatty Acid Methyl Esters*:

13. Add 0.1 mL iodomethane and 20 mg ground NaOH to 0.5 mL dimethyl sulfoxide with sample and sonicate at room temperature for 30 min.
14. Add distilled water and dichloromethane to it and collect the non-polar phase. Evaporate the collected dichloromethane layer under nitrogen gas and then synthesize the fatty acid methyl esters.

    *This method results in isolation of ALA with 81.75% of purity.

3.3 Isolation of α-Linolenic Acid from Flaxseed [16]

(a) *Extraction of Oil*:

1. Crush the flaxseed in a coffee grinder.
2. Extract the oil using a mixture of chloroform and methanol (2:1 v/v).
3. Remove the solvents in a rotary evaporator.

(b) *Saponification of Oil*:

4. Take 130 g of flaxseed oil in 2 L of methanol containing 0.3 M of potassium hydroxide (KOH).
5. Allow it to saponify at 90 °C for 2 h under a gentle stream of nitrogen.
6. On completion of saponification remove the methanol in a rotary evaporator.

(c) *Extraction of Linolenic Acid*:

7. Acidify the mixture to pH 1.0 using 1 M sulphuric acid (H2SO4).
8. Separate the top layer (layer containing free fatty acids) and wash it thrice with the same volume of distilled water.
9. Dissolve the mixture of free fatty acid in three volumes of methanol.
10. Crystallize at 4 °C to remove saturated fatty acids.
11. Carry out a second crystallization at −18 °C to obtain linolenic acid.
12. Repeat the process twice or thrice to obtain the linolenic acid with 80% purity.

3.4 Separation and Purification of ω-6 Linoleic Acid from Crude Tall Oil [12]

(a) *Pre-treatments*:

1. Pretreat the oil sample with 1% w/v NaCl solution to remove unwanted impurities. To achieve this, add 40 mL of NaCl solution to 60 mL of crude oil and mix vigorously in a separating funnel for 10 min (*see* **Note 9**).
2. Discard the aqueous layer and centrifuge the upper layer for 10 min at 10,000 rpm.

(b) *Distillation*:

3. Take 200 g of pre-treated oil sample in a 5000 mL round-bottom flask.
4. Cover with glass wool to prevent heat dissipation.
5. The temperature should be maintained between 130 and 265 °C during distillation.
6. Maintain a pressure of 10 mm Hg in the flask by air compressor.
7. Collect the produced fatty acid vapors in a condenser.

(c) *Urea Complexation of PUFAs*:

1. Mix extracted fatty acids with ethanoic: Urea solution at a 20:80 volume ratio.
2. Keep the mixture at 60 °C and stir continuously until a clear homogeneous mixture is formed.
3. Bring down the mixture to room temperature for the precipitation of urea crystals (PUFAs remain soluble in ethanol while urea crystals get settled at bottom).
4. Filter the solution with Buchner funnel using 70 mm Whatman filter paper.
5. Remove the ethanol at 60 °C to recover the PUFAs.
6. The SFAs and MUFAs are recovered by extracting the solid urea crystal residue retained on the Buchner funnel with

hexane. The urea can be solubilized with water and can be reused for urea complexation.

(d) *Separation of PUFAs by Low Temperature Crystallization*:

1. Mix the sample of PUFAs with acetone in the ratio 1:9.
2. Carry out the crystallization in the temperature range of −7 to −15 °C for 24 h (Note: The freezing point of linoleic acid is −6.5 °C and for gamma linolenic acid and alpha linolenic acid is −11 °C.).

   *With this method 95.2% purity of linoleic acid can be achieved.

3.5 Optimization of Novel Method for Isolation of High Purity Food Grade α-Linolenic Acid from Linum usitatissimum Seeds [17]

(a) *Extraction of Oil*:

1. Grind the flaxseeds and sieve through a 16–32 mesh to obtain a particle size of 0.5 mm.
2. Mix the ground material with gas chromatography (GC) grade hexane and petroleum ether in an amber colored container with a tight lid.
3. Keep the mixture in the dark for 28 days. In between the mixture must be shaken gently on alternate days.
4. Recover the filtrate on the 28th day and discard the solid residue.
5. Remove the solvents on a hot plate or in a rotary evaporator.
6. Fill the oil up to brim (leaving no oxygen in the bottle to prevent oxidation) and store at 4 °C.

(b) *Saponification of Oil*:

1. Saponify one part of oil with 2 parts of saponification solution. Carry out this operation in a temperature-controlled reactor at 60 °C for 1 h with continuous agitation.
2. On completion of the saponification add 250–300 mL water per liter of mixture.
3. Remove the unsaponifiable matter by extracting in a separating funnel using 250–300 mL of hexane per liter of the mixture.
4. Acidify the aqueous alcohol phase containing the soaps with HCl (1.19 g/mL, pH-1).
5. Recover the free fatty acids by extraction with hexane in a separating funnel.
6. Remove the excess amount of acid from extract by washing it twice or thrice with distilled water.

(c) *Urea Fractionation*:

1. Dissolve 4 g of free fatty acids in 1:4 (w/w) urea solution prepared in methanol.
2. Allow urea and urea complexes to crystallize overnight using an orbital agitator set to 100–800 rpm (20 °C).
3. Filter the solution using a vacuum pump and store the filtrate at 4 °C for 6 h before being filtered again.

(d) *Extraction of Omega-3 PUFAs*:

1. Mix the total filtrate obtained from urea fractionation with equal amounts of HCl (15 mL or 17.85 g) and hexane (15 mL).
2. Stir the mixture thoroughly for 1 h.
3. Separate the hexane layer with the help of a separating funnel.
4. Repeat the process to remove the excess amount of acid in the environment.

(e) *Esterification of Omega-3 PUFAs*:

1. Mix 1 g of filtered material with 50 mL absolute methanol, 2.5 mL acetyl chloride, and 100 mL of hexane.
2. Transfer the solution to a pressure vessel, boil for 20 min, and cool to ambient temperature in a cold-water bath.
3. Evaporate the recovered solution in a rotary evaporator and add 10 mL of hexane to it.

(f) *Fractionation of Fatty Acids by Column Chromatography*:

1. *Prepare Argentation Silica (Ag-silica) Gel*:
   (i) Dissolve 10 g of silver nitrate in 27.615 g or 35 mL of ethanol.
   (ii) Make a slurry of 100 g of silica (pore size range: 60 nm to 230 meshes ASTM) by adding to 200 mL of silver nitrate ethanol mixture with constant shaking.
   (iii) Evaporate the ethanol under vacuum using a rotary evaporator at 60 °C.
   (iv) Heat it overnight at 120 °C to activate silver impregnated silica gel.
   (v) Chill the activated silver-impregnated silica gel and store in a desiccator under dark conditions.
2. *Preparation of Column*:
   (i) Take a glass chromatography column having a diameter of 0.5 cm and a packed height of 7–9 cm.
   (ii) Block the chromatography column's outlet with glass wool to prevent the particles from escaping.

(iii) Make a slurry of 20 mL Ag-silica gel in 40 mL hexane and place it into the chromatography column that is half-filled with solvent (hexane).

3. *Extraction*:

   (i) Add the methyl PUFA solution to the chromatographic column at the rate of 0.05–0.1 mL extract-to-stationary phase loadings (*see* **Notes 10** and **11**).

   (ii) Elute the column with a solvent containing 0.784 g of acetone per 100 g of hexane. The rate of elution is set to 1–2 mL/min.

   (iii) Collect the solvent fractions.

3.6 Acid Ether and Supercritical Extraction of Docosahexaenoic Acid from Schizochytrium limacinum Using Vegetable Oils as Entrainer [18]

(a) *Acid Ether Extraction*:

1. Dry the sample, ground it, and pass through a 1 mm sieve.
2. Take 0.5 g of the sieved sample and wet it with 1 mL of anhydrous ethanol.
3. To this sample, add 5 mL of concentrated HCl and place in a water bath maintained at 70 °C for 20 min (This will rupture the cells of biomass).
4. On reaching the room temperature add 5 mL of anhydrous ethanol.
5. Extract the mixture thrice with 10 mL of anhydrous ether.
6. Combine the ether layers and evaporate the ether by heating in a vacuum evaporator.
7. Store the recovered lipid in a glass tube at −20 °C until analysis.

(b) *Supercritical CO_2 Extraction*:

1. Take the dried sample, grind it, and mix with the entrainer (*see* **Note 12**).
2. The optimum conditions for the SC-CO_2 extraction are a temperature of 50 °C, pressure of 28 MPa, and a constant flow rate of 50 mL/h.
3. Carry out the extraction for 4 h.
4. After 3 h, collect the liquid product at an interval of every 30 min.
5. The residue from the extraction can be ground again, mixed with the same entrainer at the same ratio, and loaded into the extraction vessel.
6. Store the extracted lipids in a glass tube at −20 °C until analysis.

   *The extraction efficiency of this method for the DHA is 71%–80%.

3.7 Isolation of High Purity Alpha-Linolenic Acid from Perilla Seed Oil by Counter Current Chromatography (CCC) [19]

(a) *Production of Free Fatty Acids*:

1. Take 20 g of Perilla seed oil in a 500 mL flask.
2. Add 200 mL of 1 N KOH-70% aqueous ethanol solution.
3. Reflux for 1 h at 70 °C under stirring.
4. Cool the refluxed solution to room temperature, add 100 mL water to the solution, and extract thrice with 100 mL n-hexane (total volume of n-hexane: 300 mL) to remove unsaponifiable matter.
5. Adjust the pH of the remaining aqueous phase to 1–2 with 10% HCl and extract again thrice with 100 mL n-hexane.
6. Combine the extracted phases (n-hexane phases) and dry with anhydrous sodium sulphate.
7. Finally evaporate under vacuum to remove moisture.

(b) *Isolation of Alpha-Linolenic Acid*:

1. The selected solvent is first equilibrated in a separatory funnel and the upper and lower phases are prepared.
2. Fill the coil column with the upper phase stationary phase.
3. Pump the lower phase in a head to tail direction at a flow rate of 1.0 mL/min for analytical column, 6.0 mL/min for semi-preparative column, and 45.0 mL/min for preparative column.
4. The analytical column is rotated at 1600 rpm and the preparative column is rotated at 1400 rpm for establishing the equilibrium.
5. Inject 20 mg/0.5 mL free fatty acids sample solutions in lower phase, i.e., 120 mg/3 mL for semi-preparative and 900 mg/25 mL for preparative separation.
6. Monitor the eluents with a UV detector at 205 nm and collect the fractions.
7. High yield, i.e., more than 50% of high purity (more than 80%) ALA can be obtained by this method in 1 h.
8. The sample size and flow rates can be scaled up by 6–45 times.
9. The further characterization of fatty acids can be achieved by GC/MS methods.

(c) *pH-Zone Refining CCC Separation by Spectrum HPCCC*:

1. Fill the column with a stationary phase containing TFA.
2. Load the sample through a sample loop.
3. Pump the mobile phase containing NH4OH into the column from head to tail at the rate of 1 mL/min. Rotate the apparatus during this operation at 1600 rpm.

4. Monitor the effluent continuously using a UV detector at 205 nm followed by the online monitoring of pH value.
5. Collect the effluent into test tubes using a fraction collector at 3 min intervals.

3.8 Study on Extraction and Separation of Butyric Acid from *Clostridium tyrobutyricum* Fermentation Broth in PEG/Na_2SO_4 Aqueous Two-Phase System [20]

1. Centrifuge the fermentation broth at 5000 rpm for 10 min. This will remove cell debris and other solid contents.
2. Collect the upper liquid layer (supernatant).
3. Add 9.2% aqueous solution of sodium sulfate (Na_2SO_4) into butyric acid containing fermentation broth to salt out cell protein, sugar, and nitrogen compounds.
4. Filter the broth by a microporous barrel type filter. Wash the filter residue twice to collect all butyric acid.
5. Add the filtrate dropwise to 26% PEG 6000 (w/w), and mix uniformly to make aqueous two-phase solution (ATPS) mix uniformly (*see* **Note 13**).
6. Adjust pH to 3.0, temperature to 30 °C, and let it stand for 1.5 h for phase separation.
7. After partition/phase separation, most butyric acid, acetic acid, and butanol are collected into a PEG rich layer, so separate the PEG-rich layer.
8. Add 0.01 mol/L of iodine solution to the PEG-rich phase to precipitate PEG.
9. Filtrate the precipitated PEG.
10. The butyric acid can be separated from the filtrate by atmospheric distillation at 118 °C as the boiling point of butanol, acetic acid, and butyric acid are 117.7 °C, 117.9 °C, and 163.5 °C, respectively.

3.9 Isolation of Fatty Acids and Fatty Alcohols from Wax Esters in a Wax-Ester Rich Marine Oil Without Formation of Fatty Acid Methyl Esters (FAMEs) [21]

(a) *Isolation of Wax Esters from Calanus Oil*:

1. Dissolve 322 mg of calanus oil in 6 mL chloroform.
2. Mount the column on visiprep vacuum manifold.
3. Condition the SPE column with 20 mL hexane at a flow rate of 1 mL/min.
4. Add the mixture of calanus oil and chloroform to the column.
5. Elute the neutral lipids with 30 mL chloroform: Isopropanol (2:1 v/v) solvent and dry under nitrogen gas.
6. Dissolve the dried neutral lipids in 5 mL chloroform.
7. Add this to a new SPE column preconditioned with 20 mL heptanes.

8. Elute wax esters with 50 mL heptanes, again dry under nitrogen gas, and dissolve in 5 mL chloroform.
9. The isolation process can be repeated 2–3 times for complete extraction of wax esters.

(b) *Saponification of Wax Esters and Isolation of Fatty Acids and Fatty Alcohols*:

1. Dissolve the extracted wax esters with 5 mL of chloroform, evaporate to dryness under nitrogen, and dissolve in 50 mL 1 M ethanolic NaOH to a concentration of 5 mg/mL.
2. Transfer the dissolved wax esters to five GL 18 Duran Schott test tubes (containing 10 mL ethanolic NaOH mixture each), cap, and place on a heating block for 90 min at 90 °C.
3. Cool the hydrolyzed wax esters on ice.
4. Pool the material from Duran Schoff tubes, add 50 mL MilliQ water, and 25 mL 6 M HCl.
5. Mix the acidified hydrolyzed wax esters with 25 mL heptanes to form two phases.
6. Collect the upper phase containing lipids, evaporate under nitrogen, and dissolve in 5 mL chloroform.

(c) *Separation of Fatty Acids*:

1. Condition one SPE column with 20 mL heptanes.
2. Add the chloroform containing fatty acids and fatty alcohols to the column.
3. Elute the fatty alcohols with 30 mL chloroform: Isopropanol (2:1) v/v.
4. Condition the same column again with 20 mL heptanes.
5. Elute the free fatty acids with 50 mL diethyl ether: Acetic acid (98:2) v/v.
6. Evaporate the solvents under nitrogen gas and dissolve each in 1 mL chloroform.

3.10 Characterization of Fatty Acids by TLC [11]

(a) *Preparation of TLC Plate*:

1. TLC plate can be developed on any flat, nonreactive surface of uniform thickness like glass, plastic, or aluminium varying in size from microscopic slides to larger plates (20 cm × 20 cm × 20 cm).
2. Clean the plate with detergent to remove adhering particles and grease.
3. Rinse the cleaned plate with distilled water, place in a metal rack, and dry in an oven for glass and aluminium plates (*see* **Note 14**).

(b) *Selection of Adsorbent*:

Adsorbent is selected based on the compound to be separated. Silica gel is preferred for the separation of polar compounds, while alumina is preferred for the weakly polar compounds. Silica or alumina with a mean pore diameter of 60 A° is most commonly used adsorbent in TLC. Gypsum (Calcium sulphate, $CaSO_4$) is used as a binder in adsorbent at the rate of 10%–15% to prevent flaking.

(c) *Preparation of Slurry*:

To prepare the slurry slowly add the adsorbent while stirring to the suitable solvent (water, dichloromethane, etc.). A wide mouthed bottle with a cap can be used for the preparation of slurry (*see* **Note 15**).

(d) *Coating of Plates*:

1. A thin layer of adsorbent may be applied by pouring, dipping, or spreading the slurry over the plate.
2. The use of commercial spreaders gives more uniform coatings (*see* **Note 16**).
3. Next, keep the plate horizontally set for 10 min and remove the excess adsorbent from the plate using a razor blade or glass rod.

(e) *Activation of Plate*:

1. Heat the plates at 110 °C for 1 h in an oven for activation.
2. Cool the plates and store in desiccators until used.

(f) *Sample Preparation*:

1. Dissolve the sample to be analyzed in a suitable solvent to a desired concentration.

(g) *Sample Application*:

1. The sample is applied at least 1–2 cm above the base of the plate. Care should be taken not to immerse the sample spots in the solvent (*see* **Note 17**).
2. The sample should be applied in very small quantities using a melting point capillary tube, micropipette, or microliter syringe.
3. The plate is carefully dried after spotting.

(h) *Development of Chromatogram*:

TLC can be developed in a wide variety of chambers such as small cylindrical glass jar or wide mouthed screw capped glass bottle for microscopic slides and rectangular glass tank with airtight lid for larger plates.

1. Line the inside of the developing tank with filter paper leaving a gap for viewing chrome plate.
2. Saturate the filter paper with developing solvent and pour down the solvent carefully from the side of the tank. The level of the solvent should be managed in a manner so that the spotting line (base line) on the TLC plate does not get soaked in solvent. The level should be usually 0.5–1 cm below the line of spotting.
3. Close the tank with lid and let it stand for about 5 min to get saturated with solvent vapors.
4. Place the spotted and dried plate in the chamber containing the mobile phase, apply the cover, and allow the sample to ascend by capillary action to the finishing line. The solvent front is marked once the solvent moves around 2/3rd of the plate and the plate is immediately removed from the development tank (Fig. 1).
5. Dry the plate in a fume cupboard and stain the plate with suitable developer for the visualization of colorless organic compounds (*see* **Note 18**).
6. The results of TLC are represented by retention factor (R_f value). R_f value is migration of solute relative to the solvent front.

$$R_f \text{ value} = \frac{\text{Distance travelled by the solute}}{\text{Distance travelled by the solvent front}}$$

*R_f value aids in the identification of a substance by comparison with standard. The standard must be run on the same TLC (*see* **Note 19**).

**Preparative chromatography is used to isolate and purify a particular substance from a sample. Whereas, analytical chromatography is used to separate the components of a sample.*

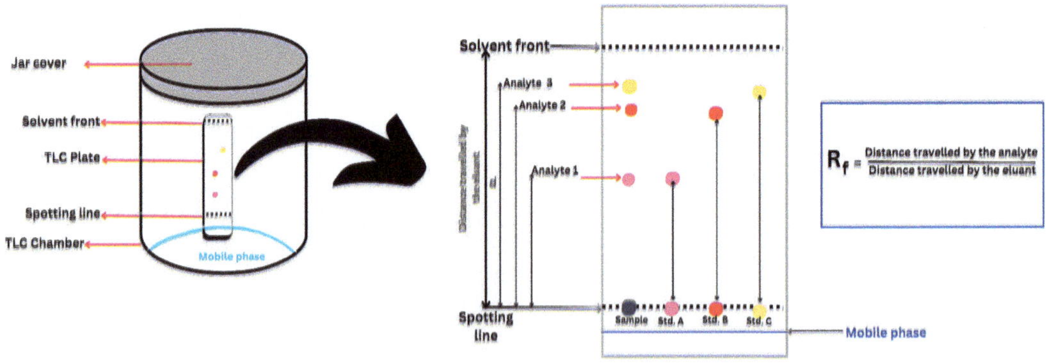

Fig. 1 TLC experimental setup

(i) *TLC for the Oil Samples* [5]:

1. Prepare TLC plates (5 cm × 2 cm) by titurating 15 g of silica gel in 30 mL of distilled water.
2. Activate the plate at 110 °C for 30 min.
3. Spot TLC plate with isolated fatty acids against their respective standards.
4. Allow the plate to dry and elute in a TLC chamber with a solvent mixture of hexane: Diethyl ether: acetic acid I the ratios of 80:20:1.
5. Once the solvent moves around 2/3rd of the plate, mark the solvent front, and remove the plate from the development chamber.
6. Dry the plate and spray bromothymol blue as a developing agent (*see* **Note 20**).
7. Calculate the R_f value using the formula given above.

3.11 Characterization of Fatty Acids by GC

1. Gas is set to flow at a constant rate.
2. Sample is injected into the sample port after careful extraction, isolation, and dilution (*see* **Note 21**).
3. The sample is carried by the carrier gas to the column and the separation takes place in the column (*see* **Note 22**).
4. The components are separated according to the desorption rate and the partitioned components come out of the column with carrier gas.
5. The carrier gas is delivered to the detector, and separation of the components is detected, and displayed on a displaying unit like computer screen in the form of a chromatogram.
6. The composition of the sample can be analyzed and quantified from a chromatogram. The optimized conditions or the quantification of fatty acids by gas chromatography are provided in Table 1. The fatty acids are identified by comparison with known standards.
7. *Calculations*:

$$\text{Concentration of sample}\left(\frac{\mu g}{mL}\right) = \frac{\text{Area under the curve of sample}}{\text{Area under the curve of standard}} \times \text{Concentration of standard}$$

3.12 Characterization of Fatty Acids by HPLC

1. Reconstitute the purified fatty acids in a suitable solvent and derivatize to fatty acid methyl esters (FAME's) to enhance detection. Also derivatize the standard fatty acids and run to serve as reference for the identification and quantification.

Table 1
Optimized conditions for GC analysis of fatty acids [21, 22]

For GC-FID	For GC-MS
GC make and model: Agilent GC6820	*Make and model*: Agilent 7890 gas chromatogram
Column type: HP-INNOWAX capillary column	*Column*: DB-5 MS capillary column coated with 5% diphenyl cross-linked with 95% dimethylpolysiloxane (30 m length × 250 μm inner diameter × 0.25 μm film thickness)
Dimensions of column: 30 m × 0.323 mm × 0.50 μm	*Sample size*: 1 μL (injected in split less mode)
Detector type: FID	*Temperature of injection*: 280 °C
Carrier gas: *Pure nitrogen*: 1.8 mL/min *Hydrogen flow rate*: 30 mL/min *Air flow rate*: 300 mL/min	*Temperature of transfer line*: 280 °C
Diversion ratio: 60:1	*Temperature of ion source*: 280 °C
Injection port temperature: 250 °C	*Energy*: −70 eV in electron impact
FID detector temperature: 250 °C	*Carrier gas*: Helium
Injection volume: 0.4 μL	*Front inlet purge flow*: 3 mL/min *Gas flow rate through column*: 1 mL/min *Column temperature*: Initial temperature was set to 50 °C for 1 min, then raised to 310 °C at a rate of 20 °C/min and finally kept at this temperature for 6 min *m/z range*: 50–500 at a rate of 12.5 spectra per second after a solvent delay of 4.78 min *Data acquisition*: Full scan mode

2. Select a suitable mobile phase and set it to flow at a constant rate through the HPLC system.
3. Inject the derivatized sample into the port of HPLC system.
4. The mobile phase carries the sample to the column, where the extraction takes place based on interaction with the column matrix.
5. The components separated in columns are detected using a suitable detector. The optimized HPLC conditions for the characterization of fatty acids are given in Table 2.
6. The fatty acids are identified by comparing retention times with those of known standards.
7. The quantity of each fatty acid can be determined by comparing peak areas with standard curves.

Table 2
Optimized HPLC conditions for the characterization of fatty acids [12]

Make and Model: Agilent Technology 1260 Infinity series
Sample size: 10 μL (100 μL of sample was diluted in 900 μL of HPLC grade hexane and out of this10 μL sample was taken for analysis)
Column: C18 (ZORBAX Eclipse XDB-C-18) (4.6 mm × 150 mm × 5 μm)
Column temperature: 50 °C
Mobile phase: 100% ACS grade hexane
Flow rate: 0.8 mL/min
Total processing time: 20 min
Detector: Refractive index

8. *Calculations*:

$$\text{Concentration of sample}\left(\frac{\mu g}{mL}\right) = \frac{\text{Area under the curve of sample}}{\text{Area under the curve of standard}} \times \text{Concentration of standard}$$

4 Notes

1. Molecular weight of potassium hydroxide is 56.105 g/mol. So, to prepare 1.75 M of ethanolic potassium hydroxide dissolve 98.184 g of anhydrous potassium hydroxide in the minimum quantity of distilled water and make volume 1 L in volumetric flask with 95% aldehyde free ethanol.
2. To prepare 90% ethanol *v/v* take 90 mL absolute ethanol and make the final volume 100 mL in a volumetric flask by adding distilled water.
3. Add 525.44 mL of concentrated HCl (purity-36%, specific gravity-1.14) in a small amount of distilled water and make the final volume to 1 L by adding more distilled water in a volumetric flask.
4. To prepare saponification solution dissolve 150 g KOH in a solution containing 500 mL distilled water and 500 mL ethanol (density = 0.789 g/mL).
5. To prevent partial deprotonation of –COOH in fatty acids to COO-, 0.2% acetic acid can be added to the solvent making the new solvent composition as 10:9:1:0.4, v/v.
6. In the extraction of high purity α-linolenic acid from perilla seed oil by Han et al. [19], the retainer used was 30 mM TFA in

upper stationary phase and the eluter was 10 mM ammonium hydroxide (NH_4OH) in the lower mobile phase.

7. Lee et al. [15] have optimized dichloromethane as the most suitable solvent for the purification of α-linolenic acid from *Perilla frutescens* var. japonica oil.
8. The quantities of methanol and urea should be increased proportionately as per the quantity of mixture used.
9. Two layers are formed in a separating funnel. The top layer contains the oil while the impurities like dust matter are collected in the bottom layer of salt solution.
10. In the study by Kumar et al. [17], 4 g loading of fatty esters on the Ag-silica gel stationary phase was added.
11. During the column chromatography the traces of saturated fatty acids/saturated esters in the mixture are eluted first followed by unsaturated fatty acids.
12. He et al. [18] have recommended using soybean or canola oil as entrainer at the rate of 50%–100% of sample as it has been found to increase the extraction efficiency by 48%–56% compared to supercritical CO_2 extraction.
13. Wu et al. [20] added pretreated butyric acid fermentation broth to PEG to make the final volume of ATPS 10 mL. In this study, 35:65 of fermentation broth and ATPS produced a good yield of butyric acid.
14. The plates must be handled carefully from the sides or from the base as any direct contact with the side to be coated with coating material (stationary phase) can cause grease contamination and may result in the formation of mechanically unstable layer that is liable for flaking.
15. The slurry should be uniform. Too thick or too thin slurries must be avoided.
16. The thickness of coating should be 150 mm for analytical TLC and 2 mm for preparative TLC.
17. The spot must be as small as possible for better separation and minimum tailing.
18. The various visualization and detection techniques for colorless organic substances are UV light and iodine vapor.
19. The standard value of R_f factor is in the range of 0–1. The value close to 0 refers to good interaction of the sample with the stationary phase. A value close to 1 means a weak or low interaction between stationary phase and sample components.
20. In the study by Kumar et al. [5], bromothymol blue was prepared in ethanol (79 g/100 g) and made alkaline with NH_4OH.

21. The fatty acids are first converted to fatty acid methyl esters (FAMEs) by the process known as methyl esterification. For this, 1 part of fatty acids/oil sample is mixed with 15 parts of hexane. To this, 2 parts of 2 N methanolic KOH are added and vortexed for 15 min. The upper layer is collected after the settling of phases, and filtered with 0.45 μm syringe filters. Other methods of FAMEs production are also discussed by Guo et al. [9] and Yang et al. [16]. The sample may be diluted with a solvent before injection. The amount of sample generally injected is 0.1–20 μL.
22. Total time through the column varies. Islam et al. [12] reported a column run time of 24 min, while Kumar et al., [5] reported a column run time of 36 min. Final column temperature and the pattern of increase of temperature of column oven also vary. Islam et al. [12] increased the temperature at 20 °C/min up to 240 °C and then held it at 240 °C for the remainder of the study. Guo et al. [9] maintained the column oven temperature at 180 °C for 3 min, increased to 200 °C at 6 °C/min, and the final temperature was maintained 240 °C. The final temperature of 240 °C was maintained for 40 min. Kumar et al. [5] maintained oven temperature 150 °C for 1 min, raised to 200 °C at 5 °C/min, hold at 200 °C for 5 min, temperature again raised to 230 °C at 10 °C/min, hold at 230 °C for 5 min and finally raised to 260 °C at 15 °C/min and held for 5 min at this temperature.

References

1. Rustan AC, Drevon AC (2005) Fatty acids: Structures and Properties. Encylopedia of Life Sciences. John Wiley and Sons, USA. https://doi.org/10.1038/npg.els.0003894
2. Swanson D, Block R, Mousa SA (2012) Omega-3 fatty acids EPA and DHA: health benefits throughout life. Adv Nutr 3:1–7. https://doi.org/10.3945/an.111.000893
3. Hackett MJ, Zaro JL, Shen WC et al (2013) Fatty acids as therapeutic auxiliaries for oral and parenteral formulations. Adv Drug Deliv Rev 65:1331–1339. https://doi.org/10.1016/j.addr.2012.07.012
4. Mukherjee S, Munshi B (2024) Reactive extraction of butyric acid from effluent streams and fermentation broth by using tri-n-octyl amine in decanol/natural oils. Biomass Conv Bioref 14:3439–3451. https://doi.org/10.1007/s13399-022-02504-2
5. Kumar A, Nirmal P, Kumar M et al (2023) Major phytochemicals: recent advances in health benefits and extraction method. Molecules 28:887. https://doi.org/10.3390/molecules28020887
6. Sharma AK, Chintala V, Ghodke P et al (2020) Extraction and purification of PUFA from microbial biomass. In: Patel AK, Matsakas L (eds) Nutraceutical fatty acids from oleaginous microalgae: a human health perspective, 1st edn. Wiley, Massachusetts
7. Sündermann A, Eggers LF, Schwudke D (2016) Liquid extraction: Bligh and Dyer. In: Wenk M (ed) Encyclopedia of lipidomics. Springer, Dordrecht
8. Zhou W, Wang Z, Alam MA et al (2019) Repeated utilization of ionic liquid to extract lipid from algal biomass. Int J Polym Sci 1: 9209210. https://doi.org/10.1155/2019/9209210
9. Guo JX, Wang CL, Wu ZJ (2013) Purification of α-linolenic acid from flaxseed oil by silver-silica gel chromatography column. Adv Mater Res 610:3580–3586. https://doi.org/10.4028/www.scientific.net/AMR.610-613.3580

10. Kumari VC, Patil SM, Ramu R et al (2022) Chromatographic techniques: types, principles, and applications. In: Egbuna C, Patrick-Iwuanyanwu KC, Shah AM, Ifemeje JC, Rasul A (eds) Analytical techniques in biosciences, 1st edn. Academic Press, London
11. Thin layer chromatography (TLC)-principle and procedure – Online Biology Notes. https://www.onlinebiologynotes.com/. Accessed 12 Nov 2024
12. Islam MS, Christopher LP, Alam MN (2020) Separation and purification of ω-6 linoleic acid from crude tall oil. Separations 7:9. https://doi.org/10.3390/separations7010009
13. Song J, Yan Y, Wang X et al (2021) Characterization of fatty acids, amino acids and organic acids in three colored quinoas based on untargeted and targeted metabolomics. LWT 140: 110690. https://doi.org/10.1016/j.lwt.2020.110690
14. Japir AAW, Salimon J, Derawi D et al (2018) A highly efficient separation and physicochemical characteristics of saturated fatty acids from crude palm oil fatty acids mixture using methanol crystallisation method. OCL 25:A203. https://doi.org/10.1051/ocl/2018003
15. Lee J, Lee MH, Cho EJ et al (2016) High-yield methods for purification of α-linolenic acid from Perilla frutescens var. japonica oil. Appl Biol Chem 59:89–94. https://doi.org/10.1007/s13765-015-0136-5
16. Yang L, Leung KY, Cao Y et al (2005) α-Linolenic acid but not conjugated linolenic acid is hypocholesterolaemic in hamsters. Br J Nutr 93:433–438. https://doi.org/10.1079/BJN20041365
17. Kumar A, Yadav RK, Shrivastava NK et al (2023) Optimization of novel method for isolation of high purity food grade α-linolenic acid from Linum usitatissimum seeds. LWT 189: 115466
18. He B, Wang Y, Dou X et al (2017) Supercritical CO2 extraction of docosahexaenoic acid from Schizochytrium limacinum using vegetable oils as entrainer. Algal Res 21:58–63. https://doi.org/10.1016/j.algal.2016.10.007
19. Han T, Yang G, Cao X et al (2019) Preparative and scaled-up separation of high-purity α-linolenic acid from *perilla seed oil* by conventional and pH-zone refining counter current chromatography. J Sep Sci 42:2360–2370. https://doi.org/10.1002/jssc.201900189
20. Wu X, Li G, Yang H et al (2015) Study on extraction and separation of butyric acid from clostridium tyrobutyricum fermentation broth in PEG/Na2SO4 aqueous two-phase system. Fluid Phase Equilib 403:36–42. https://doi.org/10.1016/j.fluid.2015.05.047
21. Schots PC, Edvinsen GK, Olsen RL (2023) A simple method to isolate fatty acids and fatty alcohols from wax esters in a wax-ester rich marine oil. PLoS One 18:e0285751. https://doi.org/10.1371/journal.pone.0285751
22. Hage DS (2018) Chromatography. In: Rifai N, Horvath RA, Wittwer TC, Hoofnagle NA (eds) Principles and applications of clinical mass spectrometry, 6th edn. Elsevier, Oxford

Chapter 12

Isolation and Characterization of Health Promoting Phytosterols

Pragya Pandey, Pintu Choudhary, Simardeep Kaur, Kuldeep Pandey, Jagveer Singh, and Amarjeet Kaur

Abstract

Phytosterols are plant-derived sterols known for their cholesterol-lowering and therapeutic effects. This chapter focuses on the downstream processes for isolating, purifying, and analyzing phytosterols from crude extracts. Extraction protocols are described in Chapter 4. The methods covered include recrystallization, urea complexation, molecular distillation, membrane filtration, and preparative HPLC for purification, along with TLC, GC-MS, HPLC, NMR, and FTIR for structural characterization. These methods facilitate the production of high-purity phytosterols for food, pharmaceutical, and analytical applications.

Key words Phytosterols, Purification, Recrystallization, Urea complexation, Molecular distillation, Membrane filtration, Preparative HPLC, Structural characterization, GC-MS, NMR, FTIR, TLC

1 Introduction

Phytosterols, including sterols and stanols, are cholesterol-like compounds naturally present in vegetable oils, cereals, nuts, legumes, and seeds [1–3]. They are structurally similar to cholesterol and this enables them to compete with dietary cholesterol during intestinal absorption, thereby reducing serum low-density lipoprotein cholesterol (LDL-C) [4, 5].

A daily intake of 2–3 g of phytosterols has been shown to lower LDL-C levels by approximately 10%, supporting their use in dietary strategies for cardiovascular risk reduction [4–6]. Beyond lipid-lowering effects, phytosterols exhibit antioxidant, anti-inflammatory, anticancer, and immunomodulatory activities, making them valuable in food, nutraceutical, and pharmaceutical applications [2].

Vegetable oil by-products such as deodorizer distillates are widely used as commercial sources of phytosterols, although cereal

Ashwani Kumar, Vidisha Tomer and Piyoosh Kumar Babele (eds.), *Phytochemicals: Extraction, Quantification, Isolation, and Characterization*, Methods and Protocols in Food Science, https://doi.org/10.1007/978-1-0716-4840-7_12,

brans, oilseeds, and microalgae are also gaining attention due to their sterol content and sustainability advantages [7, 8].

Extraction protocols from raw matrices are discussed in Chapter 4. This chapter begins from the point of crude extract acquisition and presents laboratory-scale methods for the purification and characterization of phytosterols. Crude unsaponifiable fractions often contain pigments, waxes, and fatty alcohols that must be removed before sterols can be used in analytical or product development workflows [9].

To achieve higher purity, several downstream purification techniques are described. Recrystallization using methanol or methanol-acetone solvent systems remains a cost-effective and scalable approach [10]. Urea complexation offers selective removal of saturated components and waxes through inclusion complex formation [8]. Molecular distillation provides thermal separation of non-sterol volatiles under vacuum conditions [11].

Membrane-based techniques such as ultrafiltration allow non-thermal separation of low-molecular weight impurities [12]. Preparative HPLC using C18 columns enables the isolation of individual sterols with high purity, particularly for analytical reference or formulation needs [13].

For purity assessment and identity confirmation, multiple analytical tools are required. These include GC-MS for quantitative and qualitative profiling [14], HPLC for sterol quantification [15], FTIR for functional group analysis, and NMR for structural validation [13].

This chapter outlines all relevant purification and characterization protocols in a step-by-step format, designed to support reproducibility in academic and industrial settings.

2 Materials

This section lists the laboratory equipment and reagents required for the purification and characterization of phytosterols from pre-extracted, unsaponifiable plant fractions. Extraction procedures are described in Chapter 4.

2.1 Purification Techniques

2.1.1 Recrystallization

(a) *Apparatus*:
- Two-neck round-bottom flask (250 mL).
- Water bath with temperature control.
- Magnetic stirrer with hotplate.
- Vacuum filtration setup (Büchner funnel, filter paper).
- Deep freezer (−10 °C to −20 °C).
- Desiccator.

(b) *Reagents*:
- Crude phytosterol extract.
- Methanol (HPLC grade).
- Acetone (analytical grade, optional).
- Anhydrous sodium sulfate.

2.1.2 Urea Complexation

(a) *Apparatus*:
- Hot plate or water bath.
- Magnetic stirrer.
- Refrigerator (4–8 °C).
- Vacuum filtration unit.
- Separating funnel.
- Rotary evaporator.
- Desiccator.

(b) *Reagents*:
- Crude unsaponifiable extract.
- Methanol (analytical grade).
- Urea (AR grade).
- Distilled water.
- Hexane (ACS grade).
- Hydrochloric acid (6 N).
- Anhydrous sodium sulfate.

2.1.3 Molecular Distillation

(a) *Apparatus*:
- Short-path molecular distillation unit.
- Cold trap.
- Vacuum pump (0.001 mbar).
- Heating mantle with controller.
- Receiving flask.

(b) *Reagents*:
- Crude sterol-rich fraction.
- Anhydrous nitrogen (optional, for inert atmosphere).

2.1.4 Membrane-Based Separation

(a) *Apparatus*:
- Cross-flow membrane filtration system.
- Nanofiltration or ultrafiltration membranes (MWCO 300–500 Da).
- Pressure pump (3–5 bar).

(b) *Reagents*:
- Methanol-dissolved sterol-rich fraction.
- Distilled water.

2.1.5 Preparative HPLC

(a) *Apparatus*:
- Preparative HPLC system with UV detector.
- C18 reverse-phase preparative column (250 mm × 10 mm, 5 μm).
- Fraction collector.

(b) *Reagents*:
- Methanol (HPLC grade).
- Acetonitrile (HPLC grade).
- Deionized water.
- Crude or semi-purified sterol extract.

2.2 Characterization Techniques

2.2.1 Thin-Layer Chromatography (TLC)

(a) *Apparatus*:
- TLC plates (Silica gel 60 F_{254}).
- Developing chamber.
- UV lamp.

(b) *Reagents*:
- Solvent system: Hexane and acetone.
- Anisaldehyde-sulfuric acid reagent.

2.2.2 Gas Chromatography-Mass Spectrometry (GC-MS)

(a) *Apparatus*:
- GC-MS system (e.g., Agilent 7890B/5977A).
- Capillary column (HP-5MS, 30 m × 0.25 mm, 0.25 μm).

(b) *Reagents*:
- BSTFA + 1% TMCS (derivatization reagent).
- Anhydrous hexane.

2.2.3 High-Performance Liquid Chromatography (HPLC)

(a) *Apparatus*:
- HPLC system with UV detector.
- C18 reverse-phase column (250 mm × 4.6 mm, 5 μm).

(b) *Reagents*:
- Methanol: Acetonitrile (80:20, v/v).
- Standard phytosterol mix.

2.2.4 Nuclear Magnetic Resonance (NMR) Spectroscopy

(a) *Apparatus*:
- NMR spectrometer (400 MHz or higher).

(b) *Reagents*:
- Deuterated chloroform ($CDCl_3$).
- NMR tubes.

2.2.5 Fourier-Transform Infrared (FTIR) Spectroscopy

(a) *Apparatus*:
- FTIR spectrometer.
- ATR accessory or KBr pellet press.

(b) *Sample*:
- Phytosterol sample (solid or film form).

3 Methods

3.1 Purification Techniques

3.1.1 Recrystallization (Adapted from Bian et al. [8], Srividya et al. [15], and Kasim et al. [16])

1. Weigh 5–10 g of crude phytosterol extract and transfer to a 250 mL round-bottom flask.
2. Add 100–150 mL of hot methanol (pre-heated to ~65 °C) and stir for 15–20 min using a magnetic stirrer to ensure complete dissolution (*see* **Note 1**) [16].
3. Cool the solution gradually to room temperature, then place it in a deep freezer at −10 °C to −15 °C for 24 h to allow crystallization [8].
4. Filter the resulting precipitate under vacuum using a Büchner funnel.
5. Wash the collected crystals with a small volume of ice-cold methanol to remove surface impurities.
6. Dry the purified phytosterol crystals in a desiccator over anhydrous sodium sulfate for 12–24 h (*see* **Note 2**) [15].
7. *(Optional)* Repeat recrystallization using acetone: Methanol (1:1 v/v) if higher purity is required (*see* **Note 1**) [15, 16].

The purified phytosterols appear as white to off-white crystalline solids, typically achieving >92% purity as confirmed by HPLC or GC-FID.

3.1.2 Urea Complexation (Adapted from Bian et al. [8], Sujith Kumar and Mawlong [9])

1. Dissolve 2–3 g of crude unsaponifiable matter (obtained from prior extraction and saponification) in 40–50 mL of hot methanol (~60 °C).
2. Add urea to the methanolic solution in a 1:4 (w/w) ratio (urea: sample) and stir until completely dissolved (*see* **Note 3**) [9].
3. Cool the mixture to room temperature, then refrigerate at 4 °C for 12–16 h to allow formation of urea complexes.

4. Filter the solid urea-fatty acid complex using vacuum filtration and collect the filtrate, which contains free phytosterols.
5. Acidify the filtrate to pH 2–3 using 6 N HCl to release any sterols potentially bound with residual urea (*see* **Note 4**) [8].
6. Extract sterols from the acidified mixture using hexane (3 × 30 mL) in a separating funnel.
7. Wash the hexane layer with distilled water to neutral pH and dry it over anhydrous sodium sulfate.
8. Evaporate hexane using a rotary evaporator at ~35 °C to obtain purified sterols.
9. Store the final product in a desiccator until further analysis.

The resulting fraction is typically rich in phytosterols and free from saturated fatty acids, with purity levels often exceeding 90% as confirmed by HPLC or GC-FID.

3.1.3 Molecular Distillation (Adapted from Nadeem et al. [10] and Seifollahi et al. [11])

1. Transfer 5–10 g of the crude phytosterol-rich extract into a molecular distillation unit.
2. Apply a vacuum of ≤0.001 mbar using a high-performance vacuum pump (*see* **Note 5**).
3. Heat the feed material in a short-path evaporator to 180–220 °C using a controlled heating mantle.
4. Maintain a cold trap (≤5 °C) and rotate the wiper system to ensure uniform film formation and prevent degradation.
5. Collect the distillate containing volatile impurities; the residue contains enriched phytosterols.
6. Repeat distillation if needed to further purify the sterol-rich fraction.

The residue obtained after distillation appears as a viscous, semi-solid material enriched in phytosterols. Typically, sterol content increases significantly while removing low-boiling-point contaminants.

3.1.4 Membrane-Based Separation (Adapted from Abd-Razak et al. [12])

1. Dissolve the crude or semi-purified phytosterol extract in methanol to prepare a feed solution.
2. Filter the solution through a 0.45 μm membrane to remove suspended solids (*see* **Note 6**).
3. Transfer the solution into the feed tank of a cross-flow membrane filtration system equipped with nanofiltration (NF) or ultrafiltration (UF) membranes (MWCO: 300–500 Da).
4. Maintain operating pressure between 3 and 5 bar and control the flow rate as per membrane specifications.
5. Collect the permeate and retentate separately.

6. Recover the purified sterol-rich fraction from the retentate by evaporating methanol under reduced pressure.
7. Store the dried product in a desiccator until further use.

The purified fraction appears as a dry, light-yellow solid after solvent removal. Membrane filtration enhances selectivity and reduces processing time for sterol enrichment.

3.1.5 Preparative HPLC (Adapted from Schröder and Vetter [17], Bian et al. [8])

1. Dissolve the semi-purified sterol fraction in methanol or methanol: Acetonitrile (80:20 v/v) to make a clear solution (*see* **Note 7**).
2. Filter the solution through a 0.22 μm syringe filter to remove particulates.
3. Inject the sample into a preparative HPLC system equipped with a C18 reverse-phase preparative column (250 mm × 10 mm, 5 μm) and UV detector set at 210 nm.
4. Run the column under isocratic conditions using methanol or a methanol: Acetonitrile mobile phase.
5. Collect fractions manually or using an automated fraction collector.
6. Identify sterol-rich fractions using UV absorbance and pool relevant ones.
7. Evaporate solvents from pooled fractions using a rotary evaporator and store the purified sterol in a desiccator.

Fractions rich in β-sitosterol, stigmasterol, or campesterol appear as clear peaks with distinct retention times, enabling separation of individual sterol types for use in analytical or pharmaceutical applications.

3.2 Characterization Methods

3.2.1 Thin Layer Chromatography (TLC) (Adapted from Salgado et al. [14] and Zhang et al. [13])

1. Prepare TLC plates (Silica gel 60 F_{254}) by cutting to appropriate dimensions and marking the baseline 1.5 cm from the bottom.
2. Spot 2–5 μL of sterol sample (dissolved in hexane) along with a standard sterol mixture for comparison.
3. Develop the plate in a TLC chamber pre-saturated with hexane: Acetone (9:1 v/v) as the mobile phase.
4. Allow the solvent front to reach ~80% of the plate height before removing the plate (*see* **Note 8**).
5. Dry the plate in air and visualize under a UV lamp (254 nm).
6. For enhanced visualization, spray the plate with freshly prepared anisaldehyde-sulfuric acid reagent and heat at 110 °C until spots appear (*see* **Note 9**).

Phytosterol spots typically appear violet or bluish, with Rf values ranging between 0.3 and 0.6 depending on the sterol type and mobile phase.

3.2.2 Gas Chromatography-Mass Spectrometry (GC-MS) (Adapted from Salgado et al. [14])

1. Accurately weigh 2–5 mg of the dried phytosterol extract and dissolve it in 0.5–1.0 mL of anhydrous hexane (*see* **Note 10**).
2. Transfer the solution to a sealed glass vial.
3. Add 100 μL of BSTFA + 1% TMCS (derivatization reagent) to the vial (*see* **Note 11**).
4. Heat the vial at 70 °C for 30–45 min to complete the silylation reaction.
5. Cool the solution to room temperature and inject 1 μL into the GC-MS system equipped with an HP-5MS column (30 m × 0.25 mm, 0.25 μm).
6. Use helium as the carrier gas. Maintain the injector temperature at 280 °C and apply a temperature gradient starting at 200 °C (held for 2 min), then increasing to 310 °C at 10 °C/min and held for 10 min.
7. Identify peaks by comparing retention times and mass fragmentation patterns with standard sterol spectra.

Sterol peaks (e.g., β-sitosterol, campesterol, stigmasterol) appear as distinct TMS-derivatized fragments in the chromatogram, with total ion current (TIC) confirming purity.

3.2.3 High-Performance Liquid Chromatography (HPLC) (Adapted from Zhang et al. [13] and Bian et al. [8])

1. Dissolve 2–5 mg of purified or crude phytosterol sample in 1 mL of methanol. Filter through a 0.22 μm syringe filter before injection.
2. Use an HPLC system equipped with a UV detector and a C18 reverse-phase analytical column (250 × 4.6 mm, 5 μm).
3. Set the mobile phase to methanol: Acetonitrile (80:20, v/v) at a flow rate of 1.0 mL/min (*see* **Note 12**).
4. Detect phytosterols at 210 nm. Inject 20 μL of the sample.
5. Identify peaks by comparing retention times with certified phytosterol standards (e.g., β-sitosterol, campesterol, stigmasterol).
6. Quantify sterols using external calibration curves prepared from standard sterol solutions (*see* **Note 13**).

Clear separation of sterol peaks is achieved under these isocratic conditions, and sample sterol content can be accurately quantified.

3.2.4 Nuclear Magnetic Resonance (NMR) Spectroscopy (Adapted from Wang et al. [18] and Bian et al. [8])

1. Dissolve 10–20 mg of the purified phytosterol sample in 0.6 mL of deuterated chloroform ($CDCl_3$) (*see* **Notes 14-15**).
2. Transfer the solution into a clean NMR tube (*see* **Note 13**).
3. Acquire spectra using a 400 MHz or higher NMR spectrometer.
4. Record both ^1H and ^{13}C NMR spectra for structural analysis.

5. Analyze the chemical shifts of olefinic protons (~5.3 ppm), methyl groups (~0.7–1.0 ppm), and the characteristic C-3 proton (~3.5 ppm) to confirm sterol structure.
6. Compare obtained spectra with published data for β-sitosterol, stigmasterol, and campesterol to confirm identity.

NMR confirms the structural integrity and substitution pattern of phytosterols and is especially useful for distinguishing closely related sterols.

3.2.5 Fourier-Transform Infrared (FTIR) Spectroscopy (Wang et al. [18])

1. Weigh a small amount (1–2 mg) of the dried phytosterol sample.
2. For ATR mode, place the powder directly onto the crystal and apply uniform pressure (*see* **Note 16**).
3. Alternatively, prepare a thin film by dissolving the sterol in a volatile solvent and allowing it to dry on the ATR crystal surface (*see* **Note 17**).
4. Collect spectra in the range of 4000–400 cm^{-1} at a resolution of 4 cm^{-1}.
5. Identify key functional groups based on characteristic absorption bands:
   - O-H stretching near 3400 cm^{-1}.
   - C-H stretching (CH_3, CH_2) near 2950–2850 cm^{-1}.
   - C=C stretching (alkenes) around 1650 cm^{-1}.
   - CH_3 bending around 1375 cm^{-1}.

FTIR provides a rapid and non-destructive means to verify the functional groups and structural integrity of phytosterol samples.

4 Notes

1. Methanol is preferred for recrystallization due to its good sterol solubility. A 1:1 methanol-acetone mix can improve purity. Avoid ethanol early on, as it may cause fatty acids to co-precipitate.
2. Store dried sterols in amber vials under nitrogen or vacuum at 4–8 °C. Avoid repeated freeze-thaw cycles to maintain stability.
3. Urea absorbs moisture easily. Keep it in a desiccator and prepare fresh urea-methanol solutions each time to avoid inconsistent results.
4. Acidifying the solution helps release sterols from urea complexes and improves the final yield.
5. During distillation, keep vacuum levels stable. Pressure changes can reduce sterol yield or damage heat-sensitive compounds.

6. Filter samples before membrane separation to reduce clogging. Clean membranes right after use with methanol and water.
7. Always use clean, filtered solvents and fresh mobile phases in HPLC to ensure reliable results and prevent blockages.
8. Apply small, narrow spots on TLC plates. Too much sample or wide bands can cause smearing and poor separation.
9. Prepare the TLC detection reagent fresh and store in amber bottles. Add sulfuric acid slowly and use the mix within 1–2 weeks for best color results.
10. Use 2–5 mg of sample for GC-MS. Use less for qualitative work; more may be needed for quantification. Don't exceed 5 mg to avoid overload.
11. Make sure the sample is fully dry before derivatization. Even small amounts of water can prevent proper reaction.
12. Always filter and degas HPLC samples and solvents to keep baselines stable and avoid damaging the column.
13. Make new standard curves every few days and store solutions at 4 °C in the dark to protect them.
14. Use dry, high-purity $CDCl_3$ for NMR. Moisture can interfere with the signal and reduce clarity.
15. For good NMR signals, use at least 10 mg/mL concentration of high-purity CDCl3. Weaker samples may need longer scans.
16. Use ATR for FTIR analysis. Avoid KBr pellets for sterols that absorb moisture.
17. For film-based FTIR, spread the sample evenly. Uneven films can affect peak accuracy and baseline quality.

References

1. Piironen V, Lindsay DG, Miettinen TA, Toivo J, Lampi AM (2000) Plant sterols: biosynthesis, biological function and their importance to human nutrition. J Sci Food Agric 80(7):939–966. https://doi.org/10.1002/(SICI)1097-0010(20000515)80:7
2. Shahzad N, Khan W, Md S, Ali A, Saluja SS, Sharma S (2023) Phytosterols as a natural therapeutic option for cardiovascular diseases: pharmacological insights and future prospects. Molecules 28(5):2214. https://doi.org/10.3390/molecules28052214
3. Valsta LM, Lemström A, Ovaskainen ML, Lampi AM, Toivo J, Korhonen T, Piironen V (2021) Estimation of plant sterol and stanol intake in Finland. Eur J Clin Nutr 75(7):1037–1045. https://doi.org/10.1038/s41430-020-00820-y
4. Demonty I, Ras RT, van der Knaap HC, Duchateau GS, Meijer L, Zock PL, Trautwein EA (2009) Continuous dose-response relationship of the LDL-cholesterol–lowering effect of phytosterol intake. J Nutr 139(2):271–284. https://doi.org/10.3945/jn.108.095125
5. Ras RT, Geleijnse JM, Trautwein EA (2021) LDL-cholesterol lowering of plant sterols and stanols—an updated meta-analysis of 124 randomized controlled trials. Food Funct 12(7):2993–3002. https://doi.org/10.1039/D0FO02498A
6. Poli A, Marangoni F, Corsini A, Manzato E, Marrocco W, Martini D, Visioli F (2021) Phytosterols, cholesterol control, and

cardiovascular disease. Nutrients 13(8):2810. https://doi.org/10.3390/nu13082810

7. Zhang Y, Wang Z, Xu J (2022) Novel sources and sustainable extraction methods for phytosterols: current trends and future perspectives. J Clean Prod 370:133407. https://doi.org/10.1016/j.jclepro.2022.133407
8. Bian Y, Zhang Y, Ruan LY, Feng XS (2024) Phytosterols in plant-derived foods: recent updates in extraction and analysis methods. Crit Rev Anal Chem:1–19. https://doi.org/10.1080/10408347.2024.2427128
9. Sujith Kumar MS, Mawlong I (2017) Phytosterol recovery from oilseeds: recent advances. J Food Process Eng 40(4):e12466. https://doi.org/10.1111/jfpe.12466
10. Nadeem M, Anjum FM, Arshad MU (2017) Comparative evaluation of purification techniques for phytosterol enrichment from deodorizer distillate. J Food Sci Technol 54(3): 784–790. https://doi.org/10.1007/s13197-017-2507-6
11. Seifollahi F, Eikani MH, Khandan N (2024) Optimization of operating conditions for lampante olive oil deacidification by short path molecular distillation. J Clean Prod. https://doi.org/10.1016/j.jclepro.2024.142304
12. Abd-Razak N, Zairossani M, Chew Y, Bird M (2020) Fouling analysis and the recovery of phytosterols from orange juice using regenerated cellulose ultrafiltration membranes. Food Bioprocess Technol 13:2012–2028. https://doi.org/10.1007/s11947-020-02541-7
13. Zhang X, Guo Z, Liu Y, Sun W (2020) Simultaneous analysis of free and esterified phytosterols in rice bran oil by GC-MS and HPLC. J Food Compos Anal 92:103560. https://doi.org/10.1016/j.jfca.2020.103560
14. Salgado J, Pereira MA, Silva CM (2022) Advances in derivatization strategies for the GC–MS analysis of phytosterols. J Sep Sci 45(12):2030–2040. https://doi.org/10.1002/jssc.202101234
15. Srividya AR, Vani S, Chandrasekhar P (2014) A rapid and cost-effective method for the purification of β-sitosterol using fractional crystallization and chromatography. J Am Oil Chem Soc 91(8):1273–1280. https://doi.org/10.1007/s11746-014-2477-9
16. Kasim NS, Gunawan S, Yuliana M, Ju YH (2010) A simple two-step method for simultaneous isolation of tocopherols and free phytosterols from soybean oil deodorizer distillate with high purity and recovery. Sep Sci Technol 45(15):2437–2446. https://doi.org/10.1080/01496391003789171
17. Schröder M, Vetter W (2012) High-speed counter-current chromatography for the isolation of phytosterols with high purity. J Chromatogr A 1252:44–52. https://doi.org/10.1016/j.chroma.2012.06.019
18. Wang J, Li X, Zhou L (2023) Application of FTIR and NMR for structural elucidation of phytosterols. Food Chem 398:134032. https://doi.org/10.1016/j.foodchem.2023.134032

Chapter 13

Isolation and Characterization of Health-Promoting Alkaloids

Vigneshwaran Subramani and Vidisha Tomer

Abstract

Alkaloids, a diverse group of naturally occurring nitrogen-containing compounds, are renowned for their wide-ranging health-promoting properties, including anticancer, anti-inflammatory, antimicrobial, and neuroprotective effects. Owing to their therapeutic potential, the efficient isolation and accurate characterization of alkaloids from plant and microbial sources are of critical importance in drug discovery and nutraceutical development. This chapter presents a comprehensive overview of classical and advanced methodologies for alkaloid isolation and purification by column chromatography and high-speed counter-current chromatography. Furthermore, it highlights the characterization strategies employed through various chromatographic techniques like TLC, HPLC, UPLC, and GC-MS. Emphasis is placed on the analytical precision, instrumentation setups, and procedural optimizations required for the structural elucidation and quantification of bioactive alkaloids. By bridging phytochemical isolation techniques with modern analytical chemistry, this chapter aims to facilitate the application of health-benefiting alkaloids in the biomedical and pharmaceutical sectors.

Key words Alkaloids, Extraction, Isolation, Purification, Quantification

1 Introduction

Alkaloids are naturally secondary metabolites predominantly found in plants, and trace amounts in animals and microorganisms [1]. Some well-known alkaloids are caffeine in coffee, cocaine in cocoa, quinine in cinchona bark, and nicotine in tobacco plants. These are bioactive molecules with low molecular weight and are primarily composed of nitrogen, carbon, hydrogen, and oxygen, with minor constituents of sulphur, phosphorus, and chlorine [2]. A distinguishing characteristic of alkaloids is the existence of the nitrogen atom inside the heterocyclic ring structure. This nitrogen atom plays a crucial role in their diverse biological activities [3].

Figure 1 illustrates the classification of alkaloids based on their ring structure and their biosynthetic origin. Alkaloids have been used therapeutically in numerous studies for their anticancer,

Ashwani Kumar, Vidisha Tomer and Piyoosh Kumar Babele (eds.), *Phytochemicals: Extraction, Quantification, Isolation, and Characterization*, Methods and Protocols in Food Science, https://doi.org/10.1007/978-1-0716-4840-7_13,

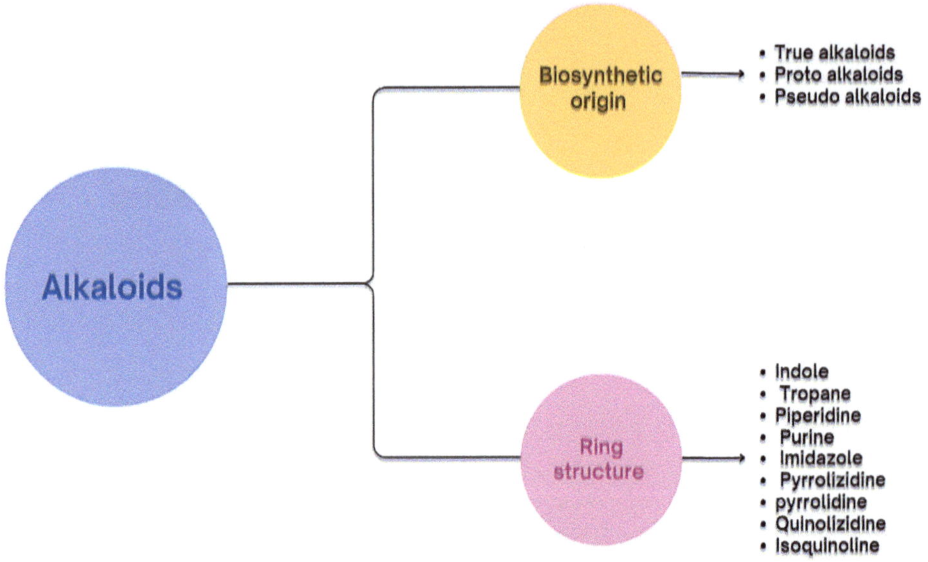

Fig. 1 Classification of alkaloids

anti-inflammatory, antiproliferative, and other health-promoting properties. Vincristine and vinblastine are potential alkaloids derived from *Catharanthus roseus* utilized in treatment of cancer. Certain alkaloids, such as vinblastine and vincristine function as anticancer agents by interfering with cellular calcium signalling cascade. Aporphine and Isoquinoline, both alkaloids have shown comparable antitumor efficacy. Phenoxazone alkaloids derived from *Pycnoporus cinnabarinus* (fungus) demonstrated anti-inflammatory, antimicrobial, and antiviral activities [4]. Research studies also showed that isoquinoline alkaloids are effective against inflammation [5]. Alkaloids also possess an antiproliferation effect by inhibiting cell growth and division. Alkaloids like zanthoxylum and evodiamine act as effective antiproliferative [6]. Most of the significant studies reported that Berberine was effective in treating neurodegenerative and inflammatory diseases. The clinical study on the effect of berberine on transgenic mice with Alzheimer's disease showed that berberine significantly reduced the amyloid-beta (Aβ) plaque aggregation and inhibited the amyloid precursor protein (APP) phosphorylation. The significant improvement in the neural function was observed in the study. This shows the therapeutic activity of berberine [7]. In addition to berberine, other alkaloids like aloperine and sinomenine were found to be effective against neurodegenerative diseases and treating inflammation. The alkaloids are often found in complex mixtures within plant matrices therefore to unlock their potential therapeutic benefits, their isolation and purification are essential. The subsequent sections will detail the isolation and characterization techniques employed for alkaloids.

2 Materials

2.1 Column Chromatography (CC) [8]

(a) **Apparatus**

- Glass column.
- Percolator.
- UV detector.

(b) **Reagents**

- Methanol.
- 1% of Acetic acid (*see* **Note 1**).
- Sodium hydroxide.
- 0.1 M HCl (*see* **Note 2**).

(c) **Standard Preparation**

- Standard alkaloids like coptisine, corysamine, magnoflorine, chelerythrine, sanguinarine, chelilutine, chelirubine, and berberine are prepared by dissolving in methanol (Concentration of 1 mg/mL).

2.2 High Speed Counter Current Chromatography (HSCCC) [9]

(a) **Apparatus**

- Reflux apparatus.
- Rotary vacuum evaporator.
- Separatory funnel.
- Sonicator.
- Macroporous resin column.
- UV-Vis Detector.

(b) **Reagents**

- 95% Ethanol (Prepared by adding 95 mL ethanol to 5 mL of distilled water).
- Methylene chloride (analytical grade).
- Methanol (analytical grade).
- 0.1 M HCl (*see* **Note 2**).
- 1% HCl (*see* **Note 3**).
- 0.3 mol/L of HCL (*see* **Note 4**).

2.3 Thin Layer Chromatography [8]

(a) **Apparatus**

- Precoated TLC plates (particle size of 0.25 mm).
- UV detector.

(b) Reagents

- Methanol.
- Acetic acid.
- 0.1 M hydrochloric acid (*see* **Note 2**).
- Sodium hydroxide.

(c) Mobile Phase

- An organic aqueous mobile phase is prepared by the addition of modifiers like mixture of methanol and 10% of 0.1 M hydrochloric acid followed by adjusting the mixtures to two different pH values—5 and 6 using acetic acid and NaOH.

2.4 High Performance Liquid Chromatography—Mass Spectrometry [10]

(a) Apparatus

- Homogenizer.
- Centrifuge.
- Cartridge.
- HPLC column (10 cm × 2.1 mm).
- Mass spectrometry.

(b) Reagents

- 60% of Methanol (prepared by adding 60 mL of methanol to 40 mL of distilled water).
- 0.1% of Hexafluoro butyric acid (*see* **Note 5**).
- Acetonitrile (HPLC grade).

2.5 Ultra Performance Liquid Chromatography—Mass Spectrometry [11]

(a) Apparatus

- Shaker.
- Sonicator.
- Rotary vacuum evaporator.
- UPLC column (100 mm × 3 mm).
- Mass spectrometry.

(b) Reagents

- Methanol (HPLC grade).
- Acetonitrile.
- 5 mM of ammonium acetate buffer (*see* **Note 6**).

2.6 Gas Chromatography—Mass Spectroscopy [12]

(a) Apparatus

- Hot air oven.
- TLC plates.
- GC column.
- Mass spectrometry.

(b) Reagents

- Petroleum ether.
- Toluene.
- Acetone.
- Ammonia.
- Ethanol.
- Dichloromethane.
- Methanol.
- Acetonitrile.

3 Methods

3.1 Column Chromatography [8]

(a) Sample Preparation

1. Sample is dried at room temperature; 200 g of the dried sample is treated with 1% cold acetic acid (2 L) using the percolation method.
2. The above mixture is adjusted to pH 6 by using NaOH incubated at ambient temperature for 2 days.
3. Further, the resultant mixture is filtered and used for isolation.

(b) Column Chromatography

1. The filtered mixture is passed through a glass column of 20 × 3 cm is packed with 100 g of Kiesel gel 60 (particle size of 0.06–0.1 mm).
2. Eluate no. 1 contains tertiary alkaloids like chelidonine, homochelidonine, protopine, and allocryptopine.
3. Similarly, eluate no. 2 contains quaternary alkaloids retained on the column with methanol and 10% of 0.1 M HCl collected and used for further analysis (Subheading 3.3 explains the characterization of alkaloids by TLC).

3.2 High Speed Counter Current Chromatography

(a) **Sample Preparation**

1. 40 g of dried sample is mixed with 400 mL of 95% ethanol and the extraction is carried out twice using the refluxing method for 2 h each time.
2. The extract mixture is filtered, followed by concentration using rotary vacuum evaporator at 45 °C.
3. The final residue is dissolved in 300 mL of 1% HCl and the pH is adjusted to 10 with ammonium hydroxide.
4. The resulting above mixture is further treated thrice with methylene chloride using 1:1 (v/v) ratio.
5. The combined organic layers of *M. cordata* are evaporated using a rotavapor at 45 °C, resulting in a brown yellow residue stored for further analysis.
6. Two-phase solvent system preparation: To isolate alkaloids, a 4: 2:2 (v/v/v) of dichloromethane: methanol: aqueous mixture of hydrochloric acid* is created and then mixed with an ionic liquid**.
   (a) (*0.3 mol/L of HCL, **([C_4mim] [BF_4])-1-butyl-3-methyllimidazolium tetrafluoroborate).
7. Each solvent system passes through a separatory funnel and solvents are stabilized by shaking at room temperature.
8. After settling down, separation of upper and lower phase occurs, which are separated, degassed and sonicated.
9. The crude extract (200 mg) is dissolved in 1.5 mL of both upper and lower phase.

(b) **Chromatographic Separation**

1. Fill the multilayer column with upper phase (stationary) at 9 mL/min followed by addition of lower phase (mobile) at 2 mL/min followed by rotating the column at 800 rpm.
2. Each of the columns are allowed to equilibrate and the sample is injected into the sample loop.
3. The column effluents are constantly recorded and detected by UV-Vis at 254 nm.
4. The separated components are collected individually and evaporated under rotary vacuum conditions to remove the solvents; however ionic liquid modifiers are removed by macroporous resin column.
5. A macroporous resin column (D-101, 400 mm × 25 mm of internal diameter) prepared with 70% of aqueous ethanol in step wise fractionation mode.
6. The effluent is recorded by UV-Vis detector at 210 nm.
7. The collected residues are evaporated under reduced pressure to obtain the purified alkaloids.

3.3 Thin Layer Chromatography [8]

(a) TLC Plate Preparation

1. A precoated TLC plates with Kiesel gel Si 60 (particle size of 0.25 mm) is used for analysis.

(b) Chromatographic Development

1. For characterization of alkaloids isolated via column chromatography (refer to Subheading 3.1), the following methods are employed.
2. Apply small volume of standard alkaloids and *Chelidonium majus* L. root extract as spot on TLC plates immersed in mobile phase.
3. Through capillary movement, mobile phase ascends the sample spot on plate until it reaches the predetermined distance from the origin.

(c) Detection and Visualization

1. After development of chromatograph, TLC plates are removed from development chamber and the plates are dried.
2. The separated spots are detected by UV at a wavelength of 366 nm.
3. Scanning of the chromatograph using densitometer in reflectance mode at wavelength of 280 nm favors quantitative analysis of alkaloid compounds.

3.4 High Performance Liquid Chromatography—Mass Spectrometry [10]

(a) Sample Preparation

1. 200 mg of homogenized lupin sample is dissolved in 1 mL of 60% of aqueous methanol followed by homogenization and centrifugation.
2. The supernatant from the mixture is collected and 50 μL of supernatant is dissolved in 1 mL of 90:10 (v/v) mixture of water and methanol.

(b) Clean-Up Process

1. The conditioning of solid phase extraction cartridge involves the addition of 1 mL of methanol.
2. The cartridge equilibrates with addition of 1 mL of 90:10 (v/v) mixture of water and methanol.
3. After diluting the sample with 1 mL of equilibrated 90:10 (v/v), mixture of water and methanol is introduced in the cartridge.
4. Allow the sample to pass through the cartridge which provides maximum interaction between analytes and stationary phase.

5. 1 mL of water: Methanol in the ratio of 90:10 (v/v) is introduced into the cartridge to collect the eluents for HPLC-MS/MS analysis.

(c) HPLC Analysis

1. The column of C18-PFP with internal diameter of 10 cm × 2.1 mm packed with 2 μm particles and equipped with security guard column is used to develop the chromatogram.
2. Mobile phase (A) consists of 0.1% of Hexafluoro butyric acid.
3. Mobile phase (B) consists of a 50:50 (v/v) mixture of acetonitrile and methanol with subsequent addition of 0.1% of Hexafluoro butyric acid.
4. The gradient elution program is set to the following conditions:
   (a) 0.0 to 0.1 min—10% B.
   (b) 0.1 to 3.1 min—Linear increase to 50% B.
   (c) 3.1 to 7.1 min—Isocratic at 50% B.
   (d) 7.1 to 10.1 min—Linear increase to 90% B.
   (e) 10.1 to 10.6 min—Isocratic at 90% B.
   (f) 10.6 to 11.1 min—Linear decrease back to 10% B.
   (g) 11.1 to 12 min—Isocratic at 10% B (re-equilibration).
5. The flow rate and injection volume of the sample are set to 0.300 mL/min and 6 μL.

(d) Mass Spectrometry

The following conditions of mass spectrometry are employed for analyzing the alkaloids. All analytes are identified in electrospray ionization positive mode operating in scheduled ionization mode:

1. Nebulizer gas pressure (air)—40 psi.
2. Capillary voltage—5500 V.
3. Source temperature—500 °C.
4. Turbo gas (nitrogen)—40 psi.
5. Source temperature—500 °C.
6. Set the ionization window to 1.5–4.5 min during the chromatographic run.

(e) Analyte—Specific Parameter Optimization

1. For each target alkaloid, addition of single standard methanolic solution of 10 ng/mL (nanograms per milliliter) (*see* **Note 7**) into mass spectrometry at flow rate of 10 μL/min.

2. In same way, two precursor ion/product ion transitions are chosen for target alkaloid for quantification and confirmation.
3. The acquisition and quantification of ions are processed in multiple reaction monitoring mode.

(f) Data Processing

1. Data processing is carried out by Multiquant 3.0.3 for identification and quantification of target alkaloids.

(g) Calculations

Recovery (%) = Area of the spiking analyte/sample area.

3.5 Ultra Performance Liquid Chromatography—Mass Spectrometry [11]

(a) Sample Preparation

1. 500 mg of powdered *Murraya koenigii*is mixed with 10 mL of methanol and then placed on shaker for 48 h.
2. The mixture is treated under rotary vacuum evaporator at reduced pressure of 35–40 °C.
3. Mix the extract residue with 1 mL of 100% methanol using sonication.
4. 5 μL of the sample is injected into UPLC system for further analysis.

(b) Preparation of Standard Solutions

1. Working standard solutions are prepared for each alkaloid (except MK279B and MK293B) by serially diluting a 1 mg/mL stock solution with methanol to create a concentration range spanning 0.003–50 μg/mL.
2. A stock solution containing all nine analytes (including mk279A and mk293A) is prepared in methanol. This mixed stock solution is then diluted to create a series of working standard solutions that cover suitable concentration ranges for each of the nine analytes.
3. For each analyte, determine the average integrated peak area.
4. Construct calibration curves for all nine analytes by plotting average peak areas against their corresponding concentrations and performing linear regression analysis for each.

(c) Chromatographic Conditions

1. In the UPLC system, the analysis is carried out in Multiple Reaction Mode (MRM), followed by connecting with a mass spectrometer.
2. The Electrospray Ionization (ESI) source on the mass spectrometer operates in negative ionization mode.

3. The chromatographic column of C18 (100 mm × 3 mm) with an average particle size of 2.6 μm used for the purification of analytes.
4. UPLC system is equipped with an autosampler in which 10 μL of sample is injected to perform analysis.
5. The mixture of 5% acetonitrile (A) and 5 mM of ammonium acetate buffer (B) combine to form the mobile phase.
6. The separation program employs time segmented elution gradient; 0–1 min: 60–70% A, 1–3 min: 70–85% A, and 3–6 min: 85–60% A at a constant flow rate of 0.5 mL/min.
7. Column temperature is set to 35 °C and the autosampler temperature to 25 °C.
8. Dwell time is set to 0.01 or 0.02 s for multiple reaction mode data collection.
9. The ESI source is configured to the following conditions:
   (a) Source temperature—25 °C.
   (b) Desolvation temperature—350 °C.
   (c) Capillary voltage—3.5 kV.
   (d) Dissolution gas flow rate of 950 L/h.
10. The collision gas, argon, flow at the rate of 0.1 mL/min whereas the flow of cone gas maintained at 60 L/h.
11. The activation energy ranges from 18 to 28 eV and inlet voltage varies between 38 and 68 V, depending on the specific analyte being analyzed.
12. Quan Lynx V4.1 SCN 714 software utilized for processing of data.

3.6 Gas Chromatography—Mass Spectroscopy [12]

(a) Sample Preparation

1. The stem of *Anabasis articulata* is dried and coarsely ground into powder with a particle size of not more than 0.5 mm.
2. Mix 65 g of coarsely ground dried stem of *Anabasis articulata* in 133 mL of petroleum ether and placed at ambient temperature for 24 h.
3. The mixture is filtered, and the resulting filtrate is dried in a hot air oven at 37 °C.
4. Further, the dried filtrate is subjected to extraction by polar and apolar solvents.

(b) Recovery of Alkaloids

1. Two extraction methods are used: (a) E1 for purification of overall alkaloid content and (b) E2.1 and E2.2 for separation of basic and quaternary ammonium alkaloids using Stas-otto method by using polar and apolar solvent.

(c) Separation of Alkaloids

1. Alkaloids are separated using a silica gel plate (60F254) or silica gel (G 60) treated sheets of aluminium.
2. For isolating E2.1 alkaloids, a solvent system of toluene/acetone/ethanol/ammonia is used in a ratio of 4:4:6:2.
3. Similarly, for isolating E2.2 alkaloids, a solvent mixture of dichloromethane/methanol is used in a ratio of 9:1.
4. Samples are spotted on TLC plates, and the resulting chromatograms are visualized under ultraviolet (UV) light at wavelength of 254 and 360 nm.
5. Fractions are collected, and then dissolved in a toluene/acetone/ethanol/ammonia system before being subjected to GC-MS analysis.

(d) GC-MS Analysis

1. The method includes resuspending each powder in 50 μL of acetonitrile, followed by constant vortexing for 15 min at room temperature.
2. The gas chromatography system consists of a DB-5 column with dimensions of 30 m × 0.25 mm internal diameter and 0.25 μm film thickness.
3. The column temperature of the gas chromatography is set to 70 °C for 2 min.
4. Gradually raise the column temperature to 200 °C at a rate of 25 °C/min for 5 min.
5. Further, the column temperature is raised to 300 °C at a rate of 3 °C/min for 10 min.
6. In addition to the temperature, other processing parameters are set to the following conditions.
   (a) Helium, the carrier gas flowing at a rate of 1.0 mL/min.
   (b) The ion source temperature—230 °C.
   (c) Temperature of interface—250 °C.
   (d) Two minutes of solvent cut time.
7. The compounds are detected in scan modality mode ranging from 100 to 1000 m/z.
8. For GC analysis, 2 μL of the developed *A.articulata* solution is injected into the column, where they interact with the mobile phase, and elution occurs.
9. Eluted compounds enter the mass spectrometer, where ionization takes place, and the compounds are measured.

(e) **Calculations**

Further the composition of the compounds are calculated by the below formula:

1. Concentration of the compound = (peak area/response factor).
2. Response factor = (calibration concentration/peak area).

4 Notes

1. 1% of acetic acid prepared by adding 10 mL of acetic acid to 990 mL of distilled water by continuous stirring.
2. For preparing 250 mL of 0.1 M of HCl, 247.92 mL of distilled water is taken followed by addition of 2.08 mL of concentrated HCl by constant stirring.
3. For preparation of 1% of HCl, 3 mL of concentrated HCl is slowly added to 297 mL of distilled water with constant stirring.
4. For preparation of 1000 mL of 0.3 M of HCl, 500 mL of distilled water is taken followed by addition of 25 mL of concentrated HCl by constant stirring and then more distilled water is added until the final volume reaches 1000 mL.
5. For preparing 0.1% of HFBA solution, 0.1 mL of HFBA is added to 70 mL of distilled water followed by constant stirring and the water is added until the final volume reaches to 100 mL.
6. For preparation of 5mM of ammonium acetate buffer, 0.096 g of ammonium acetate is dissolved in 200 mL of distilled water and then add water until the final volume reaches 250 mL.
7. To prepare 10ng/mL of methanol,a stock solution(1mg/ml) of methanol is prepared and then 10 μL of this is taken and dissolved in 9990 μL of methanol.

References

1. Umer SM, Solangi M, Khan KM, Saleem RSZ (2022) Indole-containing natural products 2019–2022: isolations, reappraisals, syntheses, and biological activities. Molecules 27:7586. https://doi.org/10.3390/molecules27217586
2. Dey P et al (2020) Analysis of alkaloids (indole alkaloids, isoquinoline alkaloids, tropane alkaloids). In: Recent advances in natural products analysis. Elsevier, Amsterdam, pp 505–567. https://doi.org/10.1016/B978-0-12-816455-6.00015-9
3. Heinrich M, Mah J, Amirkia V (2021) Alkaloids used as medicines: structural phytochemistry meets biodiversity—an update and forward look. Molecules 26:1836. https://doi.org/10.3390/molecules26071836
4. Dias DA, Urban S (2009) HPLC and NMR studies of phenoxazone alkaloids from *Pycnoporus cinnabarinus*. Nat Prod Commun 4:4. https://doi.org/10.1177/1934578X0900400409

5. Barbosa-Filho JM et al (2006) Anti-inflammatory activity of alkaloids: a twenty-century review. Rev Bras Farmacogn 16:109–139. https://doi.org/10.1590/S0102-695X2006000100020
6. Hu X et al (2018) Antiproliferative effects of alkaloid evodiamine and its derivatives. Int J Mol Sci 19:3403. https://doi.org/10.3390/ijms19113403
7. Durairajan SSK et al (2012) Berberine ameliorates β-amyloid pathology, gliosis, and cognitive impairment in an Alzheimer's disease transgenic mouse model. Neurobiol Aging 33:2903–2919. https://doi.org/10.1016/j.neurobiolaging.2012.02.016
8. Meyers CLF (2000) Column chromatography. Curr Protoc Nucleic Acid Chem Appendix 3: Appendix 3E. https://doi.org/10.1002/0471142700.nca03es03
9. Liu Y et al (2020) Isolation and purification of alkaloids from the fruits of *Macleaya cordata* by ionic-liquid-modified high-speed counter-current chromatography. J Sep Sci 43:2459–2466. https://doi.org/10.1002/jssc.201901242
10. Eugelio F (2023) Development of an HPLC-MS/MS method for the determination of alkaloids in lupins. Molecules 28:1531. https://doi.org/10.3390/molecules28041531
11. Nandan S et al (2021) Quantitative analysis of bioactive carbazole alkaloids in *Murraya koenigii* (L.) from six different climatic zones of India using UPLC/MS/MS and their principal component analysis. Chem Biodivers 18: e2100557. https://doi.org/10.1002/cbdv.202100557
12. Belyagoubi-Benhammou N, Belyagoubi L, Gismondi A, Di Marco G, Canini A, Atik Bekkara F (2019) GC/MS analysis, and antioxidant and antimicrobial activities of alkaloids extracted by polar and apolar solvents from the stems of Anabasis articulata. Med Chem Res 28:754–767. https://doi.org/10.1007/s00044-019-02332-6

Chapter 14

Isolation and Characterization of Health-Promoting Polypeptides

Poorva Sharma and Piyush Kashyap

Abstract

Polypeptides are short chains of amino acids. Peptides are released from the proteins by enzymatic hydrolysis, microbial fermentation, or food processing. Proteolytic enzymes are used to break down the native protein into smaller peptides. Characterization of these peptides is done either based on molecular weight (MW) or by identifying the sequence of amino acids. Functional properties such as solubility, water binding capacity, emulsion capacity, dispersibility and foaming capacity decide the final application of these peptides in the food industry.

Key words Microbial fermentation, Enzymatic hydrolysis, SDS-PAGE, Functional properties

1 Introduction

Proteins are one of the essential components of human diets. These bio macromolecules are formed of amino acids which are linked together by peptide bonds between the amino group of one amino acid and the carboxyl group from the following amino acid in the chain [1]. Proteins participate in a large number of physiological and functional processes. Many of these properties of proteins are attributed to biologically active peptides encrypted in the parent protein molecules [2, 3]. Peptides are released from various food sources (rich in protein) using a range of methods such as in vitro enzymatic hydrolysis (gastrointestinal digestion, in vitro hydrolysis using proteolytic enzymes of plant, animal, or microbial origin) and also during food processing (cooking, fermentation, ripening) [4]. In vitro enzymatic hydrolysis of food proteins is performed by using proteolytic enzymes such as pepsin [5], bromelain [6], trypsin [7], papain [8], alcalase [9], pancreatin [10], chymotrypsin [11], thermolysin [12], flavorzyme, neutrase [13], and protamex

Ashwani Kumar, Vidisha Tomer and Piyoosh Kumar Babele (eds.), *Phytochemicals: Extraction, Quantification, Isolation, and Characterization*, Methods and Protocols in Food Science, https://doi.org/10.1007/978-1-0716-4840-7_14,

[14]. During the action of proteolytic enzymes, proteins get fragmented at particular catalytic sites to form bioactive peptides which possess the properties like tissue affinity, specificity, and efficiency [4]. Alternatively, peptides can also be produced by fermentation method. In fermentation method, fermentation is usually carried out using different generally recognized as safe (GRAS) microbial strains which release the peptides from various food proteins by the action of microbial proteases [15]. These peptides are appraised as bioactive if they possess bio-functionalities including immunomodulatory, antidiabetic, antimicrobial, antihypertensive, and antioxidative amongst several others.

For the fractionation of peptides, the hydrolysates are subjected to fractionation on the basis of their molecular weights. Peptides fraction(s) with maximum bioactivity are purified by using various techniques such as reversed phase high pressure liquid chromatography (RP-HPLC), affinity chromatography, ion exchange chromatography, size exclusion chromatography, hydrophilic interactions liquid chromatography (HILIC), ultra-high pressure liquid chromatography (UPLC), electrospray ionization mass spectrometry (ESI-MS), and isoelectric focusing [16]. Characterization of bioactive peptides is done through a new field of study known as nutriproteomics. These techniques are gaining momentum due to their rapid and precise identification method. High-resolution mass spectrometry perhaps represents the leading method for this purpose because it is sufficiently sensitive to detect and identify different types of analytes present in complex mixtures, even when they are available at low concentrations [17].

2 Materials

2.1 Isolation of Bioactive Peptides

Prepare all solutions using distilled water and store them in refrigerated conditions. Diligently follow all waste disposal regulations when disposing of waste materials.

2.1.1 Solvent Extraction [18]

(a) Apparatus

- Vacuum centrifuge.
- Centrifuge.
- HPLC.

(b) Reagents

- *Extracting solvent*: 60% (v/v) acetonitrile (MeCN) containing 0.1% trifluoroacetic acid (TFA).
- *Resuspended solution*: 0.05% TFA/H_2O (v/v) to a concentration of 100 mg/mL.

- Reversed phase C18 35cc Sep-Pak Vac cartridges equilibrated with 0.05% TFA/H20 (v/v).
- *Eluted solution*: 10%, 20%, 30%, 40%, and 80% (v/v) MeCN containing 0.05% TFA (v/v).

2.1.2 Enzymatic Hydrolysis [5]

(a) Apparatus

- pH meter.
- Centrifuge.

(b) Reagents

Chemical Required for Protein Isolation

- 0.1 M NaOH (*see* **Note 1**).
- 6 M HCl (*see* **Note 2**).

Chemicals Required for Protein Hydrolysates Preparation

- Protein isolates.
- Protease enzymes: papain, pepsin, trypsin, and alcalase.
- 0.1 M glycine-HCl buffer, pH 2 (*see* **Note 3**).

2.1.3 Microbial Fermentation [19]

(a) Apparatus

- Incubator.
- Centrifuge.
- Water bath.

(b) Reagents

- Deionized water.
- 0.5 N NaOH.
- Whatman filter paper No. 1.
- Microbial culture for fermentation.
- Medium for microbial growth (*see* **Note 4**).

2.2 Characterization of Peptides

2.2.1 SDS Polyacrylamide Gel Electrophoresis [20]

(a) Gel Components

- Acrylamide/Bis-acrylamide solution (30% w/v) (*see* **Note 5**).
- Resolving gel buffer: 1.5 M Tris–HCl, pH 8.8.
- Stacking gel buffer: 0.5 M Tris–HCl, pH 6.8.
- Sodium dodecyl sulfate (SDS, 10%).
- Ammonium persulfate (APS, 10% w/v solution).
- Tetramethylethylenediamine (TEMED).

(b) Sample Preparation

- Protein or peptide sample.
- *SDS loading buffer (Sample buffer)*: Contains Tris–HCl, SDS, glycerol, β-mercaptoethanol or dithiothreitol (DTT), and bromophenol blue (*see* **Note 6**).

(c) Running Buffer:

25 mM Tris–HCl, 192 mM glycine, 0.1% SDS (pH ~8.3) (*see* **Note 7**).

(d) Staining and Destaining

- Coomassie Brilliant Blue R-250 or silver stain.
- Destaining solution: methanol, acetic acid, and water.

2.2.2 Amino Acid Composition

(a) Apparatus

- LCMS.
- Oven.

(b) Reagents

- 6 N Hydrochloric acid.
- Phenol (two to three drops of 0.1% phenol).
- Buffer for HPLC or amino acid analyzer (*see* **Note 8**).
- Amino acid standards (*see* **Note 9**).
- *Derivatization reagent*: o-Phthalaldehyde (OPA) for fluorescence detection.
- Dansyl chloride or PITC (phenylisothiocyanate).
- Organic solvents: Acetonitrile or methanol (HPLC grade).

2.3 Functional Properties [21]

2.3.1 Solubility

(a) Apparatus

- Analytical balance.
- Magnetic stirrer.
- Centrifuge.
- Spectrophotometer.
- pH meter.

(b) Reagents

- Sodium chloride.
- Lowry reagent.
- Distilled water.

(c) **Others**

- Whatman No. 1 filter paper.

2.3.2 Water Holding Capacity (WHC)

(a) **Apparatus**

- Analytical balance.
- Thermometer.
- Centrifuge.

(c) **Others**

- Filter paper.

2.3.3 Emulsion Capacity and Stability

(a) **Apparatus**

- Analytical balance.
- Thermometer.
- Centrifuge.
- pH meter.
- Homogenizer.

(b) **Reagents**

- Distilled water.
- Vegetable oil.
- Phosphate buffer (pH 7.0) (*see* **Note 10**).

(c) **Others**

- Filter paper.

2.3.4 Dispersibility

(a) **Apparatus**

- Analytical balance.

(b) **Reagents**

- Distilled water.

(c) **Others**

- Whatman filter paper No. 1.

2.3.5 Foam Capacity and Stability

(a) **Apparatus**

- Homogenizer.
- Graduated cylinder.

- Analytical balance.
- pH meter.

(b) Reagents

- Distilled water.
- Phosphate buffer.

2.4 Biofunctional Activity [22]

2.4.1 Antioxidant Activity

(a) Apparatus

- Spectrophotometer.

(b) Reagents

- DPPH solution (0.1 mM in methanol) (*see* **Note 11**).
- Methanol.

2.4.2 Antimicrobial Activity

(a) Apparatus

- Biosafety cabinet.
- Sterile plates.
- Micropipettes.
- Incubator.

(b) Reagents

- Test microorganisms.
- Growth media.
- Phosphate buffer saline.
- Positive control: Known antibiotic (e.g., ampicillin or gentamicin).
- Negative control: Sterile water or solvent used to dissolve the test sample.

3 Methods

3.1 Isolation of Bioactive Peptides

(a) Solvent Extraction [18]

1. Take samples (100 g) and extract twice with five volumes of extracting solution for 24 h at 4 °C. The combined supernatants were incubated at −20 °C for approximately 1 h allowing two liquid phases, an organic, MeCN-rich phase and an aqueous, salt-rich phase to be formed and separated.
2. Dry aqueous phase by using vacuum centrifuge and resuspend in solution (0.05% TFA/H_2O (v/v) to a concentration of

100 mg/mL). Remove the salt from the aqueous phase by solid phase extraction (SPE).

3. For solid phase extraction, load extracts onto reversed-phase C18 35cc Sep-Pak Vac cartridges equilibrated with 0.05% TFA/H_2O (v/v). After washing with acidified water, elute analytes with 10%, 20%, 30%, 40%, and 80% (v/v) MeCN containing 0.05% TFA (v/v).
4. Dry eluate in a vacuum centrifuge and resuspend in MQ-H_2O.

(b) Protein Isolate Preparation

1. Sample is blended in 0.1 M NaOH (1:5 w/v) and stirred for 4 h at 4 °C. Centrifuge this mixture at 4000 rpm for 20 min.
2. Precipitate the supernatant using 6 M HCl to pH 4.0 and further centrifuge at 3500 rpm for 15 min.
3. Discard the supernatant, collect the pellet, lyophilize, and store at −4 °C as protein isolate.

(c) Protein Hydrolysates Preparation

1. Protein isolates (4 mg/mL) were hydrolyzed with the enzymes pepsin and trypsin separately (1 mg/mL) in 0.1 M glycine–HCl buffer, pH 2 and 0.1 M phosphate buffer, pH 8, respectively, while varying enzyme: Substrate ratio at 1:8, 1:16, and 1: 32 (v/v).
2. The mixtures were incubated for 4 h at 37 °C with incessant mixing at intervals, and the pH was monitored throughout hydrolysis.
3. At the end of the fourth hour, the mixtures were boiled at 100 ° C for 10 min to terminate the reaction, promptly placed on ice, and centrifuged at 4000 rpm for 20 min. The supernatants were collected, freeze-dried, and stored at −20 °C.

(d) Microbial Fermentation [19]

1. For the isolation of peptides, substrates are defatted using hexane (*see* **Note 12**). Extract the defatted sample with deionized water after centrifugation (2000 *g*) to remove the extraneous raw materials.
2. During extraction the pH of the suspension is maintained (7.5–11.5) by using 0.5 N NaOH and the temperature was regulated at 50 °C by a water bath.
3. Filter the supernatant through Whatman filter paper No. 1 and store at freezing conditions until it is used in the experiment.
4. For the preparation of microbial cells for fermentation, the cells are grown till early stationary phase (*see* **Note 13**) and then

harvested by centrifugation at (6000 *g*, 15 min, 4 °C). Wash the cells twice with deionized water and resuspend in deionized water. The bacterial suspension which is used to inoculate protein rich substrates should contain 6 log CFU/mL.

5. Fermentation is carried out in 250 mL Erlenmeyer flasks with working volumes of 100 mL of protein rich isolate (heat treated for 5 min at 100 °C).
6. Fermentation time depends upon the growth kinetics of microorganisms in protein rich substrates and the desirable degree of hydrolysis.

3.2 Characterization of Peptides

(a) SDS Polyacrylamide Gel Electrophoresis

1. Assemble gel plates with spacers to create a leak-free setup.
2. Mix acrylamide/bis-acrylamide, Tris–HCl (pH 8.8), SDS, and water. Add APS and TEMED last to initiate polymerization.
3. Pour the resolving gel solution (12%) to the desired height in the gel plates. Overlay with water or isopropanol to prevent air contact. Let it polymerize for 20 min.
4. For stacking gel (4%) mix acrylamide/bis-acrylamide, Tris–HCl (pH 6.8), SDS, and water. Add APS and TEMED.
5. Remove the overlay on the resolving gel, pour the stacking gel, and insert the comb. Allow it to polymerize for 15 min.
6. The sample buffer was prepared using 5% β-mercaptoethanol, 0.01% bromophenol blue, 2% SDS, 10% glycerol, and 0.0625 mol/L Tris–HCl (pH 6.8).
7. The running buffer comprises 3 g/L Tris base, 14 g/L glycine, and 1 g/L SDS.

(b) Sample Preparation

8. Mix the protein/peptide sample with SDS loading buffer in a 1:1 ratio.
9. Heat the mixture at 95 °C for 5 min to denature proteins and reduce disulfide bonds.
10. Cool the samples to room temperature.

(c) Gel Loading

11. Remove the comb and rinse the wells with a running buffer.
12. Assemble the gel in the electrophoresis unit and fill it with a running buffer.
13. Load the molecular weight marker and protein samples into separate wells using a micropipette (10–20 μL per well).

14. Connect the electrophoresis unit to the power supply. Set the voltage to 80 V for stacking gel (until samples migrate to the resolving gel) and 120–150 V for resolving gel.

(d) **Gel Staining**

15. Remove the gel from the plates after electrophoresis and place it in a staining tray.
16. Cover the gel with Coomassie Brilliant Blue stain. Shake gently for 30–60 min.
17. Destain the gel with a destaining solution until protein bands are visible.
18. Capture an image of the gel using a gel documentation system.
19. Compare the protein bands with the molecular weight marker to determine sizes.

3.3 Amino Acid Composition

(a) **Sample Preparation**

1. Dissolve the protein or peptide sample in water or buffer.

(b) **Hydrolysis**

2. Add 5 mg of protein/peptide into a vacuum hydrolysis tube.
3. Add 1 mL of 6 N HCl to cover the sample completely.
4. Add a small amount of phenol (0.1% w/v) to prevent oxidation of amino acids.
5. Seal the hydrolysis tube under vacuum or nitrogen to remove oxygen.
6. Incubate the tube at 110 °C for 24 h in a heating block or oven.
7. After hydrolysis, cool the tube to room temperature.

(c) **Sample Processing**

8. Evaporate the acid under vacuum or nitrogen to dryness.
9. Re-dissolve the residue in the buffer or solvent suitable for HPLC or amino acid analyzer.
10. Use a 0.22 μm filter to remove particulates before injection into the instrument.

(d) **Derivatization**

11. Mix the sample with the derivatization reagent (e.g., OPA or PITC) according to reagent instructions.
12. Incubate at room temperature for the required reaction time (typically 1–5 min).

13. Inject the derivatized sample into the HPLC system immediately.

(d) Amino Acid Separation by HPLC

14. Set up the HPLC or amino acid analyzer with the reverse phase column for derivatized amino acids and ion exchange column for amino acid analyzer.
15. Prepare a gradient elution program with buffers citrate and acetonitrile.
16. Inject 20 μL of sample into the instrument.
17. Identify and quantify amino acids based on retention times and peak areas using the calibration curve.

3.4 Functional Properties

3.4.1 Solubility

1. Weight the required amount of protein sample.
2. Dissolve it in distilled water or buffer at known pH to achieve defined concentration.
3. Prepare the protein samples at different pH using suitable buffers.
4. Stir each solution for 30 min to ensure equilibrium
5. Incubate the solution for 30 min at desired temperature.
6. The polypeptide suspension underwent centrifugation at 5000 g for 15 min, followed by filtration through Whatman No. 1 filter paper.
7. Collect the supernatant without disturbing the pellet.
8. Measure the soluble protein concentration in the supernatant using Lowry assay.
9. Use a spectrophotometer to determine the absorbance at the 660 nm wavelength.
10. Calculate the soluble protein concentration from a standard curve prepared using bovine serum albumin (BSA).
11. Solubility can be measured using formula.

$$\text{Solubility (\%)} = \frac{\text{Protein concentration in supernatant}}{\text{Total initial protein concentration}} \times 100$$

3.4.2 Water Holding Capacity (WHC)

1. Weight approx. 1 g of protein powder and place the sample in a pre-weighed centrifuge tube.
2. Add 10 mL of distilled water to the centrifuge tube using a graduated cylinder.
3. Mix thoroughly to disperse the sample evenly in the water.
4. Allow the sample to hydrate and bind water by letting it stand at room temperature for 30 min. Gently shake the tube periodically to ensure thorough mixing.

5. Centrifuge the mixture at 5000 *g* for 30 min to separate unbound water.
6. After centrifugation, carefully decant the supernatant without disturbing the sediment.
7. Weigh the tube with the hydrated sediment to determine the weight of the bound water.
8. WHC is expressed as the amount of water bound per gram of the dry sample:

$$\text{Water holding capacity} = \frac{W_2 - W_1}{W} \times 100$$

Where,

W_1 = Weight of tube plus dry sample (g).

W_2 = Weight of tube plus the sediment (g).

W = Weight of the dry sample (g).

3.4.3 Emulsion Capacity

1. Dissolve the protein sample (1% w/v) in the buffer at a defined concentration and adjust the pH (7.0).
2. Gradually add vegetable oil at the ratio of 1:1 to the protein solution.
3. Homogenize the mixture using homogenizer for 5 min at consistent speed.
4. Immediately transfer the solution into measuring cylinder and measure the total volume and volume of emulsified layer.
5. Emulsion capacity is typically expressed as the maximum volume of oil emulsified per unit weight of protein:

$$\text{Emulsifying capacity (\%)} = \frac{\text{Volume of emulsified layer}}{\text{Volume of total content in the tube}} \times 100$$

3.4.4 Emulsion Stability

1. Use the above emulsion for stability test.
2. Transfer the emulsion into centrifuge tubes and centrifuge at 5000 *g* for 5 min.
3. Measure the height of emulsion layer and separated oil or water layer.
4. ES is expressed as the percentage of the emulsion layer remaining after centrifugation or incubation:

$$\text{Emulsion stability (\%)} = \frac{\text{Height of emulsified layer after test}}{\text{Initial height of emulsion layer}} \times 100$$

3.4.5 Dispersibility

1. Weigh 1 g of protein powder.
2. Measure 100 mL of distilled water in a beaker at room temperature.
3. Sprinkle the protein sample on the water and mixture underwent vigorous shaking with a glass rod and was then left undisturbed for 3 h.
4. Observe the uniformity of the dispersion:
   - Look for clumps or dispersed particles.
   - Check the clarity or turbidity of the liquid.
5. Filter the dispersed sample to separate undispersed particles, dry them, and weigh the residue.
6. Dispersibility was calculated as

$$\text{Dispersibility }(\%) = 1 - \frac{\text{Weight of residue}}{\text{Initial weight of protein sample}} \times 100$$

3.4.6 Foaming Capacity

1. Dissolve the protein sample in distilled water or buffer to prepare a 1% w/v solution and adjust the pH (7.0).
2. Pour the sample into a homogenizer and whip the solution for 5 min at constant speed.
3. Immediately transfer the foamed solution into the graduated cylinder and measure the total volume.
4. Foaming capacity is calculated as the volume of foam generated relative to the initial liquid volume.

$$\text{Foam capacity }(\%) = \frac{\text{Volume after whipping} - \text{initial liquid volume}}{\text{Initial liquid volume}} \times 100$$

3.4.7 Foaming Stability

1. Leave the graduated cylinder containing the foam at room temperature undisturbed.
2. Record the foam volume at regular time intervals, i.e., 10 min for 1 h.
3. FS is expressed as the percentage of foam volume retained over time:

$$\text{Foam stability }(\%) = \frac{\text{Foam volume tatimet}}{\text{Initial foam volume}} \times 100$$

3.5 Biofunctional Activity

3.5.1 Antioxidant Activity

1. Prepare different concentrations of peptide in methanol.
2. Mix 1 mL of each dilution with 1 mL of DPPH solution.
3. Incubate the reaction occurs in the dark for 30 min.
4. Measure the absorbance at 517 nm using a spectrophotometer.

5. Calculate the scavenging activity

DPPH scavenging activity (%)

$$= \frac{\text{Absorbance of control} - \text{absorbance of sample}}{\text{Absorbance of control}} \times 100$$

3.5.2 Antimicrobial Activity

(a) Broth Method

1. Dissolve nutrient premix powder in water as per the manufacturer's instructions.
2. Sterilize the prepared broth at 121 °C, 15 psi pressure for 20 min.
3. Freshly inoculate the selected bacterial culture into the nutrient broth.
4. Add 1 mL of the mother inoculum to 100 mL of sterile nutrient broth (1:100 ratio).
5. Place the flask in an orbital shaker incubator at 37 °C overnight.
6. Check for visible turbidity in the broth the next day, indicating bacterial growth.
7. Measure the absorbance of the culture at 600 nm using a UV spectrophotometer.

(b) Agar Well Diffusion Assay

1. Pour and solidify the nutrient agar in sterile petri dishes.
2. Inoculate the agar plate surface evenly with the microbial suspension.
3. Use a sterile cork borer to make wells (6–8 mm diameter) in the agar.
4. Fill each well with 50–100 μL of sample, positive, and negative control.
5. Incubate at 35–37 °C for 18–24 h.
6. Measure the diameter of clear zones around the wells using a ruler or caliper.

4 Notes

1. To prepare 0.1 M NaOH, dissolve 0.4 g of NaOH in 100 mL distilled water.
2. To prepare 100 mL of 6.0 M HCl from a concentrated HCl solution (11.99 M based on given specifications), take approximately 50 mL of the concentrated solution with 50 mL of water.

3. Take 800 mL of distilled water, add 7.5 g of glycine and adjust the pH 2 with HCl, and then add distilled water to make the volume 1 L.
4. Medium used for microbes growth is specific to the strains. MRS (Man, Rogosa, and Sharpe) is used for the growth of lactic acid bacteria. Generally, GRAS (generally recognized as safe) microorganisms are used for fermentation as others might cause the production of toxic compounds.
5. The concentration of the gel component may change as per peptide size. For smaller size peptide tricine, 17% Polyacrylamide gel is used.
6. To prepare a 10 mL sample buffer (4X) take 2.5 mL Tris–HCl (pH 6.8), 4 mL glycerol, 0.8 mL 10% SDS, 4 mL ®-mercaptoethanol, 2 mg Bromophenol blue.
7. To prepare a stock solution of Running/Electrophoresis buffer (5X, pH 8.3) take 15.1 g Tris–HCl, 72 g glycine, 5 g SDS, mix in 1000 mL MilliQ (Dilute to 1X before use).
8. To prepare the buffer 2 solvents (Solvent A and Solvent B) are used. Solvent A is aqueous and Solvent B is organic phase. Solvent A and Solvent B are prepared in a salt base solution. To prepare this salt base solution, 200 mM (6.31 g of ammonium formate in 500 mL) ammonium formate is prepared and pH is reduced to 3 using formic acid. Further, Solvent A is prepared by mixing 100 mL of salt solution and 900 mL distilled water. Solvent B is prepared by mixing 100 mL of salt solution and 900 mL of organic solvent, generally acetonitrile is used.
9. Selection of amino acid standards depends upon the interest of amino acid. Generally, all 20 amino acids are used as standard.
10. 0.1 M phosphate buffer, pH 8: Take 800 mL of distilled water, add 16.282 g of phosphate dibasic and 0.8878 g of phosphate monobasic and then add distilled water to make the volume 1 L.
11. For the preparation of 0.1 mM of DPPH, dissolve 39.4 mg of DPPH in 1000 mL methanol/ethanol.
12. Defatting of substrate is required as fat may hinder the isolation of peptides.
13. Time required to reach the early stationary phase is different for different microorganisms. Thus, this must be checked before starting the fermentation.

References

1. Yilmaz K, Turhan S, Saricaoglu FT et al (2020) Improvement of physicochemical, mechanical, thermal and surface properties of anchovy by-product protein films by addition of transglutaminase, and the correlation between secondary structure and mechanical properties. Food Packag Shelf Life 24:100483. https://doi.org/10.1016/j.fpsl.2020.100483
2. Shahidi F, Zhong Y (2008) Bioactive peptides. J AOAC Int 91:914–931. https://doi.org/10.1093/jaoac/91.4.914
3. Amigo L, Hernández-Ledesma B (2020) Current evidence on the bioavailability of food bioactive peptides. Molecules 25:4479. https://doi.org/10.3390/molecules25194479
4. Daliri EBM, Oh DH, Lee BH (2017) Bioactive peptides. Foods 6:32. https://doi.org/10.3390/foods6050032
5. Mazloomi SN, Mahoonak AS, Mora L et al (2021) Pepsin hydrolysis of orange by-products for the production of bioactive peptides with gastrointestinal resistant properties. Foods 10:679. https://doi.org/10.3390/foods10030679
6. Saengsuk N, Laohakunjit N, Sanporkha P et al (2022) Comparative physicochemical characteristics and in vitro protein digestibility of alginate/calcium salt restructured pork steak hydrolyzed with bromelain and addition of various hydrocolloids (low acyl gellan, low methoxy pectin and κ-carrageenan). Food Chem 393:133315. https://doi.org/10.1016/j.foodchem.2022.133315
7. Kou X, Gao J, Xue Z (2012) Purification and identification of antioxidant peptides from chickpea (Cicer arietinum L.) albumin hydrolysates. LWT—Food Sci Technol 50:591–598. https://doi.org/10.1016/j.lwt.2012.08.002
8. Wang D, Zhang M, Zou Y et al (2018) Optimization of flavourzyme hydrolysis condition for the preparation of antioxidant peptides from duck meat using response surface methodology. J Poult Sci 55:217–223. https://doi.org/10.2141/jpsa.0160155
9. Shen P, Zhou F, Zhang Y et al (2020) Formation and characterization of soy protein nanoparticles by controlled partial enzymatic hydrolysis. Food Hydrocoll 105:105844. https://doi.org/10.1016/j.foodhyd.2020.105844
10. Sitanggang AB, Sumitra J, Budijanto S (2021) Continuous production of tempe-based bioactive peptides using an automated enzymatic membrane reactor. Innov Food Sci Emerg Technol 68:102639. https://doi.org/10.1016/j.ifset.2021.102639
11. Kamal H, Mudgil P, Bhaskar B et al (2021) Amaranth proteins as potential source of bioactive peptides with enhanced inhibition of enzymatic markers linked with hypertension and diabetes. J Cereal Sci 101:103308. https://doi.org/10.1016/j.jcs.2021.103308
12. Kehinde BA, Sharma P (2020) Recently isolated antidiabetic hydrolysates and peptides from multiple food sources: a review. Crit Rev Food Sci Nutr 60:322–340. https://doi.org/10.1080/10408398.2018.1528206
13. Liu FF, Li YQ, Wang CY et al (2022) Physicochemical, functional and antioxidant properties of mung bean protein enzymatic hydrolysates. Food Chem 393:133397. https://doi.org/10.1016/j.foodchem.2022.133397
14. Kim MJ, Chilakala R, Jo HG (2022) Anti-obesity and anti-hyperglycemic effects of Meretrix lusoria protamex hydrolysate in ob/ob mice. Int J Mol Sci 23:4015. https://doi.org/10.3390/ijms23074015
15. Chalamaiah M, Ulug SK, Hong H et al (2019) Regulatory requirements of bioactive peptides (protein hydrolysates) from food proteins. J Funct Foods 58:123–129. https://doi.org/10.1016/j.jff.2019.04.050
16. de Castro RJS, Sato HH (2015) Biologically active peptides: processes for their generation, purification and identification and applications as natural additives in the food and pharmaceutical industries. Food Res Int 74:185–198. https://doi.org/10.1016/j.foodres.2015.05.013
17. Abril AG, Calo-Mata P, Villa TG et al (2024) High-resolution comparative and quantitative proteomics of biogenic-amine-producing bacteria and virulence factors present in seafood. J Agric Food Chem 72:4448–4463. https://doi.org/10.1021/acs.jafc.3c06607
18. Purohit K, Reddy N, Sunna A (2024) Exploring the potential of bioactive peptides: from natural sources to therapeutics. Int J Mol Sci 25:1391. https://doi.org/10.3390/ijms25031391
19. Mechmeche M, Kachouri F, Ksontini H et al (2017) Production of bioactive peptides from tomato seed isolate by Lactobacillus plantarum fermentation and enhancement of antioxidant activity. Food Biotechnol 31:94–113. https://doi.org/10.1080/08905436.2017.1302888
20. Sharma P, Trivedi N, Gat Y (2017) Development of functional fermented whey–oat-based product using probiotic bacteria. 3 Biotech 7:

272. https://doi.org/10.1007/s13205-017-0906-3

21. Massoud R, Hosseini AH, Massoud A (2017) Functional properties of food proteins; gelation and stable foam. In: Proceedings of the 6th international conference on science and engineering (ICES), Paris, France, pp 1–8

22. Sharma P, Sharma D, Kaur S et al (2021) Optimization of flaxseed milk fermentation for the production of functional peptides and estimation of their bioactivities. Food Sci Technol Int 27:585–597. https://doi.org/10.1177/1082013220973815

Chapter 15

Isolation and Characterization of Health-Promoting Oligosaccharides

Devendra Kumar Bhatt, Ashwani Kumar, Aditi Jain, Pragya Pandey, and Ankur Tripathi

Abstract

Oligosaccharides are the group of carbohydrates containing 2–10 sugar units. These are nondigestible in the small intestine and are utilized as a food source by gut microflora in the colon. They also help in restoring a healthy gut by selectively supporting the growth of one or more health-promoting microorganisms. A healthy gut improves digestion, immunity, and prevents lifestyle disorders such as diabetes, obesity, and cardiovascular diseases. This has increased the demand of oligosaccharides and they are now widely used in functional foods and nutraceuticals. This chapter discusses the types of oligosaccharides, their sources, and methods of isolation, and characterization.

Key words Inulin, Fructooligosaccharides, Galactooligosaccharides, Xylooligosaccharides, Pectic polysaccharides

1 Introduction

Oligosaccharides are intermediates between monosaccharides and polysaccharides that contain 2–10 monosaccharide units attached to each other either by an α or β glycosidic linkage [1]. They are naturally present in various food sources like chicory, wheat, rice husk, potatoes, onion, garlic, bananas, and artichokes and are categorized as digestible and functional oligosaccharides. The digestible oligosaccharides are made up by the glycosidic linkage between two monosaccharides, and the examples are sucrose, maltose, and lactose. They are digestible and are broken down into simple sugars by the human gut enzymes. The functional oligosaccharides are the nondigestible carbohydrates made up of 3–10 monosaccharide units [2]. The common functional oligosaccharides are fructooligosaccharides (FOS), galactooligosaccharides (GOS), xylooligosaccharides (XOS), maltooligosaccharides (MOS), isomalto oligosaccharides (IMO), human-milk

Ashwani Kumar, Vidisha Tomer and Piyoosh Kumar Babele (eds.), *Phytochemicals: Extraction, Quantification, Isolation, and Characterization*, Methods and Protocols in Food Science, https://doi.org/10.1007/978-1-0716-4840-7_15,

Table 1
Major oligosaccharides, their natural sources, and synthesis

Oligosaccharide type	Sources	Synthesis	Reference
Fructooligosaccharide (FOS)	Naturally present in cereals (wheat, barley), fruits (banana), vegetables (tomato, onion, garlic), and honey	Enzymatic hydrolysis of sucrose and microbial fermentation using fructosyltransferase enzyme	[3]
Xylooligosaccharides (XOS)	Found in agricultural byproducts such as corn cob, corn husk, cereal straw, sugarcane bagasse, coconut husk	Chemical, enzymatic, or autohydrolysis of agricultural byproducts	[4]
Galactooligosaccharides (GOS)	α-GOS are naturally found in beans, legumes (soybean, lupin, chickpea, cowpea) and β-GOS in dairy products	Can be extracted naturally from plant sources and industrially produced by transgalactosylation of galactose/lactose catalyzed by α-galactosidase/β-galactosidase, respectively	[5]
Lactulose derived galactooligosaccharides (LDGOS)	Synthetically produced	Chemical isomerization of lactose	[5]
Raffinose family oligosaccharides (RFOs)	Abundantly present in legumes, tubers, and leaves	Enzymatic conversion of sucrose and galactinol by the action of enzyme galactinol synthase	[6]
Isomaltooligosaccharides (IMO)	Agroindustrial wastes containing starch	Can be synthesized enzymatically using starch, sucrose, maltose, or dextran as carbohydrate source	[7]
Arabinooligosaccharides (AOS)	Present in plant cell wall components	Enzymatic hydrolysis of arabinose	[5]
Mannan Oligosaccharides (MOS)	Naturally present in vegetables, seeds, yeasts, and bacterial cell walls	Partial hydrolysis of the mannans polysaccharide	[8]
Chitin and Chitosan oligosaccharides (COS)	Naturally present in fungal cell wall and exoskeletons of crustaceans	Depolymerization of chitosan	[5]

oligosaccharides (HMOs), marine-oligosaccharides, pectic oligosaccharides (POS), lactosucrose (LS), and lactulose. One other important oligosaccharide is inulin. It differs from other oligosaccharides based on the chain length as it can also be a polysaccharide. A list of major oligosaccharides and their sources is also provided in Table 1. These oligosaccharides are not broken down by human digestive enzymes in the small intestine, but are utilized by the gut microbiota in the colon where they enhance the growth of selective health-promoting microorganisms in the gut. The supplementation of food with oligosaccharides also affects its texture and increases the digestion and absorption of nutrients. Some common health benefits derived by the intake of oligosaccharides include improved immunity, regulation of blood sugar levels, reduction in the risk of inflammatory diseases, cancer, cardiovascular diseases, and obesity, etc. More specifically, the supplementation of inulin enhances the product texture, and stimulates the growth of healthy bacteria such as Bifidobacterium spp. [9]. FOS supplementation increases the mRNA expression of genes related to fat digestion and absorption and also helps in fat metabolism through mobilizing intestinal bacteria and their metabolites [10]. The intake of GOS for over 3 weeks has been reported to lower the secretion of cortisol, a stress hormone [11]. It also supports the growth of *Lactobacillus plantarum* [12]. The intake of XOS increases the digestion and absorption of nutrients. The consumption of XOS for over 6 weeks has been reported to improve the microbiota balance, i.e., an increase in the count of *Lactobacillus* spp. and *Bifidobacterium* spp. and decrease in *Clostridium perfringens* was observed [13]. Another study has reported the increase in count of *L. sakei*, and *L. brevis* on XOS intake [14]. POS has an inhibitory action against enteroputrefactive and pathogenic microorganisms. They also possess bifidogenic prebiotic properties. Owing to the health benefits of oligosaccharides and their ability to support healthy microbial growth in the gut (probiotics), they are nowadays isolated from their natural sources and are fortified in functional and health-promoting foods. The commonly used isolation methods are solvent extraction methods, enzymatic methods, and microwave assisted methods. After the isolation, the oligosaccharides must be checked for their purity and composition as the purity is a deciding factor in their application in food and pharmaceutical industries. The most commonly used characterization techniques are High Performance Liquid chromatography and Fourier-Transform Infrared Spectroscopy.

This chapter discusses step by step procedures for the isolation and characterization of plant-based oligosaccharides such as inulin, FOS, GOS, XOS, and POS.

2 Materials

2.1 Isolation of Inulin [15]

(a) Apparatus

- Hot air oven.
- Sieve [1.00 mm, 100 mesh (150 μm)] Hot water bath with stirrer.
- Vacuum filtration unit.

- Vacuum oven.
- Aspirator.
- Rotary evaporator.
- Freeze dryer.

(b) **Reagents**

- Water.
- Citric acid solution (0.75%, *see* **Note 1**).
- Calcium hydroxide (5%, *see* **Note 1**).
- Phosphoric acid (10%, *see* **Note 1**).
- Charcoal powder.
- Ethanol (99%, *see* **Note 1**).

(c) **Others**

- Muslin cloth.
- Whatman filter paper (*see* **Notes 1** and **4**).
- Glass rod.

2.2 Isolation of Fructooligosaccharides [16]

(a) **Apparatus**

- Drying oven.
- Grinder/blender.
- Sieve (525 μm).
- Soxhlet extraction unit.
- Water bath with shaker.
- Centrifuge (≥4000× *g*).
- Freeze drier.
- pH meter.
- Reflux setup for derivatization.
- High-Speed countercurrent chromatography (HSCCC) system.

(b) **Reagents**

- Hexane (for defatting).
- Sodium acetate buffer (0.2 M, pH 7.0, *see* **Note 2**).
- α-Amylase (1000 U) (*see* **Notes 3** and **4**).
- Glucoamylase (7000 U) (*see* **Notes 3** and **4**).
- Ethanol (99%, *see* **Note 1**).
- Acetic anhydride.
- Pyridine.

- Petroleum ether, n-butanol, methanol, water (for HSCCC solvent system).
- Sodium methoxide.
- Methyl alcohol.

2.3 Isolation of Galactooligosaccharides [17]

(a) **Apparatus**

- Hot air oven.
- Mechanical grinder.
- Centrifuge.
- Filter (0.45 μm).
- Rotary evaporator.
- Freeze-dryer.
- HPLC system with a refraction index detector.
- 0.45 μm Millipore membrane.

(b) **Reagents**

- Ethanol (100%).
- Distilled water.
- Activated charcoal (100 mesh).
- Deionized water.
- Acetonitrile.

2.4 Isolation of Xylooligosaccharides [18]

(a) **Apparatus**

- Weighing balance.
- Hot air oven.
- Electromotive mill/grinder.
- Sieve (45 mesh).
- Microwave oven.
- Shaker.
- Centrifuge (≥3000 *g*).
- Suction filtration unit.
- Pyrex crucible.
- Vacuum dryer.
- Freeze dryer.

(b) **Reagents**

- Distilled water.
- Deionized water.

- Xylananse (2500 bacterial xylanase units/100 g, *see* **Notes 3** and **4**).
- Ethanol (95%, *see* **Note 1**).
- Activated carbon powder.

2.5 Isolation of Pectic Polysaccharides [19–21]

(a) **Apparatus**

- Cabinet dryer.
- Freeze dryer.
- Grinder.
- Ultrasonic tank.
- High pressure homogenizer.
- Microwave.
- Hot air oven.
- Rotary evaporator.
- Ultrafiltration unit.
- Freeze drier.

(b) **Reagents**

- Distilled water.
- Deionized water.
- Ethanol (80%, *see* **Note 1**).
- Deep eutectic solvents (1:3 choline chloride and ethylene glycol).
- α-amylase (5 U/mL, *see* **Notes 3** and **4**).
- Glucoamylase (5 U/mL, *see* **Notes 3** and **4**).
- Pancreatin (1 U/mL, *see* **Notes 3** and **4**).

2.6 Characterization of Oligosaccharides

- High pressure liquid chromatography (HPLC).
- Fourier transform infrared spectroscopy (FTIR).
- X-ray diffraction.
- Scanning electron microscopy (SEM).
- Nuclear magnetic resonance (NMR).

3 Methods

3.1 Isolation of Inulin

1. *Cleaning of Chicory Roots*: Wash the chicory roots with tepid water to remove adhering soil and other undesirable materials.
2. *Removal of Bitterness*:
   (a) Soak the chicory roots in 0.75% citric acid solution at a ratio of 1:10 (roots: Water w/v) for 24 h.

   (b) Wash the roots thoroughly, dry in an oven at 40 °C for 10 h, and grind to a fine powder that passes through 100 mesh sieves.

3. *Extraction*:
   (a) Add water to the root powder in 1:10 ratio (root powder: Water, w/v).
   (b) For extraction, heat the mixture to 80 °C ± 2 °C and continue heating at this temperature with continuous stirring for 1 h.
   (c) Filter the obtained crude extract through muslin cloth to remove insoluble residues.

4. *Clarification*:
   (a) Mix the crude extract with 5% slurry of calcium hydroxide at 50–60 °C for 30 min to remove the impurities like protein, pectin, and cell wall.
   (b) The flocculent so formed is precipitated and a clear bright yellow supernatant is obtained.

      *This step also increases the pH to 10–12 from normal 5–6.

5. *Filtration*:
   (a) After treatment, filter the solution under vacuum using filter paper (Whatman No. 4) to remove the precipitates.

6. *Coagulation and Filtration*:
   (a) Add 10% phosphoric acid (H_3PO_4) to the filtrate with vigorous continuous stirring.
   (b) Adjust the pH to 8–9 to precipitate excessive coagulate organic material and calcium.
   (c) Allow the filtrate to remain at 60 °C for 2–3 h and re-filter through Whatman No. 4 filter paper.

7. *Removal of Coloring Substances*:
   (a) Add charcoal powder to the filtrate at 60 °C with continuous stirring with a glass rod for 15–30 min.
   (b) Filter the clarified extract again using Whatman No. 1 filter paper.

8. *Concentration*:
   (a) Concentrate the obtained filtrate using a rotary evaporator at 60 °C to obtain 40°B inulin.

9. *Precipitation*:
   (a) Add 99% ethanol to the concentrated inulin solution and store it at 20–25 °C for 4 days.

10. *Separation*:
    (a) Remove the supernatant after 4 days using an aspirator and wash the precipitated inulin with ethanol.
11. *Drying*:
    (a) Remove the excess ethanol from the precipitated inulin by drying in a vacuum oven at 40 °C for 1 h.
12. *Freezing and Freeze Drying*:
    (a) Freeze the precipitated inulin at −20 °C for 24 h and dry the frozen inulin in a freeze dryer at −30 °C for 24 h.
13. *Milling and Storage*:
    (a) Grind the dried inulin to a size to pass through a 1.00 mm sieve and store in an airtight container for further use.

3.2 Isolation of Fructooligosaccharides

1. *Extraction of Crude Oligosaccharides*:
    (a) Collect the sample, dry in shade or at 30 °C in a dehydrator.
    (b) Homogenize the sample into a fine powder using an experimental blender.
    (c) Defat the sample (50 g peel powder) using hexane as a solvent (100 mL) in a Soxhlet apparatus at 70 °C.
    (d) Suspend the defatted sample to 100 mL sodium acetate buffer (0.2 M, pH 7.0 ± 2).
    (e) Incubate with 1 mg of 1000 U α-amylase at 95 °C for 1 h to hydrolyze the starch.
    (f) Cool, then subject the sample to 100 mg of 7000 U glucoamylase and heat at 55 °C for 48 h.
    (g) Centrifuge the enzyme treated samples at 4000 *g* for 20 min and collect the supernatant.
    (h) Extract the de-starched samples with 1000 mL of distilled water at 25 °C for 2 h.
    (i) To obtain the oligosaccharide precipitates add three volumes of 99% ethanol.
    (j) The crude oligosaccharides can be lyophilized and stored at 4 °C until further use.
2. *Isolation of Fructooligosaccharides from Crude Oligosaccharides*:

    A. *Precolumn-Derivatization*:
    (a) Add 5 g of oligosaccharide extract to 50 mL pyridine and to this add 19 g of 0.4 M acetic anhydride to reduce polarity.
    (b) Stir the reaction mixture overnight and concentrate the solution under reduced pressure.
    (c) A yellow syrup like material will be obtained from this step that will be used for further purification.

B. *Separation of Fructooligosaccharides (Kestose)*:

(a) The separation of fructooligosaccharides is carried by High-Speed Countercurrent Chromatography Separation.

(b) Prepare a two-phase solvent system: petroleum ether: N-butanol:methanol:water (3:2:1:4, v/v).

(c) Load the acetylated extract into the HSCCC system (*see* **Note 5**).

(d) On reaching thermal equilibrium, monitor the column product continuously using an ultraviolet detector at 250 nm and collect fractions at 5 min interval.

(e) Identify the acetylated kestose rich fraction based on KD value and elution time.

(f) Collect the kestose fraction and vacuum dry it.

3.3 Isolation of Galactooligosaccharides

1. *Preparation of Plant Material*:

   (a) *Selection and Cleaning*: Select the suitable plant source and clean it thoroughly to remove any contaminants.

   (b) *Drying and Grinding*: Dry the plant material in a hot air oven and grind it after drying using a mechanical grinder.

   (c) *Defatting*: Soak 300 g of dried rhizome powder in 100% ethanol at ambient temperature for 24 h.

2. *Aqueous Ethanol Extraction*:

   (a) Afterward, filter to eliminate the solvent, then extract the defatted residue twice using 3.0 L of 10% aqueous ethanol while stirring at 60 °C for 2 h.

3. *Centrifugation*:

   (a) Centrifuge the combined extract for 15 min at 3000 rpm.

   (b) Concentrate the supernatants while maintaining a temperature below 80 °C under reduced pressure.

4. *Deproteinization*:

   (a) Next, deproteinize the extract through a freeze-thaw method, repeating this process three times.

5. *Centrifugation*:

   (a) Centrifuge it again for 15 min at 3000 rpm.

6. *Charcoal Treatment (Sugar Removal)*:

   (a) Mix the resulting supernatant with 3 g of activated charcoal (100 mesh) and stir for 30 min to eliminate mono- and disaccharides.

7. *Filtration*:
   (a) Filter this mixture under vacuum and rinse the activated charcoal with 10% aqueous ethanol. The oligosaccharides that have been adsorbed onto the activated charcoal are desorbed by stirring in 50% aqueous ethanol for 30 min.
   (b) Remove the activated charcoal by filtration, then evaporate the resulting solution under vacuum.
8. *Final Processing*:
   (a) Finally, filter the remaining sample through a 0.45 μm filter and freeze-dry it to obtain the oligosaccharides (RL-GOS) as a fine white powder.

3.4 Isolation of Xylooligosaccharides

1. *Cleaning of Raw Materials*:
   (a) Wash the raw materials with distilled water and dry at 45 ° C for 2 days.
   (b) Grind the dried sample in an electromotive mill and sieve through a 45 mesh sieve.
2. *Extraction in Microwave Oven*:
   (a) Add distilled water to the ground wheat bran so as to get 50% slurry.
   (b) Expose the slurry in the microwave oven to 1400 W for 120 s (time can vary as per the sample size), cool, and add distilled water to make 10% slurry.
3. *Enzymatic Treatment*:
   (a) Add xylanase at a rate of 0.5 g/100 g substrate, i.e., 2500 bacterial xylanase units/100 g substrate.
   (b) Place the sample in a shaker at 100 rpm and 55 °C for 24 h.
   (c) Stop the reaction by boiling.
   (d) Centrifuge the sample and collect the supernatant by centrifugation at 3000 *g* for 10 min.
   (e) Add distilled water to the sediment (insoluble wheat bran) to make 50% slurry for repeated microwave hydrolysis.
   (f) Carry out the repeated microwave hydrolysis at 1400 W for 5 min.
   (g) Cool, and make 10% slurry by adding distilled water.
   (h) Add enzyme (as per step 3(a)) to perform second time enzyme reaction and repeat the steps from 3(b) to 3(d).
4. *Recovery of XOS by Ethanol Precipitation*:
   (a) Mix the supernatants and mix 50 mL of aliquot with 150 mL of 95% ethanol.

   (b) Centrifuge the mixture at 3000 *g* for 10 min and decant the precipitate.
   (c) Vacuum concentrate the liquid to remove ethanol.
   (d) Lyophilize the concentrate and weight to evaluate the recovery rate of XOS.

5. *Recovery of XOS by Activated Carbon Adsorption Method*:
   (a) Add activated carbon powder to the supernatant at the rate of 5–10% of the liquid weight.
   (b) Homogenize the contents in flasks at room temperature using a shaker at 200 rpm for 30 min to stabilize the carbon-carbohydrate adsorption.
   (c) Suction filter the mixture with a 50 mL pyrex crucible filter and wash with distilled and deionized water.
   (d) Elute it with 50% ethanol twice.
   (e) Vacuum concentrate the mixture to remove ethanol and lyophilize.
   (f) Weight to evaluate the recovery rate of XOS.

3.5 Isolation of Pectic Polysaccharides

1. *Sample Preparation*:
   (a) Take the sample, clean it, and dry it. The various dehydration methods are sun drying, cabinet dying, and freeze drying.
   (b) Grind the sample to fine powder.
2. *Removal of Ethanol Soluble Components*:
   (a) Mix 10 g of sample powder with 80% of ethanol (1:10 w/v ratio).
   (b) Treat this mixture in an ultrasonic cleaning tank at 480 W and 25 °C for 20 min for the removal of ethanol soluble components.
3. *Extraction of Polysaccharides*:

   The polysaccharides can be extracted using various methods like, hot water extraction, deep eutectic solvent extraction, using conventional heating method, deep eutectic solvent extraction with high pressure homogenization, and deep eutectic solvent extraction with microwave assisted extraction.

   A. *Hot Water Extraction*:
      (a) Collect the residue after ethanol extraction and extract polysaccharides by adding deionized water in a material-liquid ratio of 1:30 g/mL.
      (b) Heat the mixture to 95 °C and maintain the mixture at this temperature for 4 h.

B. *Deep Eutectic Solvent (DES) Extraction*:
   (a) Prepare the DES solution by mixing choline chloride and ethylene glycol in the ratio 1:3.
   (b) Mix this solvent with extraction solvent, i.e., water in the ratio of 50% v/v of extraction solvent: DES solution.
   (c) Keep the material-liquid ratio as 1:40 g/mL.
   (d) Carry out the extraction at 90 °C for 2 h.

C. *High Pressure Deep Eutectic Solvent Extraction*:
   (a) The solvent used is 1:3 choline chloride: Ethylene glycol deep eutectic solution.
   (b) The ratio of extraction solvent: DES solution is 50% v/v.
   (c) Keep the material-liquid ratio as 1:40 g/mL.
   (d) The extraction conditions are 110 MPa for 13 min.

D. *Microwave-Assisted Deep Eutectic Solvent Extraction*:
   (a) The solvent used is 1:3 choline chloride: Ethylene glycol deep eutectic solution.
   (b) The ratio of extraction solvent: DES solution is 36: 64 v/v.
   (c) Keep the material-liquid ratio as 1:50 g/mL.
   (d) The extraction power and time are 635 W for 17 min.

4. *Concentration of the Supernatant*:
   (a) After the extraction, the supernatant is collected and concentrated to about 100 mL in a rotary evaporator.

5. *Removal of Starch and Protein Residues*:
   (a) To the supernatant, add thermostable α-amylase at a concentration of 5 U/mL and incubate this at 85 °C for 6–8 h to break down starch.
   (b) Then, add glucoamylase at 5 U/mL and incubate at 55 °C for 8–10 h to further degrade the starch.
   (c) Next, add pancreatin at 1 U/mL at 45 °C for 6 h to remove protein residues.

6. *Ethanol Precipitation*:
   (a) Add three volumes of 95% ethanol to the supernatant or adjust the ethanol concentration in the supernatant to 70% (v/v).
   (b) Keep it at 4 °C for 12 h and collect the precipitates. The precipitates can also be obtained at room temperature.

7. *Membrane Filtration*:
   (a) Redissolve the precipitates in 50 mL deionized water.
   (b) Ultrafiltrate with an ultrafiltration device with a 3.5 kDa or 10^4 Da cutoff at 4000 *g* for 15 min.
   (c) Repeat the filtration three times.
   (d) The solution obtained after filtration is freeze dried at −40 to −80 °C for 48 h.

3.6 Characterization of Oligosaccharides

3.6.1 Characterization Using HPLC

1. Dissolve the isolated oligosaccharides in the deionized water or appropriate solvent.
2. Filter the solution using a 0.22 or 0.45 μm syringe filter.
3. Select the appropriate column and detector (Table 2).
4. Connect the column and equilibrate it with the mobile phase.
5. Inject the required sample (Table 2). The optimized conditions for the characterization of oligosaccharides are given in Table 2.
6. Also run the standards for comparison.
7. *Calculations*:
   - The regression standard curves should be obtained for concentration versus area (mV × C)
   - Peaks in chromatograms can be identified by coincidence of retention times with their standards.
   - The test oligosaccharides can be quantified by comparison of their areas with the corresponding calibration curve.

Table 2
Optimized conditions for the characterization of oligosaccharides

Part/parameter	Type/value	Type/value
Detector	Refractive Index Detector (RID)	Refractive Index Detector (RID)
Column	Shodex® Sugar SP0810 (300 mm × 7 mm inner diameter) with Pb2 + and a Shodex SP-G guard column (5 μm, 6 × 50 mm)	Rezex™ ROA-Organic Acid H+ (8%), 300 mm × 7.8 mm column, protected with a Carbo-H 4 × 3.0 mm ID security guard cartridge
Column operating temperature	35 °C	25 °C
Mobile phase	Distilled water	Mili Q water acidified with 2.5 mM sulphuric acid
Flow rate	1 mL/min	0.4 mL/min
Injection volume	20 μL[a]	5 μL[a]
References	[15]	[22]

[a]The sample and standards should be filtered through a 0.45 μm syringe filter before injection

3.6.2 Characterization by FTIR

FTIR is a mid-IR (4000–400 cm^{-1}) spectroscopy that measures the functional groups in materials by using infrared radiations. The materials containing covalent bonds absorb electromagnetic radiation in the IR region and the IR spectroscopy measures the amount and intensity of absorption of IR radiation made by each bound in the molecule. A particular IR will be absorbed by a particular bond in the molecule, as every bond has their particular natural frequency. FTIR develops a plot called the infrared spectrum of the measured infrared intensity versus its energy range (wavenumber cm^{-1}) [23]. To identify the unknown molecules, a fingerprint of unknown molecules is created by analyzing the unique absorption patterns and this is matched to a database of known compounds. To characterize the oligosaccharides, a universal sampling device containing diamond or zinc selenide (ZnSe) crystal can be used. The scanning can be performed at room temperature and the scanning speed, number of accumulations, and resolution can be kept 0.20 cm/s, 30 and 4 cm^{-1}, respectively. A background spectrum of air should be scanned under the same instrumental conditions before each series of measurement [22].

3.6.3 Characterization by X-Ray Diffraction (XRD)

X-rays are used to check the quality through diffraction, imaging, or scattering modes. Among these, XRD is the most effective technique for identifying the crystal structure of materials and is broadly applied to study the structural and crystallinity changes in cereal starches and other components. A multipurpose X-ray diffractometer with 40 kV radiation and incident current of 40 mA can be used to analyze the samples in continuous and step-scan mode, diffraction angle of 2Θ ranging from 10° to 40° with a step of 0.02°/s [24, 25].

3.6.4 Characterization by Scanning Electron Microscopy (SEM)

SEM studies surface morphology of dietary fiber using electrons as an imaging medium. After interacting with test materials, electrons come out in various forms and provide a three-dimensional image. It studies the microstructures of fibers for properties like roughness, porosity, pore distribution, cracks, folds, and fibrillar structure. These properties are related to water holding capacity, oil absorption capacity, swelling capacity, etc. These properties can also be used to differentiate between cellulose, lignin, hemicellulose, pectin, etc. To scan the fibre particles, they are first coated with a fine gold layer and can be observed with 200-, 500-, and 2000-fold magnification at the accelerating voltage of 15 kV [26].

3.6.5 Nuclear Magnetic Resonance (NMR)

NMR studies the chemical structures of the materials at the molecular level. This technique can be used to study the type of linkages, branching patterns, and monomeric composition in dietary fibers. It can differentiate the soluble and insoluble fibers and can also study the chemical modifications. To study the insoluble dietary fiber 13C NMR is generally used at frequencies of about

100.13 MHz. Depending on the rotor size, 200–300 mg of the sample is packed in a 4–7 mm Zirconium dioxide rotor, compact the sample, seal the rotor, and spin the sample at 5–8 kHz, at a magic angle of about 54.7°. The pulse width, contact time, spectral width, acquisition time, time domain points, transform size, line broadening, and number of scans can be selected based on the requirement. The details can be found in the study conducted by Zhang and Wang [27]. For soluble fiber, weigh 5–10 mg of fiber in 0.5–1.0 mL deuterated water (D2O), adjust pH (5.0–7.0), transfer to 5 mm NMR tube, and stir or gently heat up to 70 °C for complete dissolution, and obtain scans using 1H NMR at frequencies 400–600 MHz [28].

4 Notes

1. To prepare percent solutions, weigh the solute in grams equal to the percentage required, dissolve in small amounts of distilled water, and make the final volume 100 mL in a volumetric flask. For example: To prepare 0.75% citric acid solution, weight exactly 0.75 g citric acid, dissolve in a small volume of distilled water, transfer to a volumetric flask, and make the final volume exactly to 100 mL. For the higher volumes, the calculations can be done accordingly.
2. To prepare 1 L sodium acetate buffer solution, weigh 27.14 g of sodium acetate (molecular weight: 136.09 g/mol), and dissolve it in 800 mL of distilled water. To this, add 0.0069 g of acetic acid (molecular weight: 60.05 g/mol) after diluting in small quantities of water. Make the final volume to 1 L by adding distilled water. Adjust the pH to 7.0 by adding sodium acetate (to increase pH) or acetic acid solution (to decrease pH).
3. 1 U = Measure of the amount of enzyme that will release 1 micromole of reducing sugars, i.e., glucose per minute under standard conditions.
4. To prepare enzyme solution, the activity of the enzyme must be known. To further calculate the amount of enzyme use the formula:

$$\text{Amount of enzyme (mg)} = \frac{\text{Desired activity (U/mL)} \times \text{volume (mL)}}{\text{Activity of enzyme (U/mg)}}$$

5. Chelliah et al. [16] used a HSCCC having a 150 m long multilayer coil with an inner diameter of 1.6 mm, and total volume of 300 mL. The β-value of the prepared column ranged from 0.8 on the outside and 0.5 m on the inside. A 30 mL loop

manual sample injection was used to inject the sample into the column. The partition coefficient-distribution constant (KD) between 0.5 and 2 was found optimum. The stationary phase was in the upper phase and the mobile phase was in the lower phase. The flow rate was 2.0 mL/min and revolution speed was 800 rpm.

References

1. Zhao C (2021) Oligosaccharides in food. In: Xiao J, Sarker DS, Asakawa Y (eds) Handbook of dietary phytochemicals, 1st edn. Springer, Singapore
2. Ibrahim OO (2018) Functional oligosaccharides: chemical structure, manufacturing, health benefits, applications and regulations. J Food Chem Nanotechnol 4:65–76. https://doi.org/10.17756/jfcn.2018-060
3. Flores-Maltos DA, Mussatto SI, Contreras-Esquivel JC et al (2016) Biotechnological production and application of fructooligosaccharides. Crit Rev Biotechnol 36:259–267. https://doi.org/10.3109/07388551.2014.953443
4. Samanta AK, Jayapal N, Jayaram C et al (2015) Xylooligosaccharides as prebiotics from agricultural by-products: production and applications. Bioact Carbohydr Diet Fibre 5:62–71. https://doi.org/10.1016/j.bcdf.2014.12.003
5. Belorkar SA, Gupta AK (2016) Oligosaccharides: a boon from nature's desk. AMB Express 6:1–11. https://doi.org/10.1186/s13568-016-0253-5
6. Elango D, Rajendran K, Van der Laan L et al (2022) Raffinose family oligosaccharides: friend or foe for human and plant health? Front Plant Sci 13:829118. https://doi.org/10.3389/fpls.2022.829118
7. Gonçalves DA, Teixeira JA, Nobre C (2022) In situ enzymatic synthesis of prebiotics to improve food functionality. In: Kuddus M, Aguilar CN (eds) Value-addition in food products and processing through enzyme technology, 1st edn. Academic Press, San Diego/Cambridge, MA
8. Faustino M, Durão J, Pereira CF et al (2021) Mannans and mannan oligosaccharides (MOS) from Saccharomyces cerevisiae–a sustainable source of functional ingredients. Carbohydr Polym 272:118467. https://doi.org/10.1016/j.carbpol.2021.118467
9. Sheng W, Ji G, Zhang L (2023) Immunomodulatory effects of inulin and its intestinal metabolites. Front Immunol 14:1224092. https://doi.org/10.3389/fimmu.2023.1224092
10. Yan Z, Tang X, Wu R et al (2025) Effect of fructo-oligosaccharides on growth performance and meat quality in broilers. Front Vet Sci 11:1485077. https://doi.org/10.3389/fvets.2024.1485077
11. Johnstone N, Milesi C, Burn O et al (2021) Anxiolytic effects of a galacto-oligosaccharides prebiotic in healthy females (18–25 years) with corresponding changes in gut bacterial composition. Sci Rep 11:8302. https://doi.org/10.1038/s41598-021-87865-w
12. Fuhren J, Schwalbe M, Peralta-Marzal L et al (2020) Phenotypic and genetic characterization of differential galacto-oligosaccharide utilization in Lactobacillus plantarum. Sci Rep 10: 21657. https://doi.org/10.1038/s41598-020-78721-4
13. Lin SH, Chou LM, Chien YW et al (2016) Prebiotic effects of xylooligosaccharides on the improvement of microbiota balance in human subjects. Gastroenterol Res Pract 2016:5789232. https://doi.org/10.1155/2016/5789232
14. Jaichakan P, Nakphaichit M, Rungchang S et al (2021) Two-stage processing for xylooligosaccharide recovery from rice by-products and evaluation of products: promotion of lactic acid-producing bacterial growth and food application in a high-pressure process. Food Res Int 147:110529. https://doi.org/10.1016/j.foodres.2021.110529
15. El-Kholy WM, Aamer RA, Ali ANA (2020) Utilization of inulin extracted from chicory (Cichorium intybus L.) roots to improve the properties of low-fat synbiotic yoghurt. Ann Agric Sci 65:59–67. https://doi.org/10.1016/j.aoas.2020.02.002
16. Chelliah R, Park SJ, Oh S et al (2023) An effective universal protocol for the extraction of fructooligosaccharide from the different agricultural byproducts. MethodsX 10: 102096. https://doi.org/10.1016/j.mex.2023.102096

17. Yang X, Zhao Y, He N et al (2010) Isolation, characterization, and immunological effects of α-galacto-oligosaccharides from a new source, the herb Lycopus lucidus Turcz. J Agric Food Chem 58:8253–8258. https://doi.org/10.1021/jf101217f
18. Wang TH, Lu S (2013) Production of xylooligosaccharide from wheat bran by microwave assisted enzymatic hydrolysis. Food Chem 138:1531–1535. https://doi.org/10.1016/j.foodchem.2012.09.124
19. Wongkaew M, Tangjaidee P, Leksawasdi N (2022) Mango pectic oligosaccharides: a novel prebiotic for functional food. Front Nutr 9:798543. https://doi.org/10.3389/fnut.2022.798543
20. Wu DT, Geng JL, Li J et al (2024) Efficient extraction of pectic polysaccharides from thinned unripe kiwifruits by deep eutectic solvent-based methods: chemical structures and bioactivities. Food Chem X 21:101083. https://doi.org/10.1016/j.fochx.2023.101083
21. Wang J, Li J, Luo X et al (2024) Structural properties and biological effects of pectic polysaccharides extracted from Tartary buckwheat sprouts by high pressure-assisted deep eutectic solvent extraction. LWT 203:116397. https://doi.org/10.1016/j.lwt.2024.116397
22. Redondo-Cuenca A, Herrera-Vázquez SE, Condezo-Hoyos L et al (2021) Inulin extraction from common inulin-containing plant sources. Ind Crop Prod 170:113726. https://doi.org/10.1016/j.indcrop.2021.113726
23. Khan SA, Khan SB, Khan LU et al (2018) Fourier transform infrared spectroscopy: fundamentals and application in functional groups and nanomaterials characterization. In: Sharma SK, Verma DS, Khan LU, Kumar S, Khan SB (eds) Handbook of materials characterization, 1st edn. Springer, Cham
24. Yu L, Wen D, Tan M et al (2024) Physicochemical and structural properties of soluble dietary fibres in navel orange peel modified by superfine grinding and their immunomodulatory activities. Food Chem Adv 5:100752. https://doi.org/10.1016/j.focha.2024.100752
25. Li X, Wang B, Hu W et al (2022) Effect of γ-irradiation on structure, physicochemical property and bioactivity of soluble dietary fiber in navel orange peel. Food Chem X 14:100274. https://doi.org/10.1016/j.fochx.2022.100274
26. Karra S, Sebii H, Yaich H et al (2020) Effect of extraction methods on the physicochemical, structural, functional, and antioxidant properties of the dietary fiber concentrates from male date palm flowers. J Food Biochem 44:e13202. https://doi.org/10.1111/jfbc.13202
27. Zhang J, Wang ZW (2011) Soluble dietary fiber from Canna edulis ker by-product and its enzymatic and antioxidant activities. Food Biotechnol 25:336–350. https://doi.org/10.1080/08905436.2011.617256
28. do Espirito Santo BLS, da Silva ÉC, Cândido CJ et al (2020) Dietary fiber chemical structures and physicochemical properties of edible Pouteria glomerata fruits, native from Brazilian Pantanal. Food Res Int 137:109576. https://doi.org/10.1016/j.foodres.2020.109576

Chapter 16

Chromatography (HPLC-DAD) Methods for Analyzing Alkaloids from Plants

Purnima Tripathi, Neha Sharma, and Anand Krishna Singh

Abstract

Phytochemicals derived from plants play a crucial role in promoting human health. Plants are rich sources of essential nutrients, including proteins, fats, vitamins, minerals, carbohydrates, and various bioactive compounds with significant nutritional and therapeutic value. Despite their importance, many plant-derived compounds remain unexplored by the scientific community. Therefore, effective methods for the screening and characterization of phytochemicals—such as phenolics, alkaloids, and flavonoids—are essential to unlock their full potential. Chromatographic techniques, particularly High-Performance Liquid Chromatography (HPLC), have proven highly effective for the qualitative and quantitative analysis of phytochemicals. A more advanced approach, HPLC coupled with a Diode Array Detector (HPLC-DAD), enhances this process by combining the efficient separation capabilities of HPLC with the ability to detect and identify compounds based on their UV-Visible absorption spectra. This method is especially effective for analyzing complex mixtures, including plant alkaloids, pharmaceuticals, and other bioactive compounds. This chapter focuses on the application of HPLC-DAD in the screening and characterization of plant-derived alkaloids.

Key words Bioactive compounds, Plant extraction, HPLC

1 Introduction

Plants play a vital role in human life, not only as a primary source of oxygen but also by providing essential nutrients such as proteins, vitamins, carbohydrates, cholesterol-lowering compounds, antioxidants, and other biologically active substances [1]. These therapeutic chemical compounds produced by plants are known as phytochemicals, synthesized through metabolic processes as primary and secondary metabolites. Plants with proven medicinal benefits are often referred to as "herbal medicinal products" or "herbal drugs," with their secondary metabolites being of particular interest. Analyzing these complex herbal matrices is challenging due to the presence of hundreds of diverse compounds with varying chemical and physical properties. Plants naturally produce a wide

Ashwani Kumar, Vidisha Tomer and Piyoosh Kumar Babele (eds.), *Phytochemicals: Extraction, Quantification, Isolation, and Characterization*, Methods and Protocols in Food Science, https://doi.org/10.1007/978-1-0716-4840-7_16,

range of secondary metabolites, including alkaloids, phenols, essential oils, flavonoids, tannins, and saponins. While not directly involved in plant growth and reproduction, these compounds enhance plant resilience against environmental stressors. Additionally, they exhibit significant therapeutic potential, offering protection against diseases like diabetes, cardiovascular disorders, cancer, and obesity, making them valuable in the pharmaceutical and food industries [2, 3].

Alkaloids are a structurally diverse group of nitrogen-containing secondary metabolites naturally produced by plants. They play a crucial role in plant defense, protecting against herbivores, pathogens, and environmental stressors, while also regulating plant growth and development. Structurally, alkaloids are basic, cyclic organic compounds with nitrogen in their ring systems and are classified based on molecular skeletons or botanical origins [4]. Due to their potent pharmacological effects, alkaloids account for nearly 60% of plant-derived drugs, with therapeutic applications such as analgesic (morphine), anticancer (vincristine), antiasthmatic (ephedrine), antihypertensive (reserpine), antipyretic (quinine), and antihyperglycemic (piperine) [5]. In animals, alkaloids like morphine, strychnine, quinine, and nicotine are clinically significant, while alkaloid-related compounds such as serotonin, dopamine, and histamine function as key neurotransmitters [6].

To address these challenges, researchers have employed a variety of chromatographic techniques, including high-performance liquid chromatography, gas chromatography, thin-layer chromatography, and their hyphenated techniques with mass spectrometry [7]. Among these, HPLC and GC have emerged as the most widely used chromatographic methods for the separation, identification, and quantification of phytochemicals [8, 9]. HPLC was first developed by Russian botanist M. S. Tswett in 1903, it has undergone numerous advancements and improvements since then. HPLC with diode array detection (DAD) is one of the most recent and preferred techniques for analysis of crude plant compounds. HPLC-DAD is an effective, flexible, and quickest chromatographic technique [3]. It is an important qualitative and quantitative technique that is commonly used for pharmaceutical and biological sample estimation. The HPLC-DAD works on the principle of differential partitioning between a stationary phase and a mobile phase. Different components in the sample interact with these phases to varying extents, causing them to move through the column at different rates, which leads to their separation. In the separation module the stationary and mobile phase consisting of opposite polarity are equipped with a high-pressure pump and the separation is achieved by their interactions. As compounds elute from the column, the DAD measures their absorbance across a broad range of wavelengths, generating both chromatograms and UV-Vis absorbance spectra. Unlike conventional UV detectors that monitor a single wavelength at a time, the DAD simultaneously scans multiple wavelengths, providing detailed spectral data for each peak. This capability enhances compound identification, especially in complex mixtures. This chapter provides descriptions about characterization of alkaloids compounds from plant extract with the use of enhanced HPLC chromatographic techniques.

2 Materials

2.1 Apparatus

- Analytical balance.
- Soxhlet extractor.
- Rotary evaporator.
- Centrifuge.
- Filter paper.
- Whatman filter paper.
- Syringe filters (0.22 μm or 0.45 μm).
- Column C18 (as per requirement).
- High-Performance Liquid Chromatography with Diode Array Detector (HPLC-DAD).

2.2 Reagents

- Plant materials in dry and powdered form.
- Solvents: Methanol, ethanol, acetonitrile, chloroform, formic acid or trifluoroacetic acid (TFA) (HPLC Grades), water (Milli-Q), ammonium acetate (optional, for buffer preparation), Alkaloids standards (as per targeted alkaloids) *see* **Notes 1** and **2**.

3 Methods: Preparation of Plant Material and Extraction of Alkaloids

3.1 Preparation of Plant Material

1. Collection and Drying: Collect the desired plant material (100 g) and dry it under shade to prevent degradation of alkaloids.
2. Powdering: Grind the dried material into a fine powder using a mechanical grinder to increase the surface area for extraction.
3. Weighing: Accurately weigh 5–10 g of the powdered sample for extraction.

3.2 Defatting (Optional But Recommended for Oily Seeds)

1. Soak the sample in petroleum ether (1:10 w/v) for 4–6 h with occasional shaking.
2. Filter with Whatman filter paper and discard the petroleum ether layer.

3.3 Extraction of Alkaloids

1. Add methanol containing 1% formic acid to the defatted sample (1:20 w/v ratio).
2. Sonicate the mixture for 30–45 min at room temperature to enhance extraction. Allow the sample to stand for 12–24 h for complete extraction.
3. Macerate or reflux the defatted plant material with acidified ethanol or methanol (using 1–2% HCl) for 24–48 h, *see* **Note 3**.
4. Filter the mixture to separate the liquid extract from the plant residue.
5. Use a rotary evaporator to concentrate extract up to 50% of the filtrate.
6. Centrifuge the concentrated extract at 5000 rpm for 10 min.
7. Collect the supernatant and repeat extraction with fresh solvent if necessary.
8. Evaporate the combined extracts under reduced pressure using a rotary evaporator at 40 °C to remove excess methanol.
9. Reconstitute the dried residue in a small volume (2–5 mL) of HPLC-grade methanol or the mobile phase used for HPLC.
10. Filter the reconstituted sample through a 0.22 μm or 0.45 μm syringe filter to remove particulate matter, *see* **Note 4**.
11. Transfer the filtrate into HPLC vials for analysis.

3.4 Isolation of Alkaloids (Liquid-Liquid Extraction)

1. Alkaloids, nitrogenous bases present in plants as organic acid salts, are extracted using liquid–liquid extraction. Acidification converts them into water-soluble salts, separating from neutral compounds. Basification regenerates free alkaloids, which are transferred into a non-polar solvent. Evaporation yields crude alkaloids, later purified by crystallization or chromatography.

3.5 Basification

1. Add ammonia solution dropwise to the aqueous extract until pH reaches 9–10. This converts alkaloid salts to their free base forms.

3.6 Extraction

1. Transfer the basified solution into a separatory funnel.
2. Extract with chloroform or dichloromethane (3 × 50 mL).
3. Collect the organic layers containing the alkaloids.

3.7 Drying

1. Dry the combined organic layers over anhydrous sodium sulphate.
2. Filter and evaporate the solvent under reduced pressure to obtain crude alkaloids.

3.8 Sample Filtration for HPLC

1. Reconstitute the crude alkaloid extract in HPLC-grade methanol or the mobile phase.
2. Filter through a 0.22 μm or 0.45 μm syringe filter.
3. Transfer to HPLC vials for analysis.

3.9 Centrifugation

1. Centrifuge the extract at 5000 rpm for 10 min.
2. Collect the supernatant and repeat extraction with fresh solvent if necessary.

3.10 HPLC-DAD Method for Alkaloid Identification

Alkaloids are nitrogen-rich compounds present in a variety of plants, including morphine, caffeine, and nicotine. Due to their diverse chemical structures, alkaloids can be difficult to analyze. However, HPLC offers an effective and reliable method for their separation and identification.

3.11 Step 1: HPLC System Setup

1. Column: C18 reverse-phase column (e.g., 4.6 × 250 mm, 5 μm particle size), *see* **Notes 5** and **6**.
2. Detector: Diode Array Detector (DAD).
3. Wavelength Range: 200–400 nm (monitoring alkaloid-specific absorption around 254 nm or 280 nm).

3.12 Step 2: Mobile Phase Preparation

1. Mobile Phase A: Water with 0.1% formic acid.
2. Mobile Phase B: Acetonitrile with 0.1% formic acid, *see* **Note 7**.

3.13 Step 3: HPLC Conditions

HPLC conditions	
Gradient elution program	0–5 min: 95% A/5% B 5–20 min: Linear gradient to 70% A/30% B 20–30 min: Linear gradient to 50% A/50% B 30–35 min: Return to initial conditions (95% A/5% B) for re-equilibration
Flow rate: 0.5–1.0 mL/min	Run time: 35–40 min
Injection volume:10–20 μL	Column temperature: 26–30 °C

See **Notes 8** and **9**

3.14 Data Acquisition

1. Record chromatograms at multiple wavelengths (e.g., 254 and 280 nm) using DAD to detect different alkaloids.
2. Scan the full UV spectrum (200–400 nm) to identify unique absorbance peaks.
3. Peak identification: The peaks of the sample can be identified in the chromatogram by comparing retention times with those of known standards.

4. Quantification: Use calibration curves from known standards to quantify the concentration of the phytochemicals in the sample. Measure the area under the peaks for quantification.
5. Various software tools are available for analysing HPLC-DAD data, supporting data acquisition, peak identification, integration, quantification, spectral analysis, and reporting. Notable examples include Chromeleon (Thermo Fisher), Clarity (DataApex), LabSolutions (Shimadzu), and Chromatoprofiler (open-source).

3.15 Identification

1. Compare retention times and UV spectra with authentic alkaloid standards (e.g., trigonelline, diosgenin, caffeine, atropine, reserpine).
2. Use spiking techniques by adding known standards to confirm peak identities.

3.16 Quantification

1. Construct calibration curves using serial dilutions of known alkaloid standards.
2. Quantify the concentration of alkaloids in the sample based on peak area integration, *see* **Note 10**.

3.17 Quality Control and Validation

1. System suitability: Check parameters like retention time reproducibility, peak symmetry, and resolution.
2. Blank runs: Run solvent blanks to check for contamination.
3. Repeatability: Inject the same sample multiple times to ensure precision.

4 Notes

1. Use HPLC-grade solvents to avoid contamination.
2. Filter all samples and mobile phases to prevent blockages in the HPLC system.
3. Ensure proper pH adjustment during extraction to optimize alkaloid recovery.
4. The mobile phase solvent is vacuum filtered to eliminate any gases and solids that could potentially clog the chromatographic column.
5. Before injection of sample, the stationary column should be flushed and equilibrated with the mobile phase in order to ensure the column performance consistency.
6. C18 Reverse-Phase Column packed with silica particles bonded to octadecylsilane (C18) groups are most commonly used for the phytochemical analysis as it is suitable for non-polar or moderately polar compounds.

7. Ammonium acetate buffer can be used instead of formic acid as ammonium acetate has low or neutral pH, it helps to obtain better peaks or shape when sample in polar compounds. Formic acid is highly acidic so it may interfere the peaks. Alternatively, ammonium acetate buffer can be used instead of formic acid for better ionization control.
8. The pressure of the delivery pump should be ranging around 250–5000 psi.
9. It typically involves gradient or isocratic elution using a combination of water, an organic solvent (such as methanol or acetonitrile), and occasionally an acidic modifier (like formic acid or phosphoric acid).
10. Compounds are identified by comparing their retention times to those of standards, along with UV/Vis spectra and/or MS data. The concentration of each compound is quantified by comparing the peak area to calibration curves generated using known standards.

Acknowledgments

This work was supported by Madhya Pradesh Biotechnology Council, Madhya Pradesh, India (sponsored research project (File No. 1417/CST/R&D(BioSci.)/2023)).

References

1. Singh AK, Yadav D, Sharma N et al (2021) Dipeptidyl peptidase (DPP)-IV inhibitors with antioxidant potential isolated from natural sources: a novel approach for the management of diabetes. Pharmaceuticals 14:586. https://doi.org/10.3390/ph14060586
2. Parmar HS, Kar A (2008) Antiperoxidative, antithyroidal, antihyperglycemic and cardioprotective role of Citrus sinensis peel extract in male mice. Phytother Res 22:791–795. https://doi.org/10.1002/ptr.2367
3. Pan Q, Saiman MZ, Mustafa NR et al (2016) A simple and rapid HPLC-DAD method for simultaneously monitoring the accumulation of alkaloids and precursors in different parts and different developmental stages of Catharanthus roseus plants. J Chromatogr B 1014:10–16. https://doi.org/10.1016/j.jchromb.2016.01.034
4. Aryal B, Raut BK, Bhattarai S et al (2022) Potential therapeutic applications of plant-derived alkaloids against inflammatory and neurodegenerative diseases. Evid Based Complement Alternat Med 2022:7299778. https://doi.org/10.1155/2022/7299778
5. Ameyaw Y, Duker-eshun G (2009) The alkaloid contents of the ethno-plant organs of three antimalarial medicinal plant species in the eastern region of Ghana. Int J Chem Sci 7:48–58
6. Heinrich M, Mah J, Amirkia V (2021) Alkaloids used as medicines: structural phytochemistry meets biodiversity—an update and forward look. Molecules 26:1836. https://doi.org/10.3390/molecules26071836
7. Zhou JL, Qi LW, Li P (2009) Herbal medicine analysis by liquid chromatography/time-of-flight mass spectrometry. J Chromatogr A 1216:7582–7594. https://doi.org/10.1016/j.chroma.2009.05.054
8. Dhooghe L, Mesia K, Kohtala E et al (2008) Development and validation of an HPLC-method for the determination of alkaloids in the stem bark extract of Nauclea pobeguinii. Talanta 76:462–468. https://doi.org/10.1016/j.talanta.2008.03.036
9. Mradu G, Saumyakanti S, Sohini M et al (2012) HPLC profiles of standard phenolic compounds present in medicinal plants. Int J Pharmacogn Phytochem Res 4:162–167

Chapter 17

In Vitro Evaluation of Phytochemicals for Their Health-Promoting Potential

Purnima Tripathi, Anand Krishna Singh, and Neha Sharma

Abstract

Phytochemicals, or bioactive compounds, are naturally occurring plant constituents with significant health benefits. Although not classified as essential nutrients, they possess powerful antioxidant, anti-inflammatory, and immune-modulating properties. Major groups, including flavonoids, polyphenols, alkaloids, carotenoids, and vitamins C and E, help reduce oxidative stress, slow aging, and prevent chronic conditions such as cardiovascular diseases, diabetes, obesity, cancer, and neurodegenerative disorders. Compounds like curcumin, resveratrol, and quercetin enhance heart health by lowering cholesterol, improving circulation, and preventing arterial plaque accumulation. Additionally, phytochemicals promote gut microbiota balance, and enhance cognitive function. Several analytical assays assess the health benefits of phytochemicals. The DPPH, ABTS, and FRAP assays evaluate antioxidant potential, essential for neutralizing oxidative stress and supporting cardiovascular and neurological health. The MTT, NRU, and LDH assays measure cytotoxicity and cytoprotective effects, aiding in the identification of anticancer and neuroprotective agents. The gut microbiota modulation assay examines their role in maintaining microbial balance, which influences immunity, metabolism, and disease prevention. These analytical methods provide valuable insights into the therapeutic potential of phytochemicals, supporting their application in functional foods, nutraceuticals, and pharmaceuticals. By assessing their antioxidant, cytoprotective, and gut-modulating properties, these assays contribute to the development of strategies for disease prevention and overall health enhancement.

Key words Phytochemicals, Antioxidant, Anti-inflammatory, Cytoprotective, Oxidative stress

1 Introduction

Bioactive compounds, often known as phytochemicals, are naturally occurring plant components that provide a number of health advantages. Though not classified as essential nutrients, they have potent immune-modulating, antibacterial, anti-inflammatory, and antioxidant qualities. By neutralizing free radicals, they lower oxidative stress, slow aging, and help prevent chronic diseases [1]. Key groups like flavonoids, polyphenols, alkaloids, carotenoids, and vitamins C and E contribute to these effects. Phytochemicals also

Ashwani Kumar, Vidisha Tomer and Piyoosh Kumar Babele (eds.), *Phytochemicals: Extraction, Quantification, Isolation, and Characterization*, Methods and Protocols in Food Science, https://doi.org/10.1007/978-1-0716-4840-7_17,

reduce inflammation linked to cardiovascular diseases, arthritis, and neurodegenerative disorders [2]. Compounds such as curcumin, resveratrol, and quercetin support heart health by lowering cholesterol, improving circulation, and preventing arterial plaque buildup. Their role in cancer prevention includes inhibiting tumour growth and reducing DNA damage. Additionally, phytochemicals promote gut microbiota balance, enhance digestion, and improve cognitive function while reducing the risk of neurodegenerative diseases [3, 4]. They also support immune function and aid in diabetes management by improving insulin sensitivity. Found in fruits, vegetables, whole grains, and herbs, phytochemicals contribute to overall health and disease prevention.

The DPPH (2,2-diphenyl-1-picrylhydrazyl) antioxidant assay evaluates antioxidant potential by measuring a compound's ability to donate electrons or hydrogen atoms, causing a color change from purple to yellow. Antioxidants like flavonoids, polyphenols, and carotenoids reduce oxidative stress, inflammation, and DNA damage. They protect cardiovascular and brain health, inhibit cancer cell proliferation, and promote skin health by slowing collagen degradation [5]. The ABTS (2,2′-azino-bis{3-ethylbenzothiazoline-6-sulfonic acid}) antioxidant assay measures antioxidant capacity by assessing a compound's ability to neutralize ABTS++ radicals, causing a colour change from blue green to colorless. Unlike DPPH, it evaluates both hydrophilic and lipophilic antioxidants. Identified antioxidants reduce oxidative stress, support cardiovascular and brain health, inhibit cancer proliferation, and protect collagen, aiding in anti-aging and pharmaceutical applications. The FRAP (Ferric Reducing Antioxidant Power) assay assesses antioxidant activity by measuring the reduction of ferric (Fe^{3+}) to ferrous (Fe^{2+}) ions, indicated by a color change. Unlike DPPH and ABTS, it evaluates reducing power. Identified antioxidants combat oxidative stress, support cardiovascular and brain health, reduce cancer risk, and preserve skin elasticity, making FRAP valuable for functional foods and pharmaceuticals [6].

The MTT ([3-(4,5-dimethylthiazol-2-yl)-2,5-diphenyltetrazolium bromide]) assay evaluates cell viability and cytotoxicity by measuring mitochondrial activity, as viable cells convert MTT into formazan, causing a color change. It helps identify anticancer compounds that inhibit tumor growth while preserving healthy cells. Additionally, it assesses cytoprotective and neuroprotective effects, making it valuable for drug discovery and functional food development [7]. Similarly, the NRU (Neutral Red Uptake) assay assesses cell viability by measuring the ability of viable cells to absorb and retain neutral red dye in lysosomes [8]. This assay reflects cellular health and metabolic activity, identifying cytoprotective compounds that enhance cell survival and combat oxidative stress. It is widely used to evaluate the anticancer properties and neuroprotective potential of phytochemicals in conditions such as

Alzheimer's and Parkinson's diseases. The LDH (Lactate Dehydrogenase) assay assesses cell membrane integrity by measuring LDH enzyme release from damaged cells. Through inducing tumor cell death and protecting healthy cells, it facilitates the discovery of cytoprotective substances that maintain cell integrity, lower oxidative stress, and enhance anticancer actions. This assay is also valuable for evaluating neuroprotective compounds for conditions like Alzheimer's and Parkinson's diseases, ensuring the safety of bioactive compounds in drug and nutraceutical development [9].

Additionally, the gut microbiota modulation assay examines the impact of phytochemicals on gut microbiota, crucial for digestion, immunity, metabolism, and disease prevention. A balanced microbiota supports overall health, while imbalances contribute to obesity, diabetes, inflammation, and neurodegenerative disorders [10]. This assay identifies compounds that enhance beneficial bacteria, regulate glucose metabolism and weight, improve immune function, and support the gut-brain axis, aiding in the development of functional foods, probiotics, and therapeutic strategies for better health. Overall, these analytical methods play a key function in understanding the health benefits of phytochemicals. By evaluating their antioxidant, cytoprotective, and gut-modulating properties, these assays guide the enhancement of functional foods, nutraceuticals, and pharmaceuticals that promote good health and overall well-being.

2 Materials

2.1 Antioxidant Activity

The most commonly used methods for the determination of antioxidant activity are DPPH (2,2-diphenyl-1-picrylhydrazyl) antioxidant assay, ABTS (2,2′-azino-bis(3-ethylbenzothiazoline-6-sulfonic acid)) antioxidant assay, and FRAP (Ferric reducing antioxidant power) Assay. The DPPH assay is a generally applied technique to evaluate the antioxidant activity of compounds by measuring their capability to donate electrons or hydrogen atoms to neutralize DPPH radicals. DPPH is a stable free radical with a deep purple colour that absorbs at 517 nm in UV-Vis spectrophotometry. When reduced by an antioxidant, it forms DPPH-H, resulting in a color change from purple to yellow. The decline in optical density at 517 nm indicates antioxidant capacity, expressed as percentage (%) free radical scavenging activity or IC_{50} value. This simple, fast, and reproducible method is widely applied to assess plant extracts and synthetic antioxidants [11]. The ABTS (2,2′-azino-bis (3-ethylbenzothiazoline-6-sulfonic acid)) assay is a spectrophotometric method used to estimate the antioxidant capacity of compounds. It is based on the ability of antioxidants to neutralize the $ABTS^{\bullet+}$ radical cation, which is formed by the oxidation of ABTS using an oxidizing agent such as potassium persulfate. The ABTS•$^+$ radical has a blue-green color with strong absorbance at 734 nm. When an antioxidant donates an electron or hydrogen atom, ABTS•$^+$ is reduced to a color less or less colored form, and the decrease in absorbance is measured. The extent of absorbance reduction corresponds to the antioxidant strength of the sample. The results are often expressed as Trolox Equivalent Antioxidant Capacity (TEAC) by comparing the activity to that of Trolox, a vitamin E analog. The ABTS assay is widely used due to its versatility, as it can measure both hydrophilic and lipophilic antioxidants. It is a simple, rapid, and reproducible method for evaluating total antioxidant activity in food, biological samples, and natural products [12]. The FRAP (Ferric Reducing Antioxidant Power) assay is the spectrophotometric method used to evaluate the antioxidant capacity of compounds based on their capability to reduce ferric ions (Fe^{3+}) to ferrous ions (Fe^{2+}). The reaction occurs in an acidic medium (pH ~3.6), where the Fe^{3+}-TPTZ (2,4,6-tripyridyl-s-triazine) complex is reduced to Fe^{2+}-TPTZ, developing an intense blue/green color with maximum absorbance at 593 nm. The increase in absorbance is directly correlated with the sample of antioxidant capacity. The FRAP value is typically expressed as Fe^{2+} equivalents, comparing the activity to a standard such as ferrous sulfate or Trolox. This fast, simple, and reproducible method is widely used to evaluate antioxidants in food, plant extracts, and biological fluids. Unlike other assays, FRAP measures only the reducing power of antioxidants and does not detect radical scavenging activity [11]. FRAP assay works best with water-soluble antioxidants, but it can also be adapted for lipid-soluble antioxidants by adjusting the solvent. The samples should always be run in triplicates to ensure accuracy and reproducibility.

2.1.1 DPPH Assay

(a) Apparatus

- UV-vis Spectrophotometer.
- Cuvettes.
- Pipettes and tips.
- Incubator with shaker.

(b) Reagents

- *DPPH solution*: 0.1 mM DPPH solution is prepared in methanol or ethanol *see* **Note 1**.
- *Solvents*: Methanol, ethanol, distilled water (For dilution of stock solutions as needed).
- *Blank solution*: Typically, the solvent used (methanol/ethanol) without DPPH or test samples.
- *Control antioxidant*: A known antioxidant (e.g., ascorbic acid) to compare results.

2.1.2 ABTS Assay

(a) Apparatus

- UV-vis Spectrophotometer.
- Cuvettes.
- Pipettes and tips.
- Incubator with shaker.

(b) Reagents

- *ABTS solution*: 10 mM solution *see* **Note 2**.
- *Potassium persulfate* ($K_2S_2O_8$): 2.45 mM *see* **Note 3**.
- *Solvents*: Methanol, ethanol, or distilled water (for dissolving samples).
- *Control antioxidant*: A known antioxidant like ascorbic acid or Trolox for comparison.
- *Blank solution*: Solvent used to dissolve the test samples (without ABTS or antioxidants).

2.1.3 FRAP Assay

(a) Apparatus

- Spectrophotometer.
- Microplate reader.
- Pipettes and tips.
- Cuvettes (microplate or cuvette-based depending on the mode of assay to be performed).

(b) Reagents

- Acetate Buffer *see* **Note 4**.
- Ferric chloride solution ($FeCl_3 \cdot 6H_2O$) *see* **Note 5**.
- 2,4,6-Tris (2-pyridyl)-s-triazine (TPTZ) solution *see* **Note 6**.
- *Positive Control*: A known antioxidant like Trolox (a water-soluble vitamin E analogue) or ascorbic acid.
- *Solvents*: Methanol, ethanol, or distilled water.

2.2 Cell Viability and Cytotoxicity Assays

Cell viability and cytotoxicity assays are essential in toxicology, pharmacology, and biotechnology for assessing cellular health. These assays evaluate a cell's ability to maintain metabolic activity, membrane integrity, and proliferation when exposed to drugs, toxins, or genetic modifications. They provide both qualitative and quantitative insights into cellular responses, ensuring accurate claims about efficacy and safety. Cell viability reflects a cell's capacity to function, while cytotoxicity measures the extent to which a substance induces cell death. These assays are crucial for understanding cellular physiology and the impact of external factors on cell survival and function. Cell viability and cytotoxicity assays are based on key principles, including metabolic activity (mitochondrial function, ATP production), membrane integrity, intracellular enzyme activity (LDH release), DNA/RNA content, and morphological changes. These assays are widely used to assess drug cytotoxicity in cancer cells while evaluating potential side effects on nontarget cells. They also play an important role in environmental toxicology, studying pollutants, chemicals, and nanoparticles' impact on cellular health. Additionally, these assays aid in cellular stress analysis, probing signalling pathways, and understanding mechanisms of cell death, providing valuable insights into disease development and therapeutic interventions to [13, 14]. The various methods used to study the cytotoxic activity of phytochemicals are MTT ([3-(4,5-dimethylthiazol-2-yl)-2,5-diphenyltetrazolium bromide]) assay, NRU (Neutral Red Uptake) assay, Lactate Dehydrogenase (LDH) assay, and Gut microbiota modulation assay. MTT assay, a simple and widely accepted technique to evaluate cell viability and proliferation, is based on the reduction of yellow MTT into insoluble purple formazan by mitochondrial reductases in viable cells. Solubilization and subsequent measurement by spectrophotometry at 570 nm is performed in accordance with cell viability. The intensity of the resulting color is proportional to the number of viable cells. MTT assay is a colorimetric assay used to assess cell viability and cytotoxicity by measuring mitochondrial activity [14]. The neutral red uptake (NRU) assay known as cell viability test that detects cytotoxic effects based on the uptake and retention of neutral red (NR), a weak cationic dye, in lysosomes of living cells. NR enters cells via non-ionic diffusion, accumulates, and binds to lysosomal components, stabilized by negatively charged proteins. Viable cells retain NR, while damaged or dead cells lose this ability due to lysosomal integrity loss. Cytotoxicity is quantified by measuring NR uptake, which is inversely related to toxicity. The NRU assay is highly sensitive, making it an effective method for assessing cell membrane and lysosomal integrity [15]. The Lactate Dehydrogenase (LDH) assay is a widely used method for assessing cell viability and cytotoxicity by measuring LDH release into the culture medium following cell membrane damage. LDH, a stable cytoplasmic enzyme, catalyzes the oxidation of lactate to pyruvate, using NAD^+ as a cofactor, which is subsequently reduced to NADH and H^+. The accumulation of NADH is quantitatively measured, providing an indicator of cell membrane integrity and overall cell viability. Since LDH is only released upon membrane disruption, this assay is a reliable and sensitive method for detecting cell damage and cytotoxic effects [14, 16]. Phytochemicals can significantly influence gut health by modulating the gut microbiota, which plays a vital role in human well-being. The gut microbiome comprises diverse microorganisms, including bacteria, fungi, viruses, archaea, and protists, residing in the gastrointestinal (GI) tract. With approximately 3×10^{13} live bacteria, these microbes impact metabolism, immune function, inflammation, and the brain-gut axis. A balanced microbial diversity is crucial for maintaining host homeostasis, while disruptions can lead to dysbiosis, linked to various diseases. Prebiotics and probiotics promote microbial balance by supporting beneficial bacteria, potentially preventing or alleviating conditions like diabetes. Enhancing gut health through these substances contributes to overall physiological stability and disease resistance [17].

2.2.1 MTT Assay

(a) Apparatus

1. CO_2 incubator (37 °C, 5% CO_2).
2. Pipettes and tips.
3. Microplate reader (570 nm wavelength).
4. Sterile laminar flow hood.

(b) Cell Culture-Related Materials

- Cell lines (e.g., HeLa, MCF-7).
- 96-well tissue culture plate.
- DMEM or RPMI-1640 culture medium (with 10% FBS).
- Trypsin-EDTA solution *see* **Note 8**.
- Phosphate-buffered saline (PBS) *see* **Note 9**.

(c) Reagents

- *MTT dye (3-(4,5-dimethylthiazol-2-yl)-2,5-diphenyltetrazolium bromide)*: Dissolve 5 mg/mL in phosphate buffer saline. Protect MTT solution from light as it is light sensitive.
- Dimethyl sulfoxide (DMSO) or isopropanol (to dissolve formazan crystals).

2.2.2 NRU Assay

(a) Apparatus

- CO_2 incubator (37 °C, 5% CO_2).
- Sterile laminar flow hood.
- Pipettes and tips.
- Microplate reader (540 nm wavelength).

(b) Cell Culture Materials

- Cell lines (e.g., HeLa, MCF-7).
- Culture medium (e.g., DMEM or RPMI-1640 with 10% PBS buffer).
- Trypsin-EDTA solution *see* **Note 8**.
- Phosphate-buffered saline (PBS) *see* **Note 9**.

- Neutral red dye (3-amino-7-dimethylamino-2-methylphenazine hydrochloride) *see* **Note 10**.
- DMSO (Dimethyl Sulfoxide).
- Glacial acetic acid.
- Ethanol (96%).
- Neutral red destaining solution *see* **Note 11**.

2.2.3 LDH Assay

(a) **Apparatus**

- CO_2 incubator (37 °C, 5% CO_2).
- Sterile laminar flow hood.
- Microplate reader (490 nm or 492 nm wavelength).
- Pipettes and tips.

(b) **Reagents**

- LDH substrate solution *see* **Note 12**.
- NAD+ (Nicotinamide Adenine Dinucleotide) solution *see* **Note 13**.
- Sodium lactate solution *see* **Note 14**.
- Tetrazolium salt (INT; Iodonitrotetrazolium chloride).
- Diaphorase enzyme.
- Lysis buffer (Triton X-100 or NP-40 in PBS) *see* **Note 15**.
- LDH reaction mixture *see* **Note 16**.

2.2.4 Gut Microbiota Modulation Assay

(a) **Apparatus**

- Anaerobic chamber or jars with gas packs.
- Incubator (37 °C) for bacterial culture.
- Centrifuge.
- Spectrophotometer or microplate reader.
- qPCR machine.
- 16S rRNA sequencing platform.

(b) **Media and Cultures**

- Biological samples: Human or animal fecal samples (collected under anaerobic conditions).
- Brain heart infusion (BHI) broth or reinforced clostridial medium (RCM).
- Anaerobic basal media (for fecal fermentation).
- PBS (Phosphate-buffered saline) *see* **Note 9**.
- Probiotics (e.g., Lactobacillus, Bifidobacterium strains).
- Prebiotics (e.g., inulin, fructooligosaccharides).

- Antibiotics or antimicrobial agents (Penicillins, Cephalosporins, Tetracyclines, Macrolides, etc.).
- Phytochemical samples.

2.3 Antimicrobial Activity Assays

The antimicrobial activity assay is used to evaluate the effectiveness of antimicrobial agents (such as antibiotics, plant extracts, or nanoparticles) against bacterial or fungal strains. Phytochemicals are often assessed for their ability to inhibit the growth of pathogens, including bacteria, fungi, and viruses. The major methods for assessing antimicrobial activity are disc diffusion method, minimum inhibitory concentration (MIC). The disk diffusion method is commonly used to measure the antimicrobial activity of phytochemicals. In order to use this technique, disks impregnated with the phytochemicals are placed onto an agar plate that has been infected with the designated bacteria. After incubating the plate, a zone of inhibition surrounding the disks will appear if the phytochemical is effective in preventing bacterial growth. The effectiveness of the phytochemical is assessed based on the size of the zone of inhibition. To evaluate the results, the zone of inhibition is compared to that of a positive control, typically an antibiotic-impregnated disk placed on a separate agar plate. The minimum concentration of an antimicrobial agent that stops the growth of bacteria or pathogens is found using the minimum inhibitory concentration (MIC) test. This test is frequently utilized in clinical laboratories and research to assess the susceptibility of bacterial strains to antimicrobial compounds, including antibiotics and phytochemicals [18]. Proper aseptic conditions should be maintained to prevent contamination during these tests. The use of sterile discs avoids cross-contamination. The tests should be performed in triplicates to avoid cross-contamination, and the use of proper controls ensures reliable interpretation.

2.3.1 Disc Diffusion Method

(a) Apparatus

- Incubator (37 °C for bacteria, 28–30 °C for fungi).

(b) Reagents

- Bacterial strains (e.g., *E. coli*, *Staphylococcus aureus*, *Pseudomonas aeruginosa*).
- Growth media Mueller-Hinton agar (MHA) for bacterial testing, potato dextrose agar (SDA) for fungal testing.
- Antibiotics (e.g., ampicillin, tetracycline).
- Plant extracts or phytocompounds.

(c) Others

- Sterile cotton swabs.
- Pipettes and sterile tips.
- Agar plates (Petri dishes).
- Filter paper discs.

2.3.2 Minimum Inhibitory Concentration (MIC)—Broth Dilution Method

(a) Apparatus

- Microtiter plates (for MIC assay).
- Incubator (37 °C for bacteria, 28–30 °C for fungi).
- Spectrophotometer or microplate reader (600 nm for OD measurement).

(b) Reagents

- Mueller-Hinton broth (MHB) or Luria Bertani (LB) broth for liquid culture, nutrient broth for bacterial culture preparation.
- Antibiotics (e.g., ampicillin, tetracycline).
- Plant extracts or phytocompounds.

(c) Others

- Sterile cotton swabs.
- Pipettes and sterile tips.
- Sterile cork borer.

3 Methods

3.1 DPPH Assay

(a) Preparation of Sample, Stock, Blank, and Control

1. *Preparation of test samples*: Extract the plant samples with a suitable solvent (e.g., methanol, ethanol). Prepare stock solutions (e.g., 0.25–5 mg/mL).
2. *Preparation of stock solution*: Prepare stock solutions (e.g., 0.25–5 mg/mL).
3. *Preparation of positive control*: Use a known antioxidant such as ascorbic acid. Prepare a series of concentrations for comparison (e.g., 10, 20, 40, 50 μg/mL).
4. *Preparation of blank*: This will contain only the solvent without DPPH.

(b) Analysis

1. Pipette out 100 μL of the DPPH solution (0.1 mM) into each well of 96-well microplate.

2. Pipette out 100 μL of test sample (or control standard) at varying concentrations into corresponding wells containing the DPPH solution.
3. Gently shake or pipette the mixture to ensure proper mixing.
4. Let the reaction proceed for 30 min to 1 h in the dark at 25–30 °C (room temperature). A decrease in absorbance will result from the test sample's positive standard reducing the DPPH radical.
5. After incubation, use a spectrophotometer or microplate reader to measure each well's absorbance at 517 nm. For the control (blank), record the absorbance of the DPPH solution without test samples.

(c) Calculations

The antioxidant activity can be calculated by applying the following formula:

$$\%\text{Inhibition} = \frac{\text{O.D.of control} - \text{O.D.of test sample}}{\text{O.D.of control}} \times 100$$

(d) Interpretation of Results

1. The greater the percentage inhibition, the stronger the antioxidant activity of the test sample. Compare the results to the positive control (known antioxidant) to calculate the relative strength.
2. Plot the % inhibition against the test sample concentration. The concentration at which 50% inhibition occurs (IC_{50}) can be calculated to measure the potency of the antioxidant.

3.2 ABTS Assay

(a) Activation and Dilution of ABTS Solution

1. *Activation of ABTS solution*: Add 88 μL of $K_2S_2O_8$ solution to 1 mL of the ABTS stock solution. Allow the mixture to stand in the dark at room temperature for 12–16 h to facilitate the formation of the ABTS radical cation (ABTS++). Once this time has elapsed, the solution should exhibit a deep green/ blue color *see* **Note 7**.
2. *Dilution of ABTS solution*: Before using the ABTS++ solution for the assay, dilute it with the solvent (typically distilled water or ethanol) to an absorbance of 0.70 ± 0.04 at 734 nm, measured using a spectrophotometer. This ensures the ABTS++ solution is at the appropriate concentration for the assay.

(b) Preparation of Control and Test Samples

1. *Preparation of control*: Use Trolox or ascorbic acid as a positive control. Prepare known concentrations of solutions for comparison.
2. *Preparation of test samples*: Dissolve the test compounds (plant extracts, phytochemicals, etc.) in an appropriate solvent (methanol, ethanol, or water) to the desired concentration.

(c) Analysis

1. In a 96-well microplate or cuvette, add a fixed volume of the ABTS++ solution (e.g., 150 μL) to each well or cuvette.
2. Add test sample (e.g., 10–50 μL) to each well containing ABTS++ solution.
3. Mix the solution thoroughly by gently shaking or pipetting.
4. Allow the mixture to react at room temperature for 6 min (or until the color stabilizes). This allows the antioxidant to scavenge the ABTS++ radical.
5. After incubation, measure the absorbance at 750 nm via a spectrophotometer or microplate reader.
6. Measure the absorbance of the blank solution (solvent) in order to correct for any background interference.

(d) Interpretation and Data Analysis

1. The antioxidant activity can be estimated by using the following formula:

$$\%\text{Inhibition} = \frac{\text{O.D.of control} - \text{O.D.of test sample}}{\text{O.D.of control}} \times 100$$

2. The results can be compared to a control antioxidant such as Trolox or Ascorbic acid to evaluate the relative antioxidant activity of the test samples.

3.3 FRAP Assay

(a) Preparation of the Working Reagents

1. *Preparation of working FRAP reagent*: It must be prepared just before use. To prepare this mix 10 mL acetate buffer with 1 mL TPTZ solution and 1 mL ferric chloride solution. This must be used within 3 h. Working FRAP reagent should be pale yellow orange in color. The FRAP reagent is usually pre-warmed to 37 °C in a water bath for best results, but it should not be hot during the assay. Ensure that the reagent remains warm (37 °C) throughout the experiment.

(b) Preparation of Test Samples and Controls

1. *Preparation of test samples*: If using plant extracts, phytochemicals, or other substances, dissolve them in an appropriate solvent (e.g., methanol, ethanol, or water) to the desired concentration.
2. *Preparation of blank solution*: This should be the solvent without any sample or FRAP reagent, used to set the baseline absorbance.
3. *Preparation of positive control*: Use a known antioxidant like Trolox or ascorbic acid. Prepare standard solutions at different concentrations (e.g., 10, 20, 50 μM) for comparison.

(c) Analysis

1. Pipette 1 mL of FRAP reagent into each well of a 96-well microplate (or into cuvettes if using a spectrophotometer).
2. Add 100 μL of the test sample (plant extract, chemical, or control) to each well containing the FRAP reagent.
3. Gently mix the contents by pipetting or shaking the microplate.
4. Allow the mixture to incubate at 37 °C for 30 min. During this time, the antioxidant present in the sample will reduce the ferric ions (Fe^{3+}) to ferrous ions (Fe^{2+}), forming a blue-colored complex with TPTZ.
5. After 30 min of incubation, measure the absorbance at 593 nm using a spectrophotometer or microplate reader. The absorbance will be proportional to the antioxidant power of the sample.
6. The antioxidant activity is expressed in terms of the FRAP value, which is calculated using the standard curve. Prepare a standard curve with Trolox or ascorbic acid at known concentrations (e.g., 10, 20, 50 μM).

(d) Interpretation and Data Analysis

Plot the absorbance values against the standard's known concentrations (e.g., Trolox). This curve is used to convert the absorbance values of the test samples into corresponding antioxidant capacities.

3.4 MTT Assay

1. Prepare a single-cell suspension by trypsinizing the adherent cells or resuspending suspension cells.
2. Count cells using a hemocytometer or an automated counter.
3. Seed 5000–10,000 cells per well in a 96-well plate with 100 μL of complete culture medium.
4. Incubate the plate at 37 °C with 5% CO_2 for 12–24 h to allow cell attachment.

5. Prepare the test compounds or drugs in culture medium at different concentrations.
6. Replace the old medium with 100 μL of fresh medium containing the test compound.
7. Incubate the cells for the desired treatment time (e.g., 24, 48, or 72 h).
8. Add 10–20 μL of MTT solution to each well (final concentration: 0.5 mg/mL).
9. Incubate for 3–4 h at 37 °C in the dark. The viable cells will convert MTT into purple formazan crystals.
10. Carefully remove the medium without disturbing the formazan crystals.
11. Add 100–150 μL of DMSO to dissolve the formazan crystals.
12. Shake the plate gently for 10–15 min at 25 °C.
13. Measure the absorbance at 570 nm by using a microplate reader (reference wavelength: 630 nm).
14. Normalize the readings with blank control wells (no cells).
15. Calculate cell viability using the formula:

$$\%\text{Cell viability} = \frac{\text{Absorbance of treated cells}}{\text{Absorbance of control cells}} \times 100$$

Generate dose-response curves or IC_{50} values if required.

3.5 NRU (Neutral Red Uptake) Assay

1. Prepare a single cell suspension by trypsinizing adherent cells or resuspending suspension cells. Maintain sterility while handling cells.
2. Count the cells using a hemocytometer or automated cell counter.
3. Seed 5000–10,000 cells per well in a 96-well plate with 100 μL of complete culture medium.
4. Incubate the cells for 24 h at 37 °C with 5% CO_2 to allow attachment.
5. Prepare the test compounds in culture medium at different concentrations.
6. Replace the old medium with 100 μL of fresh medium containing the test compound.
7. Incubate the plate for the desired exposure time (e.g., 24, 48, or 72 h).
8. Remove the treatment medium and add 100 μL of neutral red solution per well. Always protect the Neutral Red dye from light as it is light-sensitive. Also ensure even distribution of neutral red solution for uniform staining.
9. Incubate for 2–3 h at 37 °C to allow live cells to uptake the dye.

10. After incubation, carefully remove the neutral red solution. Do not over-incubate with neutral red as excessive uptake may affect results.
11. Wash the cells gently with phosphate buffer saline to remove excess dye.
12. Add 100 μL of destaining solution per well.
13. Shake the plate gently for 10–15 min at room temperature to dissolve the dye.
14. Measure the absorbance at 540 nm using a microplate reader.
15. Normalize with blank control wells (no cells).
16. Calculate the percentage of cell viability using the formula:

$$\%\text{Cell viability} = \frac{\text{Absorbance of treated cells}}{\text{Absorbance of control cells}} \times 100$$

Plot dose-response curves or determine IC_{50} values if required.

3.6 LDH Assay

1. Prepare a single-cell suspension by trypsinizing adherent cells or resuspending suspension cells.
2. Count the cells using a hemocytometer or automated cell counter.
3. Seed 5000–10,000 cells per well in a 96-well plate with 100 μL of complete culture medium.
4. Incubate the plate for 24 h at 37 °C with 5% CO_2 to allow attachment.
5. Prepare the test compounds in culture medium at different concentrations.
6. Replace the old medium with 100 μL of fresh medium containing the test compound.
7. Incubate the plate for the desired exposure time (e.g., 24, 48, or 72 h).
8. Include three types of controls:
   - Low control (spontaneous LDH release): Untreated cells in normal medium.
   - High control (maximum LDH release): Cells treated with lysis buffer (Triton X-100) to release all LDH.
   - Blank control: Medium without cells.
9. After incubation in different exposed time then (*see* **step** 7) transfers 50–100 μL of culture supernatant from each well to a new 96-well plate. Avoid cell debris contamination when transferring the supernatant.
10. Avoid disturbing the cell monolayer to prevent additional LDH release.

11. Add 100 μL of LDH reaction mixture to each well containing the collected supernatant. Protect the LDH reaction mixture from light as the tetrazolium salt is light-sensitive.
12. Incubate the plate at 37 °C for 30 min in the dark. Ensure equal incubation times for all samples to maintain consistency.
13. Measure the absorbance at 490 nm using a microplate reader. Use appropriate positive and negative controls for accurate interpretation.
14. Calculate the percentage of cytotoxicity using the formula:

$$\text{Cytotoxicity (\%)} = \frac{\text{LDH}_{\text{EXrelease}} - \text{LDH}_{\text{SpoRelease}}}{\text{LDH}_{\max} - \text{LDH}_{\text{SpoRelease}}} \times 100$$

This protocol quantitatively measures cell membrane integrity, reflecting the degree of cell damage or death induced by different treatments.

Generate dose-response curves or determine IC_{50} values if required.

3.7 Gut Microbiota Modulation Assay

1. Collect fresh fecal samples from humans or animals in sterile containers. Ensure sterility while handling fecal samples to avoid contamination.
2. Store samples on ice or in anaerobic conditions to maintain microbial viability and prevent microbiota alterations.
3. Prepare fecal slurry by diluting the sample (e.g., 10% w/v) in anaerobic PBS.
4. Homogenize the slurry and filter to remove debris if needed.
5. Transfer 1–2 mL of fecal slurry into anaerobic culture tubes or microplates.
6. Add test compounds (e.g., probiotics, prebiotics, antibiotics) at desired concentrations.
7. Include controls:
   - Positive control (known gut-modulating agent).
   - Negative control (untreated fecal sample).
8. Incubate anaerobically at 37 °C for 24–48 h.
9. After incubation, collect samples for analysis:
   - pH measurement (to assess fermentation products).
   - Optical density (OD 600 nm).
10. Centrifuge the samples at 12,000 rpm for 10 min to separate supernatant and bacterial cells.
11. Extract bacterial DNA using a DNA extraction kit. The extracted DNA should be stored at Store extracted DNA at −20 °C or −80 °C for sequencing.

12. Perform 16S rRNA sequencing or qPCR for microbial identification.
13. Analyze short-chain fatty acid (SCFA) production using HPLC or GC-MS.
14. Compare microbial diversity and abundance between treated and control samples. Use appropriate negative and positive controls for accurate interpretation.
15. Analyze changes in SCFA levels, pH, and bacterial population shifts. Use bioinformatics tools (QIIME, Kraken, or MetaPhlAn) for microbiome data analysis.

3.8 Disc Diffusion Method (Kirby-Bauer Method)

1. Prepare bacterial or fungal inoculum by adjusting to 0.5 McFarland standard (~1.5×10^8 CFU/mL).
2. Spread 100 μL of bacterial suspension evenly on an MHA/SDA plate using a sterile cotton swab.
3. Lace sterile filter paper discs (6 mm) with the help of sterile cork borer on the agar surface.
4. Load 10–20 μL of test compound onto the discs and allow absorption.
5. Incubate the plates at 37 °C for 18–24 h (for bacteria) or 28–30 °C for fungi.
6. Measure the zone of inhibition (ZOI) around each disc using a ruler.

3.9 Minimum Inhibitory Concentration (MIC)—Broth Dilution Method

1. Prepare serial dilutions of the test compound in a 96-well microtiter plate using MHB.
2. Add 100 μL of standardized bacterial suspension (~5×10^5 CFU/mL) to each well.
3. Include positive control (antibiotic-treated) and negative control (untreated bacteria).
4. Incubate at 37 °C for 18–24 h.
5. Measure OD at 600 nm to determine bacterial growth inhibition.
6. The MIC is the lowest concentration of the test compound that inhibits visible growth.

4 Notes

1. *DPPH solution*: Dissolve 2.4 mg of DPPH powder in methanol or ethanol to a final volume 100 mL to get a 0.1 mM DPPH solution. Store the solution in a dark bottle and protect it from light as it is light sensitive.

2. *ABTS solution*: Dissolve 7 mg of ABTS salt in 1.0 mL of distilled water to make a 10 mM stock solution.
3. *Potassium persulfate ($K_2S_2O_8$) solution*: Dissolve 0.0662 g of $K_2S_2O_8$ in distilled water and dilute to a final volume of 100 mL with distilled water or phosphate buffer (pH -6.8) in a volumetric flask.
4. *Acetate buffer, 300 mmol/L, pH 3.6*: 3.1 g sodium acetate trihydrate is dissolved in 800 mL distilled water; 16 mL glacial acetic acid is added, and the pH measured and adjusted if necessary to pH 3.6 with glacial acetic acid or 1 mol/L sodium hydroxide. The final volume is brought to 1 L with water. The buffer can be stored at room temperature. This reagent should appear clear and colorless.
5. *Ferric chloride solution, 20 mmol/L*: 270 mg ferric chloride hexahydrate is dissolved in 50 mL of water. The reagent is stored at 4 °C in a dark bottle.
6. *TPTZ solution, 10 mmol/L in 40 mmol/L HCl*: 156 mg 2,4,6 tripyridyl-S-triazine (TPTZ) is dissolved in 50 mL of 40 mmol/L hydrochloric acid (HCl). The reagent is stored at 4 °C and can be used for at least 2 weeks after preparation.
7. ABTS++ solution should be prepared freshly to maintain its activity.
8. *Trypsin EDTA solution*: Dissolve 0.5 g trypsin in 20–30 mL PBS and 0.2 g EDTA in 20–30 mL PBS, adjusting pH to 7.2–7.4. Mix 25 mL of each, dilute to 100 mL, sterilize, store, and warm before use.
9. *Phosphate buffer saline*: Dissolve NaCl, KCl, Na_2HPO_4, and KH_2PO_4 in 800 mL distilled water, adjust pH to 7.2–7.4, and dilute to 1 L. Sterilize by filtration or autoclaving. Store at room temperature or 4 °C for long-term use.
10. *Neutral red solution*: Dissolve 0.1 g of neutral red in 100 mL of distilled water, then filter sterilize using a 0.22 μm filter or autoclave, and store in a dark bottle.
11. *Neutral red destaining solution*: 50% ethanol + 1% acetic acid + 49% distilled water.
12. *LDH substrate solution*: For the forward reaction (Lactate → Pyruvate), dissolve sodium lactate (50–100 mM) in Tris–HCl buffer (pH 7.4–8.0), add NAD^+ (0.2–1.0 mM), and adjust the final volume with distilled water. For the reverse reaction (Pyruvate → Lactate), dissolve sodium pyruvate (1–10 mM) in phosphate buffer (pH 7.0–7.5), add NADH (0.2–1.0 mM), and protect from light. Store both solutions on ice.

13. *NAD+ (Nicotinamide adenine dinucleotide) solution*: Weigh 0.663 g NAD^+, dissolve in 50–100 mM phosphate or Tris–HCl buffer (pH 7.4–8.0), adjust pH if needed, dilute to 100 mL, aliquot, and store at −20 °C or 4 °C.
14. *Sodium lactate solution*: Dissolve sodium lactate in water or neutralize lactic acid with NaOH. Adjust pH (7.4–8.0), optionally add buffer, dilute to 100 mL, and store at 4 °C for use.
15. *Lysis buffer*: Prepare a lysis buffer by mixing 1× PBS with 0.5–1% Triton X-100 and optional protease inhibitors for protein stability.
16. *LDH reaction mixture*: The LDH reaction mixture contains 50–100 mM Tris–HCl buffer (pH 7.4–8.0) to maintain pH, 0.1–0.2 mM NADH as a coenzyme, 1–2 mM pyruvate as the substrate, LDH enzyme for catalysis, and distilled water to adjust the volume.

Acknowledgments

This work was supported by Madhya Pradesh Biotechnology Council, Madhya Pradesh, India (sponsored research project (File No. 1417/CST/R&D(BioSci.)/2023)).

References

1. Zhang YJ, Gan RY, Li S et al (2015) Antioxidant phytochemicals for the prevention and treatment of chronic diseases. Molecules 20: 21138–21156. https://doi.org/10.3390/molecules201219753
2. Nash RJ (2001) In: Tringali C (ed) Bioactive compounds from natural sources: isolation, characterisation and biological properties, 1st edn. CRC Press, London
3. Cai B, Zhang Y, Wang Z et al (2020) Therapeutic potential of diosgenin and its major derivatives against neurological diseases: recent advances. Oxidative Med Cell Longev 2020: 3153082. https://doi.org/10.1155/2020/3153082
4. Aryal B, Raut BK, Bhattarai S et al (2022) Potential therapeutic applications of plant-derived alkaloids against inflammatory and neurodegenerative diseases. Evid Based Complement Alternat Med 2022:7299778. https://doi.org/10.1155/2022/7299778
5. Thaipong K, Boonprakob U, Crosby K et al (2006) Comparison of ABTS, DPPH, FRAP, and ORAC assays for estimating antioxidant activity from guava fruit extracts. J Food Compos Anal 19:669–675. https://doi.org/10.1016/j.jfca.2006.01.003
6. Re R, Pellegrini N, Proteggente A et al (1999) Antioxidant activity applying an improved ABTS radical cation decolorization assay. Free Radic Biol Med 26:1231–1237. https://doi.org/10.1016/S0891-5849(98)00315-3
7. Freimoser FM, Jakob CA, Aebi M et al (1999) The MTT [3-(4,5-dimethylthiazol-2-yl)-2,5-diphenyltetrazolium bromide] assay is a fast and reliable method for colorimetric determination of fungal cell densities. Appl Environ Microbiol 65:3727–3729. https://doi.org/10.1128/AEM.65.8.3727-3729.1999
8. Kaffy J, Brinet D, Soulier JL et al (2016) Designed glycopeptidomimetics disrupt protein-protein interactions mediating amyloid β-peptide aggregation and restore neuroblastoma cell viability. J Med Chem 59:2025–2040. https://doi.org/10.1021/acs.jmedchem.5b01629
9. Decandia D, Gelfo F, Landolfo E et al (2023) Dietary protection against cognitive impairment, neuroinflammation and oxidative stress in Alzheimer's disease animal models of

lipopolysaccharide-induced inflammation. Int J Mol Sci 24:5921. https://doi.org/10.3390/ijms24065921

10. Iatcu CO, Steen A, Covasa M (2022) Gut microbiota and complications of type-2-diabetes. Nutrients 14:166. https://doi.org/10.3390/nu14010166
11. Singh AK, Jatwa R, Joshi J (2014) Cytoprotective and dipeptidyl peptidase-IV (DPP-IV/CD26) inhibitory roles of Ocimum sanctum and Momordica charantia extract. Asian J Pharm Clin Res 7:115–120
12. Ozgen M, Reese RN, Tulio AZ et al (2006) Modified 2,2-azino-bis-3-ethylbenzothiazoline-6-sulfonic acid (ABTS) method to measure antioxidant capacity of selected small fruits and comparison to ferric reducing antioxidant power (FRAP) and 2,2′-diphenyl-1-picrylhydrazyl (DPPH) methods. J Agric Food Chem 54:1151–1157. https://doi.org/10.1021/jf051960d
13. Harput US, Genc Y, Saracoglu I (2012) Cytotoxic and antioxidative activities of Plantago lagopus L. and characterization of its bioactive compounds. Food Chem Toxicol 50:1554–1559. https://doi.org/10.1016/j.fct.2012.01.019
14. Raiola A, Del Giudice R, Monti DM et al (2015) Bioactive compound content and cytotoxic effect on human cancer cells of fresh and processed yellow tomatoes. Molecules 21:33. https://doi.org/10.3390/molecules21010033
15. Edziri H, Mastouri M, Mahjoub A et al (2011) Toxic and mutagenic properties of extracts from Tunisian traditional medicinal plants investigated by the neutral red uptake, VITOTOX and alkaline comet assays. S Afr J Bot 77:703–710. https://doi.org/10.1016/j.sajb.2011.03.007
16. Han JH, Lee EJ, Park W et al (2023) Natural compounds as lactate dehydrogenase inhibitors: potential therapeutics for lactate dehydrogenase inhibitors-related diseases. Front Pharmacol 14:1275000. https://doi.org/10.3389/fphar.2023.1275000
17. Manderson K, Pinart M, Tuohy KM, Grace WE et al (2005) In vitro determination of prebiotic properties of oligosaccharides derived from an orange juice manufacturing by-product stream. Appl Environ Microbiol 71:8383–8389. https://doi.org/10.1128/AEM.71.12.8383-8389.2005
18. Devanaboyina N, Ramalakshmi N, Sudeepthi BS et al (2013) Preliminary phytochemical screening, quantitative estimation and evaluation of antimicrobial activity of Alstonia macrophylla stem bark. IJSIT 2:31–39

Chapter 18

LC-MS/MS and GC-MS Method for the Determination of Phenolics in Plant Tissue

Shiva Kant and Piyoosh Kumar Babele

Abstract

Studies on secondary metabolites such as phenolics have gained enormous popularity due to their potential applications in agriculture, human nutrition, and health. In plants they play a critical role in plant defense and pollination. In humans, plant phenolic compounds are reported to be involved in lowering the oxidation stress diseases due to their anti-inflammatory, anti-allergic, and anti-carcinogenic activities. They are present in different types of plant tissues and foods, but extracting and characterizing them is challenging due to their presence in complex matrices. Plant metabolomics utilizing liquid chromatography (LC) and/or as chromatography (GC) coupled with tandem mass spectrometry (MS) is ideal for identifying and quantitating a wide range of primary and secondary plant metabolites, including phenolics. This chapter discusses the most relevant studies from recent years related to the analysis of polyphenols in various matrices using comprehensive LC–MS/GC-MS.

Key words Secondary metabolites, Phenolics, LC-MS, GC-MS, Derivatization

1 Introduction

Phenolic compounds (PC) are the most abundant secondary metabolites of plants, with over 8000 different phenolic structures identified to date. Phenolics are a group of metabolites derived from the secondary pathways of plants. Although there is not a single classification, phenolics could be categorized into seven groups: Flavonoids, phenolic acids, simple and condensed tannins, stilbenes, coumarins, lignans, and lignins [1, 2]. They are commonly present in various plant products such as fruits, vegetables, cereals, roots, and leaves [3]. Plant polyphenols serve multiple functions, including acting as antioxidants, forming structural polymers like lignin, serving as attractants (like flavonoids and carotenoids), providing UV protection (through flavonoids), functioning as signaling compounds (such as salicylic acid and flavonoids), and playing a role in defense responses (like tannins and phytoalexins),

Ashwani Kumar, Vidisha Tomer and Piyoosh Kumar Babele (eds.), *Phytochemicals: Extraction, Quantification, Isolation, and Characterization*, Methods and Protocols in Food Science, https://doi.org/10.1007/978-1-0716-4840-7_18,

as well as contributing to plant's color [4, 5]. For example, anthocyanins, a subgroup of flavonoids, provide the orange, red, blue, and purple colors in fruits and vegetables like apples, berries, beets, and onions [6]. Present in all plant parts, they are also crucial to the human diet. Phenolic compounds have important physiological roles, offering anti-aging, anti-inflammatory, antioxidant, and anti-proliferative benefits [5, 7]. Consuming plant foods rich in antioxidants can lower the risk of chronic diseases, such as diabetes, cancer, and cardiovascular issues, by managing oxidative stress.

It is very important to study phenolic compounds because they may help human health. These compounds have complicated chemical structures and are found in different amounts. Therefore, it is necessary to measure them correctly in food and beverages to know how they might affect our bodies. Mass spectrometry (MS) is the most common analytical tool for performing plant metabolomics [8, 9]. MS coupled with liquid or gas chromatography (LC and/or GC, respectively) is a preferred method due to its higher sensitivity and lesser amount of sample requirements [8, 10]. GC-MS analysis offers influence measurements of reproducibility, dynamic range, and universal mass spectral library for compounds with small molecular weight. On the other hand, LC–MS covers a large array of compounds predominant as secondary metabolites such as terpenoids and phenolics [9, 10]. These two methods were equally adapted to allow both inclusive impression and full analysis of phenolics in plants [11, 12]. LC–MS-based approaches are expected to be of particular importance in plants, owing to the highly rich biochemistry of plants and variations in the range of physicochemical properties, molecular masses, and structure among classes of phenolics [11]. Separation and identification of several phenols simultaneously are highly challenging and often need optimization and development of suitable methods [13, 14]. LC-MS and GC-MS can be sensitive, fast, and reproducible platforms for the qualitative and quantitative estimation of phenols and are described in this chapter.

2 Materials

2.1 Apparatus

- Mixer and grinder.
- Vortex mixer.
- Refrigerated centrifuge.
- Ultrasonic bath.
- Analytical and precision balance.
- 0.45 μm nylon-membrane syringe filter.

- Liquid chromatography-mass spectrophotometer (LC-MS) machine with required columns and software for data analysis.
- Gas chromatography-mass spectrophotometer (LC-MS) machine with required columns and software for data analysis.

2.2 Reagents

- Methanol (LC-MS grade).
- Formic acid (analytical grade).
- Acetic acid (analytical grade).
- Standards of phenolic compounds (depends on the target phenolics).
- Deionized water (18.6 MΩ cm resistivity).
- Ribitol.
- Methyl tert-butyl ether (MTBE).
- Pyridoxine hydrochloride.
- Methoxamine hydrochloride (MeOX).
- N-Methyl-N-trimethylsilyl trifluoroacetamide (MSTFA).
- O-bis(trimethylsilyl)trifluoroacetamide (BSTFA).
- Trimethylchlorosilane (TMS).

3 Methods

3.1 Stock and Preparation of Working Standard Solutions

1. Prepare stock solutions at a concentration of 1 mg/mL of desired standards of phenolic compounds individually by dissolving accurately weighed standards in pure HPLC grade methanol.
2. To ensure the complete dissolution of standards, apply ultrasonication for 10 min.
3. Filter all standards through 0.45 μm nylon-membrane syringe filter assembly.
4. Prepare a mixture of stock solution containing all the desired standards at a concentration of (0.1 mg/mL) in methanol.
5. Dilute the stock standard mixture with the methanol to obtain reference working standard solutions with different concentrations ranging from 0.05 to 20 μg/mL for plotting standard curves.
6. Keep all the standards at −20 °C and thaw them properly before analysis.

3.2 Extraction and Fractionation Process of Phenolic Compounds

1. Phenolic compounds can be extracted using different liquid-liquid and solid liquid extraction protocols. The extraction techniques have been discussed in detail in Chapter 1. A dry extract rich in phenolic compounds should be dissolved in methanol to a concentration of 5 mg/L.

2. Filter the samples through a membrane filter with porosity of 0.45 μm, collect in airtight LC vials and keep them away from the moisture, *see* **Note 1**.
3. Before analysis, dilute the samples with methanol and either subjected directly to LC–MS analysis or stored at −20 °C until analysis.

3.3 LC-MS Analysis

1. Configure the HPLC instrument following standard operating procedures. Make sure that the equipment is correctly calibrated (according to the manuals of each LC-MS instrument) before use, *see* **Note 2**.
2. Use a sequence consisting of multiple blanks at the beginning and end of the analysis. Ideally, inject at least three replicates per sample to ensure reproducibility of the analyses.
3. If different types of extracts are going to be analyzed, run blank samples between them to avoid possible carryover peaks, *see* **Note 3**.
4. Table 1 details the HPLC and MS parameters.

Table 1
HPLC and MS parameters

Chromatographic conditions	MS conditions
Autosampler and injector: Flow through needle injector (FTN) with 15-μL needle	Ionization mode: Negative ion mode
Column: C18 column (250 mm × 4.6 mm, 5 μm)	Acquisition mode: Negative selected ion monitoring (SIM)
Mobile phase A: 0.2% acetic acid in 95% water and 5% MeOH Mobile phase B: 0.2% acetic acid in 50% water and 50% acetonitrile	Capillary voltage: −3.5 kV
Elution: linear gradient elution: 0–45 min, 10–20% B; 45–85 min, 20–55% B; 85–97 min, 55–100% B; 97–110 min, 100% B	Nebulizing gas flow: 1.5 L/min
Column temperature: 40 °C Sample temperature: 10 °C	Dry gas flow: 12 L/min
Injection volume: 5 μL	Curved desolvation line (CDL) temperature: 250 °C
Flow rate: 0.50 mL/min	Block source temperature: 400 °C
Needle wash solvent: water: methanol: acetonitrile: isopropanol (20:40:20:20; v/v) Seal wash solvent: water: methanol: (80:20; v/v)	Mass acquisition: 50–2000 Da

3.4 Gas Chromatography—Mass Spectroscopy (GC-MS) for Profiling of Plant Phenolics

3.4.1 Sample Derivatization for GC-MS

1. Completely dry the extracted samples using the rotary evaporator under vacuum.
2. MeOX-TMS derivatize all the extracted samples and standards using pyridine, methoxamine hydrochloride, and N-Methyl-N-trimethylsilyl trifluoroacetamide, *see* **Note 4**.
3. For MeOX-TMS derivatization add 35 μL of pyridine containing methoxyamine hydrochloride (20 mg/mL) to the dried sample. Incubate the samples at 37 °C, 900 rpm for 2 h, then add 49 μL of MSTFA (N-methyl-N-(trimethylsilyl)-trifluoroacetamide) to the sample and re-incubate for 30 min.
4. Centrifuge the samples at 13,000 *g* for 10 min and transfer the supernatant to new inserts and GC-vials for GC-MS data acquisition.

3.4.2 GC-MS Analysis

1. Configure the GC instrument following standard operating procedures. Make sure that the equipment is correctly calibrated (according to the manuals of each GC-MS instrument) before starting its use, *see* **Note 5**.
2. Use a sequence consisting of multiple blanks at the beginning and end of the analysis. Ideally, inject at least three replicates per sample to ensure reproducibility of the analyses.
3. If different types of extracts are going to be analyzed, run blank samples between them to avoid possible carryover peaks.
4. Table 2 details the GC and MS parameters.

Table 2
GC and MS parameters

Chromatographic conditions	MS conditions
Injection source: GC ALS Liner: Split, taper, wool, low pressure drops	Ionization mode: Negative ion mode
Syringe size: 10 μL	Acquisition mode: Negative selected ion monitoring (SIM)
Column: HP-5MS 5% phenyl methyl silox (30 m × 250 μm × 0.25 μm)	Capillary voltage: −3.5 kV
Injection volume: 2 μL	Nebulizing gas flow: 1.5 L/min
Mode: Splitless	Dry gas flow: 12 L/min
Carrier gas flow (initial): 0.6 mL/min, gas flow (post run): 1 mL/min	MS source: 230 °C (maximum 250 °C) MS quad: 150 °C (maximum 200 °C)
Column temperature range: 0–325 °C	Block source temperature: 400 °C
Max temperature: 450 °C; initial temperature: 50 °C	Mass acquisition: 50–2000 Da

3.5 Data Analysis

1. The acquired raw file for each condition should be checked for reproducibility and baseline correction using MS data processing software.
2. Different commercial software of the instrument's own brand e.g., Mass Hunter and Chemstation (Agilent) software can be used for profiling and data analysis (https://www.agilent.com/en/product/software-informatics/mass-spectrometry-software).
3. To convert the raw MS data and its pre-processing use open-source software such as msConvert (https://bio.tools/msconvert), MZmine 2.0 (https://mzmine.github.io/features.html), and SIRIUS (https://boecker-lab.github.io/docs.sirius.github.io/) [14].
4. Transfer the full-scan spectra data file in supported file format to the Automatic Mass Spectral Deconvolution and Identification System (AMDIS), (http://www.amdis.net/) which deconvolutes the spectra and searches for compounds using the deconvoluted full spectra.
5. Compare the retention times and mass ion fragments (m/z) identifier ions and match against the NIST (https://chemdata.nist.gov/) and Fiehn (https://fiehnlab.ucdavis.edu/projects/fiehnlib) metabolomics library to identify the unknown phenolic compounds.
6. Consider the metabolites having more than 70% probability score in NIST and Fiehn metabolomics library and having peaks closely related with those of authentic standards.
7. Normalize the peaks with internal standard and quantify the unknown phenolic compounds using the calibration curves of corresponding standard solutions.
8. Prepare the final metadata excel file for statistical analysis and principal component analysis (PCA) using Metaboanalyst 6.0 software (https://www.metaboanalyst.ca/).

4 Notes

1. Purification, fractionation, and concentration of the sample are of great importance for polyphenol analysis [15]. Generally, the solvent extraction implies the co-extraction of other non-phenolic substances, such as sugars, glycosides, organic acids, fats, alkaloids, terpenoids, waxes, and pigments [15]. Hence, one additional step of clean up prior to liquid chromatography analysis is necessary, with the aim of removing these substances and avoiding possible interferences.

2. Ensure that the LC-MS and GC-MS system is operational. Start with system suitability tests as given in your instrument's manuals, especially for tuning the instrument. Tuning parameters and tuning results will readily indicate if the mass spectrometer portion of your MS system is in control, yielding high quality, sensitive measurements. Performing an autotune followed by a manual tune on your analytes will yield the best signal quality. Keep separate tune files for each compound group.
3. Ionization in mass spectrometry is generally influenced by the specific compounds being analyzed. As a result, it is essential to conduct an infusion and fine-tune the liquid chromatography (LC) and mass spectrometry (MS) parameters during method development to achieve the best signal for your analyte. This includes adjusting factors such as the pH of the mobile phase, the mode and polarity of ionization, and the settings for the MS source, which encompass voltages, flow rates, and temperatures.
4. Derivatization is a crucial step in GC-MS analysis, especially when dealing with polar, labile, or poorly volatile compounds. The goal of derivatization is to make an analyte more volatile, less reactive, and thus improve its chromatographic behavior. Derivatization typically involves silylation, alkylation, or acylation reactions. The reagents used for silylation react with active hydrogen, whereas the reagents for acylation react with polar functional groups. Silylation is a very common derivatization technique, and is useful for a wide variety of compounds. Silylation replaces active hydrogens (like those in OH, SH, NH, or COOH groups) with trimethylsilyl (TMS) groups. This reduces polarity, enhances volatility, and improves stability, making the compound more suitable for GC analysis. N, O-bis(trimethylsilyl)trifluoroacetamide with 1% trimethylchlorosilane (BSTFA + TMCS) are commonly used for this reaction.
5. In case of GC-MS, at least 80% of all problems you may encounter are due to the "injector" (i.e., gas flows, gas leaks, pressure control, loose fittings, and liner incompatibilities). Ensure that you have all necessary parts ready to maintain your instrument, including gas leak detectors, nuts, ferrules, spare columns, filaments, and other consumables.

References

1. Al Mamari HH (2021) Phenolic compounds: classification, chemistry, and updated techniques of analysis and synthesis. In: Badria FA, Blumenberg M (eds) Phenolic compounds – chemistry, synthesis, diversity, non-conventional industrial, pharmaceutical and therapeutic applications, 1st edn. IntechOpen, London

2. Pratyusha S (2022) Phenolic compounds in the plant development and defense: an overview. In: Hasanuzzaman M, Nahar K (eds) Plant stress physiology – perspectives in agriculture, 1st edn. IntechOpen, London
3. Naczk M, Shahidi F (2006) Phenolics in cereals, fruits and vegetables: occurrence, extraction and analysis. J Pharm Biomed Anal 41: 1523–1542. https://doi.org/10.1016/j.jpba.2006.04.002
4. Cheynier V (2012) Phenolic compounds: from plants to foods. Phytochem Rev 11:153–177. https://doi.org/10.1007/s11101-012-9242-8
5. Luna-Guevara ML, Luna-Guevara JJ, Hernández-Carranza P et al (2018) Phenolic compounds: a good choice against chronic degenerative diseases. In: Rahman AU (ed) Studies in natural products chemistry, 1st edn. Elsevier, Oxford
6. Cappellini F, Marinelli A, Toccaceli M et al (2021) Anthocyanins: from mechanisms of regulation in plants to health benefits in foods. Front Plant Sci 12:748049. https://doi.org/10.3389/fpls.2021.748049
7. Oluwole O, Fernando WB, Lumanlan J et al (2022) Role of phenolic acid, tannins, stilbenes, lignans and flavonoids in human health – a review. Int J Food Sci Technol 57:6326–6335. https://doi.org/10.1111/ijfs.15936
8. Babele PK, Srivastava A, Selim KA, Kumar A (2023a) Millet-inspired systems metabolic engineering of NUE in crops. Trends Biotechnol 41:701–713. https://doi.org/10.1016/j.tibtech.2022.10.008
9. Babele PK, Srivastava A, Young JD (2023b) Metabolic flux phenotyping of secondary metabolism in cyanobacteria. Trends Microbiol 31:1118–1130. https://doi.org/10.1016/j.tim.2023.05.005
10. Babele PK, Young JD (2020) Applications of stable isotope-based metabolomics and fluxomics toward synthetic biology of cyanobacteria. Wiley Interdiscip Rev Syst Biol Med 12:e1472. https://doi.org/10.1002/wsbm.1472
11. Castro-Moretti FR, Cocuron JC, Castillo-Gonzalez H et al (2023) A metabolomic platform to identify and quantify polyphenols in coffee and related species using liquid chromatography mass spectrometry. Front Plant Sci 13:1057645. https://doi.org/10.3389/fpls.2022.1057645
12. Abd-El-Aziz NM, Hifnawy MS, Lotfy RA et al (2024) LC/MS/MS and GC/MS/MS metabolic profiling of Leontodon hispidulus, in vitro and in silico anticancer activity evaluation targeting hexokinase 2 enzyme. Sci Rep 14:6872. https://doi.org/10.1038/s41598-024-57288-4
13. Jin HF, Shi Y, Jiao YH, Cao J (2024) Separation and determination of phenolic compounds using novel deep eutectic solvent-in-water microemulsion electrokinetic chromatography. Anal Chim Acta 1297:342359. https://doi.org/10.1016/j.aca.2024.342359
14. Mallmann LP, Rios AO, Rodrigues E (2023) MS-FINDER and SIRIUS for phenolic compound identification from high-resolution mass spectrometry data. Food Res Int 163: 112315. https://doi.org/10.1016/j.foodres.2022.112315
15. López-fernández O, Domínguez R, Pateiro M (2020) Determination of polyphenols using liquid chromatography–tandem mass spectrometry technique (LC–MS/MS): a review. Antioxidants 9:479. https://doi.org/10.3390/antiox9060479

Index

Ashwani Kumar, Vidisha Tomer and Piyoosh Kumar Babele (eds.), *Phytochemicals: Extraction, Quantification, Isolation, and Characterization*, Methods and Protocols in Food Science, https://doi.org/10.1007/978-1-0716-4840-7,

MIX
Papier aus verantwortungsvollen Quellen
Paper from responsible sources
FSC® C105338

If you have any concerns about our products,
you can contact us on
ProductSafety@springernature.com

In case Publisher is established outside the EU,
the EU authorized representative is:
Springer Nature Customer Service Center GmbH
Europaplatz 3, 69115 Heidelberg, Germany

Printed by Libri Plureos GmbH
in Hamburg, Germany